Chemical and Biological Hazards in Food

Chemical and Biological
HAZARDS IN FOOD

Edited by

J. C. AYRES, A. A. KRAFT, H. E. SNYDER, H. W. WALKER

DEPARTMENT OF DAIRY AND FOOD INDUSTRY
Iowa State University

(Facsimile of the 1962 Edition)

HAFNER PUBLISHING COMPANY
NEW YORK AND LONDON
1969

Printed and Published by
HAFNER PUBLISHING COMPANY, INC.
31 East 10th Street
New York, N.Y. 10003

Liibrary of Congress Catalog Card Number: 70-93999

Proceedings of the International Symposium
on Food Protection held at Iowa State University
Ames, Iowa, May 10-12, 1962

.

Printed in U.S.A. by
NOBLE OFFSET PRINTERS, INC.
NEW YORK 3, N. Y.

AUTHORS

M. T. Bartram   Ph.D.
  Chief, Bacteriological Branch
  Division of Microbiology
  Bureau of Biological and
    Physical Sciences
  Food and Drug Administration
  Department of Health, Education,
    and Welfare
  Washington, D. C.

Paul R. Cannon   M.D.
  Department of Pathology
  University of Chicago
  Chicago, Illinois

Gail M. Dack   Ph.D.
  Director, Food Research Institute
  University of Chicago
  Chicago, Illinois

Francis A. Gunther Ph.D.
  Professor of Entomology and
    Insect Toxicologist
  Citrus Experiment Station
  University of California
  Riverside, California

Wayland J. Hayes, Jr. M.D., Ph.D.
  Chief, Toxicology Section
  Communicable Disease Center
  Department of Health, Education,
    and Welfare
  Atlanta, Georgia

Betty C. Hobbs   Ph.D.
  Director of Food Hygiene
    Laboratory
  Central Public Health Laboratory
  Colindale, London
  England

Leon Jacobs   Ph.D.
  Chief, Laboratory of Parasitic
    Diseases
  National Institute of Allergy and
    Infectious Diseases
  Bethesda, Maryland

Joe Kastelic   Ph.D.
  Professor of Animal Nutrition
  University of Illinois
  Urbana, Illinois

C. J. Kensler   Ph.D.
  Vice President, Life Sciences
    Division
  Arthur D. Little, Inc.
  Cambridge, Massachusetts

Arnold J. Lehman   M.D.
  Director, Division of Pharmacology
  Bureau of Biological and Physical
    Sciences
  Food and Drug Administration
  Department of Health, Education,
    and Welfare
  Washington, D. C.

Karl F. Mattil   Ph.D.
Associate Director of Research
Swift and Co.
Chicago, Illinois

D. A. A. Mossel   Ph.D.
Head, Laboratory of Bacteriology,
Central Institute for Nutrition
and Food Research
Utrecht, The Netherlands

Emil M. Mrak   Ph.D.
Chancellor, University of
California
Davis, California

C. F. Niven   Ph.D.
Scientific Director
American Meat Institute
Foundation
Chicago, Illinois

Hans Riemann   D.V.M.
Assistant Director
Danish Meat Institute
Roskilde, Denmark

Harold W. Schultz   Ph.D.
Head, Department of Food and
Dairy Technology
Oregon State University
Corvallis, Oregon

Nevin S. Scrimshaw M.D., Ph.D.
Head, Department of Nutrition, Food
Science and Food Technology
Massachusetts Institute of
Technology
Cambridge, Massachusetts

P. M. F. Shattock   Ph.D.
Reader, Department of
Microbiology
The University
Reading, Berkshire
England

L. B. Sjöström
Vice President
Arthur D. Little, Inc.
Cambridge, Massachusetts

George F. Stewart   Ph.D.
Chairman, Department of Food
Science and Technology
University of California
Davis, California

H. L. A. Tarr   Ph.D.
Director, Pacific Fisheries
Experimental Station
Fisheries Research Board
of Canada
Vancouver, British Columbia

A. J. Vlitos   Ph.D.
Director of Research
Tate and Lyle
Central Agricultural Research
Station
Trinidad, West Indies

PREFACE

THE CONSUMER has become acutely aware of the fact that additives, intentional or otherwise, profoundly influence the desirability and edibility of his food products. Many added substances such as the vitamins, amino acids, and trace minerals with which foods are fortified not only improve the nutritional character of the products into which they are incorporated, but additionally often satiate the appetite and enhance gustatory appeal. Formerly, the principal role of added substances in foods was that of preservative agents, i.e., as bacteriostatic agents or inhibitors of chemical deterioration. More recently there have been many reasons for the presence, intentional or otherwise, of chemical additives in foods. While some of these materials are still used to retard or arrest decomposition, instability and other changes, others are specifically employed as nutrients, coloring and flavoring agents, plasticizers, neutalizers, humectants and coating agents. Also, a number of chemical agents serving as pesticides, plant growth regulators or synthetic animal estrogens, migrate into portions of products and so must likewise be considered as additives.

Growth of the food industry has been marked by rapid development of convenience items for the consumer, e.g., precooked frozen foods. Increasing awareness of the need for control of microorganisms having public health significance has accompanied this expansion. Microorganisms and their toxic products are, on occasion, associated with foods that have not been adequately protected during preparation, processing, or subsequent handling and storage.

During this century preparation of foods has changed from custom production in the home to mass production in processing plants. These developments have created great demands for people trained in food technology. In recognition of this demand, curricula in food technology were set up in several universities. Iowa State University was one of the pioneers in this area. As interest in this program increased at I.S.U., it became apparent that new facilities were needed. Funds were provided by the Iowa legislature and the National Institutes of Health for constructing and equipping a food preservation laboratory. Upon completion of the new building, staff members of the Department of Dairy and Food Industry thought it appropriate to honor the occasion with an international conference on an important topic relating to food preservation. With this in mind, a symposium was organized to focus attention on various chemical and biological problems involving environmental hazards to the food supply.

Presentations by several of the world's foremost authorities on chemical and biological hazards in foods have been incorporated into the present volume. This book includes not only the papers presented but also a resumé of the thoughts and ideas exchanged by the several participants whose names are given in the accompanying list (vide infra).

The editors are indebted to all of the participants for the enlightening and spirited discussions which added significantly to the value of the symposium. Particular thanks are due the speakers for their fine cooperation in preparing their papers for publication. In addition, we wish to thank many on the staff at I.S.U. for their generous cooperation and enthusiastic help with the organization and execution of the symposium — without their support the task would have been far more difficult. We wish particularly to thank Dr. V. H. Nielsen, Head of the Department of Dairy and Food Industry for his support, encouragement and help with many of the problems that arose during this time.

Finally, this conference was supported, in part, by a Public Health Service research grant, EF-39, from the Division of Environmental Engineering and Food Protection, Public Health Service and, in part, by funds received from the I.S.U. Graduate College, the Dairy Creamery Fund and the Short Course Fund. We are most grateful for these sources of support and to the Iowa State University Press through whom publication was effected.

August 1962   John C. Ayres
                                       A. A. Kraft
                                       H. E. Snyder
                                       H. W. Walker

CONTENTS

V. MICROBIAL TOXINS

VI. SUMMARY

Introduction

HAROLD W. SCHULTZ
OREGON STATE UNIVERSITY

I N THESE DAYS when there are so many meetings, conferences, and symposia, there must have been some very good
reasons why this symposium was planned. Symposia, when
properly planned and conducted, contribute substantially to scientific progress because they are an important means for determining the present status of knowledge in a particular field, for
evaluating the progress of research, and for stimulating and di-
recting greater effort to increase knowledge and solve problems.

In research, appraisals or evaluations must be carried out
constantly. The independent scientist must evaluate and reevalu-
ate his research continuously, not only in terms of what he himself has done, but also in relation to what others have done or are
doing. This evaluation or appraisal provides a basis for determining what findings need to be reaffirmed and how this might be
done, and provides the guidelines to delve deeper in search of
new knowledge. Research directors, those who determine research policy, as well as those who wish to apply or make practical use of research results, also continuously evaluate.

Obviously, there are other ways than the symposium by which
the value of research and the progress made in solving problems
can be judged. The independent scientist is aided in appraising
his own research efforts by having a thorough acquaintance with
the scientific literature with which he is most concerned. Another way is through personal conferences with his laboratory or
institutional associates, even though they may not be working in
the same specific field. Such conferences are highly desirable in
judging techniques, results, and theories which are being developed. They are also usually stimulating to the researcher because responses are immediate and on a personal basis. Communication with scientists who may be working in the same or

related fields elsewhere can provide assistance and guidance with excellent results. Getting a few persons of similar interests together in conferences is another means now rather widely used.

Meetings of scientific societies provide opportunities for presentation of the scientist's findings and exposing them to analysis by others, as well as for learning from presentations of others. Publication of research findings exposes the scientist's procedures and results to many more scientists and may evoke immediate or delayed confirmation, refutation, or suggestions for new research.

These, as well as other methods, can be applied to individual studies in determining their value, merit, or excellence in themselves. But when there must be judgment as to the contribution of these studies to solving a rather well defined yet broader problem, the situation becomes more difficult because there are more scientific disciplines to consider. The symposium has great value in this situation by bringing together scientists who work on separate phases of a complex problem. Each can place his contributions before the others so that all can evaluate the total accomplishments toward solving the whole problem. This also reveals phases which need more attention or may expose newly recognized problems or phases.

The Symposium on Food Protection permits an appraisal of the present status of knowledge, and of the progress being made in solving chemical and biological problems involving environmental hazards to the food supply. Each participant reports progress to date in a phase in which he is a highly competent scientist. At the conclusion of the symposium we will have a clearer picture of how real these hazards are today and what may be required to remove them.

The Department of Dairy and Food Industry at Iowa State University is to be complimented for sponsoring a symposium on the subject of food protection. There has not been, to my knowledge, a symposium previously which has such broad coverage of the subject of chemical and biological hazards in food as this one. Thus we should be able to appraise possible food hazards in the broadest sense and estimate the safety of our total food supply as never before. This will make a genuine contribution in the interest of public health.

Finally, the Department of Dairy and Food Industry should be thanked for permitting the participants and those attending the symposium to become better informed on a complex subject in so short a period of time. However, it should be expected that the symposium may contribute to the dilemma of science by exposing new vistas of ignorance. Dr. Warren Weaver, Vice-President for the Natural and Medical Sciences, Rockefeller Foundation, describes our science dilemma as follows:

"We keep, in science, getting a more and more sophisticated view of our essential ignorance. Is science really gaining in its assault on the totality of the unsolved? As science learns one answer, it learns several new questions. It is as though scientists were working in a great forest of ignorance, making an ever-larger circular clearing within which, not to insist on the pun, things are clear. But, as that circle becomes larger and larger, the circumference of contact with ignorance also gets longer and longer. Science learns more and more. But there is an ultimate sense in which it does not gain; for the volume of the appreciated but not understood keeps getting larger."

PART ONE
Food Additives: Appraisal

1

Technical Benefits of Food Additives

EMIL M. MRAK

UNIVERSITY OF CALIFORNIA

THE MEANING of the term food additives seems to vary with individuals and attitudes. In this paper I have taken the definition of the Food Protection Committee which is: "A substance or mixture of substances, other than a basic food-stuff, which is present in a food as a result of any aspect of production, processing, storage, or packaging." The term, of course, does not include chance contaminants; but it does include two broad types of food additives — intentional and incidental.

Intentional additives are substances such as salt added purposely to perform specific functions. Incidental additives, on the other hand, are substances which, though they have no function in the finished food, become a part of the food product through some phase of production, processing, storage, or packaging. An incidental additive could be an agricultural chemical applied to the crop or a substance that migrates into food from a packaging material.

It is apparent, therefore, in discussing the technological benefits of food additives, that we must include chemicals used in agricultural production, processing, packaging, and even those used in the home.

In developing this paper I tried to determine when chemicals were first used in foods. The beginning is lost in antiquity; however, I believe there is some relationship to the great revolutions in the history of man as given by Darwin. For example, the first great revolution came with the discovery of fire which resulted in the process of smoking. This method of preservation, of course, meant the inclusion of certain components of smoke such as pyroligneous acid, formaldehyde, and even potential "carcinogens" in the food.

Agriculture was another revolution caused by man. It is

difficult to determine just when this involved the use of chemicals, but we know that Indians in this country used dead fish for fertilization, the Chinese used ethylene gas for ripening, and the Greeks treated raisins with alkali (ashes) before drying. In more recent times, developments have been abundant. Dusting with sulphur was started by Duchatel in 1850. Bordeaux mixture was discovered by Mallaradet in 1882 when, interestingly enough, he tried to prepare a repugnant material which, when placed on his grapes, would discourage people from stealing them. This serendipity gave man one of the best fungicides of all times.

Tremendous advances in the use of agricultural chemicals have occurred in more recent years. For example, synthetic herbicides were introduced in 1938, synthetic organic fungicides in 1940, chlorinated hydrocarbons (DDT) in 1942, synthetic rodenticides in 1944, and organic phosphate insecticides in 1947. Since then, a multiplicity of chemicals has been developed for agricultural purposes, and these have had a tremendous impact on production, quality, and distribution of foods and food products.

Another great revolution caused by man was urbanization; and this, too, had an influence on the use of chemicals; but it is difficult to be specific. As cities developed it became necessary to transport foods to them from the country. At first livestock was transported on foot, but milk fermented and formed butter as it joggled along in bags on the backs of camels. Without doubt, the situation encouraged the use of preservatives such as salt and smoke. One can hardly start reading the story of the American westward movement without finding reference to "sowbelly and beans," or in other words, salt pork and beans. Some may doubt the technological benefits of this type of preservation, but it certainly helped people travel and keep alive though they may have suffered from dyspepsia, which in turn started the "corn flakes" crusade. As time went on, of course, other means of preservation, including the use of chemicals, were developed; and these enabled an easier and better movement of foods to the centers of population.

The industrial revolution of man is the most recent and is continuing, and this is the period in which an enormous increase in the use of chemical additives has taken place. At the same time in this period, our foods have become more abundant, nutritious, convenient, varied, acceptable, and widely distributed.

Now I should like to dwell a little on the technological advantages of agricultural chemicals, or those we call incidental additives. These chemicals are used for a variety of reasons, and they have had a profound effect on the availability, quality, and cost of our food. I hardly need to mention the benefits of synthetic fertilizers, yet there are those who would outlaw all fertilizers except natural ones; and they would eliminate all fertilizer factories except the natural one. We all know that weeds compete with cultivated plants for moisture and nutrients, so the development and application of weed killers has meant better crops and cheaper production. The extensive use of insecticides and fungicides has enabled the farmer to remain in business; and he keeps many crops on the market that would have been eliminated or would have become so costly that only the wealthy could afford them. The destructive potential of agricultural pests is well indicated by the potato famines of Ireland. The potato blight (Phytophthora infestans) destroyed crops and caused thousands of people to die of starvation. Other fungi can be equally bad. In the past, ergot poisoning caused by the fungus Claviceps purpurea, which grows on cereal crops, was quite common. Only a few years ago, surprising as it may seem, several people died from ergot poisoning in Europe. Such destructive occurrences no longer take place in this country, and a substantial part of this advance can be attributed to the use of agricultural chemicals.

Food wastage during storage has been, and still is, a very serious problem. These losses caused by rodents, insects, and fungi annually amount to as much as 33 million tons of good food. Believe it or not, this is enough to feed the entire population of the United States for one year. According to Robert Brittain (1952) this means, "If one person out of every 14 or 15 in the world should die yearly from starvation because of the real lack of food to go around, we could say quite literally that he was done to death by these predators." The development of rodenticides has been important in the reduction of storage losses although we still have a long way to go.

Losses during shipment of fresh produce have been substantial. The development of "hydrocooling," which involves the use of chlorine, has greatly reduced such losses in items such as cherries and asparagus. In fact I have been told by one shipper that hydrocooling reduced his losses in fresh cherries during shipment from over 50 per cent to less than 10 per cent.

Defoliation of plants such as sorghum by use of chemicals is another procedure that has meant a reduction of the cost of production of feed crops, and hence, animal protein.

Hormone sprays are being used to prevent immature fruit from dropping (June drop) and experimentation is underway on "thinning" blossoms by use of sprays in order to produce larger and better fruit.

One rather interesting development has been the use of

ethylene as a ripening agent for certain fruits. For example, honeydew melons harvested for shipment must be treated with ethylene, otherwise they do not ripen properly and are inedible. The development of this gas has a rather interesting history. It was observed many years ago by the Chinese that if they would place persimmons in a closed room and burn oil lamps in the room, the fruit would ripen much faster. Not until this century did Denny and his associates learn that the burning process produced ethylene gas, and this in turn was responsible for the ripening. The application of this gas was questioned by the zealots who would oppose the use of any chemical, until it was learned that nature itself produces the gas in certain fruits, and this is a natural means of ripening. As a matter of fact, at one time, the canning industry in California took advantage of this by placing boxes of ripe pears near boxes of immature pears, realizing full well that the immature pears would ripen more rapidly this way than when kept next to other immature pears.

A number of chemicals have also been used to protect or enhance the development of animals. Certain hormones reduce the time required in feed lots, and antibiotics in feed enable faster growth in some instances.

Discussion on the use of agricultural chemicals may well be summarized by a statement on the situation in California. In a report of the Governor's Committee concerned with Policy on the Use of Agricultural Chemicals, it was pointed out that the cash receipts from California's combined crops, livestocks, and allied production in 1959 exceeded $3 billion. This income amounted to 9 per cent of the total cash receipts from farming in our entire country.

It has been well established that the record high production in California could not be achieved without the extensive use of agricultural chemicals. There are no accurate data on the quantities of chemicals used by California farmers, but it is estimated that about 20 per cent of all the pesticides employed in the United States are applied in California. This represents about 200 different chemicals, over 1,400 brands, and the treatment of over 7 million acres by farmers and commercial operators. In some instances, these applications have prevented the complete destruction of certain agricultural industries. For example, in recent years the Mexican bean beetle and the Khapra beetle (a type of carpet beetle from India) have been a very serious menace of stored grain material. Because of the effective use of pesticides, the Mexican beetle was eradicated and the Khapra beetle has been almost eliminated from the scene. Should any of these pests have become established, the ravages would have been reflected in substantially higher costs and lower quality of the foods affected. Specific measurements have been made by the University of California of the cost attributed to the tuber worm, just one of the pests that attack potatoes. The cost of production of potatoes would be increased from 7.3 to 39 per cent depending on the variety and where grown; but by the use of proper control measures, the losses are less than 1 per cent. Advantages of the use of agricultural chemicals can be realized from the fact that in the United States, 23 per cent of the average income is spent for food while the rest of the world spends 60 per cent. Furthermore, foods eaten in the United States are more varied and of superior quality.

Now I would like to discuss intentional additives used in processing. Several hundred chemicals are used for a variety of reasons and the benefits are great indeed. The processor has the needs of the consumer uppermost in his mind. Of particular concern is consumer acceptance and the desire to make available to the consumer the best flavor, texture, color, and other characteristics of importance; furthermore, the food should be stable so that, even after a prolonged period of storage or handling under adverse conditions, it is still acceptable and nutritious.

Ease of use (convenience) has been built into many of our foods, and in many cases this has required the use of additives. We all know how simple it is to prepare instant potatoes and coffee or use cake and other mixes without failure. New convenience items are still appearing on the market, and it appears that the "sky is the limit." Of course, the factor of safety is always uppermost in the minds of everyone, and chemicals play an important role in protecting foods against organisms that cause food poisoning.

Great care is taken to retain the nutritional values of foods during processing. In some cases, however, vitamins, minerals, or other substances may be added to fortify the food from a nutritional point of view.

Packaging is used more today than ever in the history of the food industry. This enables mobility, easy handling, and storage of the product all the way along the line, from production to consumption. All these factors have involved the increased use of chemicals. On the other hand, if chemicals had not been used intensively, it would have been impossible to produce the stable, nutritious, safe, and convenient foods with high acceptance that we have today. We would still be back in the days of "staple products," and I doubt that even the most ardent organic farmer, or perhaps even Bicknell himself, would desire to go back to those days.

The present-day use of food additives during processing has been summarized in an excellent manner by Ikeda and Crosby in the University of California Manual Number 26. They give nine basic uses for additives during the processing and handling of foods. They are: preservatives, nutritional supplements, coloring, flavoring, improved functional properties, processing aids, moisture content control, alkalinity control, and physiological activity control. Preservatives are used to prevent microbiological and chemical spoilage and insect infestation. The importance and nature of microbiological and insect spoilage are well known. Chemical spoilage may involve a variety of changes including the browning reaction, discoloration, change in texture, loss of flavor, and rancidification. A number of different chemicals are used to inhibit these processes.

A number of nutritional supplements are added to flour including vitamin B_1, riboflavin, and iron. The enrichment of bread has brought about a nutritional adequacy in this country that does not exist elsewhere in the world. Another example of supplementation is the addition of iodine to table salt, which is so important in the "goiter belt."

Synthetic and natural colors are used so the consumer may have the advantage of a food with a characteristic appealing color that is consistent from one purchase to the next. There is an abundance of evidence to indicate that color has a profound influence on acceptability.

One of the most important reasons for the use of chemical additives in foods is to improve or stabilize the flavor. Food must not only look good, it must taste good. It would be impossible to list the number of foods that contain added flavoring agents, for flavoring agents make up the largest and most diversified group of food additives. Sometimes chemicals such as monosodium glutamate are added to enhance flavor though these chemicals do not impart a flavor. These substances are used for enhancing the flavor in certain foods such as soups, salads, and fish sausage. It has been pointed out to me on a number of occasions that from a physiological standpoint, flavors are important because they induce the flow of stomach juices.

A number of agents are used to improve the functional properties of foods which are so important in many of the convenient foods we have today. These chemicals include stabilizing agents, curing and bleaching substances, emulsifiers and those used for thickening, firming or maturing. Stabilizing agents, for example, are used in chocolate milk to prevent the particles from settling to the bottom of the glass. They are also added to frozen desserts to increase the viscosity and help prevent the formation of large coarse crystals.

Curing and bleaching agents are used in the milling and baking industries. Freshly milled wheat flour has a yellowish color caused by small quantities of natural pigments. Such flour also lacks the qualities necessary to make an elastic and stable dough. When flour is stored and allowed to age for several months, it gradually becomes whiter due to oxidation and "matures" to make a good product for baking. The use of chemicals has speeded this and actually standardized the product so that we have better flour now than ever before.

Emulsifiers permit the dispersion of very small particles of one liquid within another. Without the use of such substances, it would not be possible to have the fine salad dressings, mayonnaise, and other similar products we have on the market today. They also play a role in improving the uniformity, volume, and "grain" of breads and rolls. Furthermore, they prevent the development of an unsightly greyish appearance (bloom) on the surface of chocolate candies.

Chemicals such as calcium salts are used as firming agents; for example, small quantities of this salt are added to canned tomatoes to retain form.

A number of food additives that may be classified as processing aids are: sanitizing agents, chemicals to remove extraneous materials, metal-binding compounds, and anti-foam agents. Detergents are used to facilitate the removal of dirt and insects; silicones act as anti-foam agents in fermentation and citric acid as a sequestrant for binding metals.

Chemicals are employed to increase and decrease moisture levels in a variety of commodities; glycerine is used in marshmallow candies as a humectant; calcium silicate is added to table salt and sodium stearate to garlic powder as anti-caking agents. The matter of moisture control, however, is still far from being solved; for example, it would be highly desirable to find some agent that would prevent the transfer of any moisture from raisins into dry cereals when packaged together.

Various acids, alkalies, and salts are added to food to obtain the desirable pH, which in turn may be concerned with taste, preservation, or stability.

Chemicals that control physiological activity are applied to the fresh product and act as either ripening or anti-metabolic agents. Ethylene gas, for example, hastens the ripening of honeydew melons and bananas. On the other hand, maleic hydrazide prevents the sprouting of potatoes.

It would hardly be fair to cover the subject of food additives without mentioning the wide use of chemicals in the home, even by those who would outlaw the use of these substances in agriculture and processing. For example, sugar, salt, vinegar, spices, baking soda, monosodium glutamate, and artificial food colors are used in almost every home. It is ironical that those who complain so much about the use of chemicals in foods use them in their homes without giving the matter a second thought. Tsze, a Chinese philosopher, once said, "Those who know, do not speak; those who speak, do not know." This may well be our problem.

It is my firm belief that the future will not only see the use of more chemicals, as indicated by Ikeda and Crosby, but we will be compelled to use more chemicals because of the need to increase the production of foods and to prevent losses. Our population is increasing at a fantastic rate, and it will not be long before our food requirements will exceed our present-day production. Not only must we produce more foods in the reasonably near future, we must produce them more efficiently and on less acreage. Even then we may find it difficult to maintain our present high standard of living. Today, for more and better foods, we pay about the same share of our income as in the pre-war days. In 1935-39, our food expenditures amounted to about 23 per cent of our income, whereas today it is about 21 per cent. Furthermore, if we would eat the same as we did 20 years ago, an even smaller share (16 per cent of the average consumer income) would go for food. The reason for this is that we are paying more for convenience, packaging, and parking spaces around chain stores. Since 1921, our crop production per acre has increased 47 per cent; per animal breeding unit, 74 per cent; and the output per man hour of labor, 249 per cent. According to the U. S. Department of Agriculture, if the present rate of population increase continues, by 1975 we will require increases in billions of pounds: 16.3 of red meat, 20.7 of fruits and vegetables, and 47 of milk, as well as 20 billion eggs. If our production should remain at the present level, we would need about 200 million more acres of crop land than we have at present. But we don't have 200 million more acres of crop land; and as a matter of fact, we are taking many acres out of production. This means, therefore, that we must increase our efficiency in agricultural production; and in my opinion, agricultural chemicals will play an extremely important role in this change.

It has been pointed out that cultivated crops in North America are attacked by over 3,000 economically important species of insects, many plant disease agents, and an unestimated number of nematodes, rodents, weeds, and other predators. Truly, we have a long way to go before these are controlled, and chemicals offer great opportunities for this control. If our entomologists are correct, we actually produce more insect protein than animal protein per acre of pasture land.

In 1954 the United States Department of Agriculture estimated that to offset the loss of agricultural production caused by pests, an extra 88 million acres would need to be cultivated, and the losses subsequent to harvest would be equal to the production of an additional 32.8 million acres. It is apparent that we should be conducting intensive research for the development of new chemicals to increase food production. I don't believe, however, that we should stop there, for the opportunities for use in processing are also great.

In closing I should like to say a word about food safety. Much has been said about the dangers in our modern foods, but there has been no sound basis for this. Those who would exclude the use of chemicals are uninformed, and unfortunately, they are a vocal group.

Many people, unfortunately, have forgotten that foods are chemicals. When one tries to trace the scientific knowledge on the chemistry of foods, he realizes that such knowledge is really quite recent. One of the earliest works is that of Liebig, who in the middle of the 1800's talked about nitrogenous or flesh-forming foods and non-nitrogenous or heat-generating foods. Such thinking, of course, gave way when information on the detailed compositions of foods appeared. The earliest volumes of the "Experiment Station Record" contained reference to USDA Bulletin 28 published by W. O. Atwater and C. D. Woods. This publication contains the history of food analysis and explanation of terms used in discussing the values of a large number of American foods. It gives "proximate" composition values for water, ash, fat, total nitrogen, carbohydrates, and fuel value. I mention this because I believe it is important that we realize the idea that foods are really chemicals does not go back very far. I think it is also important to point out that we have forgotten that some of the chemicals we use in foods are actually present in foods; for example, ethylene or sodium chloride. The number of chemicals occurring in food is large indeed. The confusions and delusions concerning the safety of natural foods with all their natural chemicals are extraordinary. As an example, safrole is in trouble because of its possible carcinogenic properties; yet it occurs naturally in a number of natural foods that have been used for hundreds of years. Then again, a closely related compound, apiole, has been found in celery and other common foods. Should we outlaw all these foods ? We need to be sensible; but the philosophy that foods are chemicals has gotten away from some people, and this is unfortunate.

Recently we have heard a great deal about poisoning of babies with salt, and an elaborate description of this was given in the

April 27, 1962, issue of Life magazine. This incident certainly bears out a statement by William E. Baier of Sunkist Growers in Southern California. I have quoted him on a number of occasions because I think his statement is an important one. He stated that Congressman Miller, in his amendment, recognized perhaps for the first time in all legislative history, this simple but often ignored fact: "There are no harmless substances; there are only harmless ways of using substances." This, I believe, is the important message that must be imparted to those who have lost confidence in our food supply because of the use of food additives.

LITERATURE CITED

Bicknell, Franklin. 1960. Chemicals in Your Food and in Farm Practice. Emerson Books, Inc. New York.

Brittain, Robert. 1952. Let There Be Bread. Simon and Schuster, New York.

Ikeda, R. M. and D. G. Crosby. 1960. Chemicals and the food industry. Univ. of California Press. Agr. Exp. Sta. Manual No. 26.

2

Chemicals in Food Products

PAUL R. CANNON
UNIVERSITY OF CHICAGO *

FIFTY YEARS have passed since Dr. Harvey Wiley resigned from his post as Chief of the Bureau of Chemistry of the U. S. Department of Agriculture. This resignation followed by only six years the passage of the first Pure Food and Drug Law of 1906. Although his withdrawal from governmental responsibilities signified frustration and defeat in his crusade against dishonest food processing, he at least had been able to establish guidelines for the prevention of reckless adulteration of food products by unscrupulous food processors and food vendors. To Dr. Wiley chemicals in foods meant essentially a rather limited group of deleterious adulterants, for example, coal-tar dyes, alum, borax, benzoic acid, benzaldehyde, salicylic acid, formaldehyde, sodium sulphite, saltpeter, methyl alcohol, etc. The unwarranted use of such substances brought into clearer focus some of the cruder stigmata of cupidity and greed. In combating the forces responsible for these adulterations Dr. Wiley succeeded in reducing the numbers of nutritional misdemeanors, but in doing so he made numerous enemies and aroused powerful antagonists.

Despite his vigorous efforts in the broad area of food protection he was obviously hampered by his limited knowledge of many important aspects of nutrition. For example Dr. Wiley had of course never heard of vitamins, and such chemical entities as niacin, thiamin, riboflavin and ascorbic acid were totally unknown to him. Neither had he heard about essential amino acids, essential fatty acids, commercial hydrogenation of oils, chemical carcinogens, antimetabolites, antivitamins, "running fits" in dogs, iodized salt, insulin, staphylococcal enterotoxin, DDT, antibiotics,

---

*Chairman, Toxicology Study Section, National Institutes of Health.

stress reactions, cortical steroids, heat injury to protein, the Maillard reaction, etc. Moreover, the great killer, coronary thrombosis, did not come into clinical prominence until 1912, the year that Dr. Wiley left the governmental service. It may be noted that it was also in that year that Funk announced his "vitamine hypothesis." In short, today's questions concerning the problems of chemicals in food products are vastly more varied and complex than they were in 1912. In fact, current methods for the detection of chemicals in foods have become so delicate that experts spend a great deal of time trying to establish such elusive quantities as the chemical equivalents of zero. Our committee from the Food Protection Committee, after struggling somewhat unrewardingly with this problem, began to refer to itself as the "committee on nothing."

My subject might fittingly cover a considerable array of problems, for example, such questions as the influence of agricultural chemicals upon food production and food surpluses, pesticidal residues, tolerances, enforcement measures, carcinogens in foods, Delaney clauses, intentional additives, grandfather clauses, GRAS lists, etc. Certainly these are the sorts of questions which stimulate legislative actions, promote the writing of letters to congressmen, and influence the publication of books about the "poisons in our food." However, I shall touch upon them but lightly since several of them are to be considered in detail in our symposium. Instead I propose to comment upon a few of the questions which have a direct bearing upon certain biological hazards incident to the presence or absence of specific chemicals in foods. Furthermore, I shall consider hazards which are related both to natural conditions and to man-made interventions.

It is always interesting to recall the many nutritional problems which have been associated with the absence of essential chemicals from foods. Indeed it is still a source of wonderment when one reflects upon the slow development of the concept of a metabolic disease as the resultant of a dietary lack of some specific essential nutrient. Such a lack was of course evident in states of water-deprivation or of prolonged dietary absences of protein, carbohydrate and fat, and much early history is replete with accounts of the interrelationships of wars, famines and pestilences as consequences of economic and social dislocations and their accompanying shortages of dietary essentials. But, lacking knowledge about vitamins and essential minerals, it took a long time for students of disease to realize that certain pathologic processes may be due specifically to the absence of a particular dietary element. Now the facts about scurvy, beri beri, pellagra and rickets and about such specific mineral deficiencies as iodine-deficiency goiter, iron-deficiency anemia, calcium-deficiency osteomalacia, etc. are too well-known to require comment, particularly since, with the discovery of the specific vitamins and minerals concerned, these deficiency diseases have virtually disappeared or are disappearing from many parts of the world. Similarly, specific deficiency states are being prevented by the addition to flour, cornmeal, rice, breakfast foods, milk, salt, etc. of an assortment of chemical additives, e.g., thiamin, riboflavin, ascorbic acid, vitamins A and D, iron, calcium, lysine, methionine, iodine, and indirectly, by the use of nonfat milk solids, yeast, soyflour, cottonseed meal, wheat germ concentrate, etc. I mention these well-known achievements only to reemphasize the fact that all of these remedial and preventive agents are chemical additives and that they are placed intentionally in our foods for nutritive improvement and not for pecuniary gain.

With the disappearance or lessening of these deficiency states, although not necessarily because of it, we have passed, in this country, and within a few decades, from a condition of relative food scarcity at least in respect to specific dietary elements, to one of almost embarrassing excess. Thus today we are confronted with problems which can be characterized as "problems of excess," namely, of excesses in food production, in food consumption, in agricultural management, and in an excessively apprehensive state of "chemophobia." By chemophobia I mean an exaggerated fear of toxicity brought about by the widespread use of chemicals in food production and processing. It was because of these changing conditions that the Food Protection Committee was established in 1950 and that the Toxicology Study Section of the National Institutes of Health was organized more recently in an effort to implement research in relation to many of these problems.

Accompanying such "problems of excess" is an undue interest in diseases associated with and in some instances related to our changing dietary patterns. I refer specifically to the degenerative and malignancy diseases, notably cardiovascular disease, osteoarthritis, and cancer. These diseases have come to the tops of our mortality tables mainly because of today's longer life expectancy. Such an expectancy has followed the disappearance of many of the lethal maladies of infancy and childhood. Associated with this is the diminution of important diseases of the midyears, notably tuberculosis, enteric disease and respiratory disease. Conspicuous in these changes have been the advances in sanitation and public health, in medical therapy, and especially in the availability of sulfonamides, antibiotics and other so-called "wonder drugs."

Because so many persons now live into the older age groups and in consequence acquire degenerative diseases, the idea has developed that these diseases must be direct resultants of our changing environment. Possibly in some instances they are; as, for example, in the case of cancer of the lung in its presumed relationships to excessive smoking of cigarettes or to exposure to fumes and dusts, to deafness in relation to urban and industrial noise, etc. However, most chronic diseases are the consequences of multiple contributing factors, no one of which can be called the actual or sole cause. For example, our economy of excess has contributed materially in the past two decades to the current prevalence of overnutrition and obesity. Concomitantly, however, other factors have contributed to our increasing physical immobility, as, for instance, through the growing use of automobiles and other labor-saving devices. These, too, have added their quota to the rising tide of national adiposity. Nevertheless, because of the expanding use of agricultural chemicals, particularly the insecticides and pesticides, an excessive fear of chemicals has also developed. Accompanying this chemophobia has been the assumption that, since some of these chemicals persist as residues in food products at times, their ultimate consequences must necessarily be harmful through the gradual poisoning of a vast population. Persons activated by such fears display little confidence indeed in the enormous detoxifying capacity of the liver. However, the post hoc ergo propter hoc type of illogical reasoning is not unusual. In an effort to clarify the situation and to ascertain the facts relating to such problems the U. S. Congress, the Departments of Agriculture and of Health, Education and Welfare, the Food and Drug Administration, food processors and manufacturers and others have gone to great lengths to guard and protect the public against any dangers which might be related to these changing agricultural and food processing and distributing practices. Unfortunately, it is easier to generate fears than it is to maintain confidence, particularly in a world dominated by the realities of hydrogen bombs, radioactive fallout and recurring recessions.

A fear which has been particularly dominant has stemmed from the idea that chemical residues in food products might possess carcinogenic potentialities and contribute thereby to the initiation of cancer. Inasmuch as cancer in this country ranks second as a cause of death, and because its causes are so largely unknown, it is of course possible that some of these residues might be carcinogenic. Because of such a possibility the subcommittee on carcinogenesis of the Food Protection Committee undertook its study of the problem at the request of the advisory industry committee. The subcommittee assumed that even though the likelihood is slight that chemical residues in food might initiate cancer in man, it is important to know the facts as quickly as possible. Several other organizations and groups of scientists in this country and abroad have also directed their efforts along these lines, including many in the chemical industries. It is evident, however, that a great deal of careful and time-consuming work will be required to evaluate all of the chemical compounds now in existence which might get into foods through carelessness, ignorance or accident. Among the compounds thus far scrutinized and found to be carcinogenic for certain animals are butter yellow, dulcin, thiourea, thioacetamide, certain waxes, stilbestrol, safrole, aminotriazole, aramite and acetylaminofluorine. At this time I will call attention only to Publication No. 749 of the subcommittee on carcinogenesis of the Food Protection Committee (1960) and the Fifth Report of the Joint FAO/WHO Expert Committee on Food Additives (1960). These two reports contain many references to contributions by individuals and by groups.

Conspicuous in the attempts of the U. S. Congress to forestall any hazardous consequences of potentially carcinogenic agents which might enter the food supply was the passage in the amendments to the Food, Drug and Cosmetic Act (1958 and 1960) of the so-called anticancer or Delaney clauses. In short, before a great deal was actually known about the carcinogenic hazards of chemical agents in food products the new law made it illegal for such agents to be present in food destined for interstate commerce. This legislative action has aroused much discussion, favorable and otherwise. Among other things it illustrates one of the important problems of our subject, viz., the conflict between two points of view, the scientific and the legal.

It is not my intention to discuss this legislative problem now; time and the legislators themselves will have to clarify their own turbidities. The only basic question is: How hopeful can we be that this kind of legislation can actually protect us to any significant degree against carcinogenic hazard? In essence the law states that whenever there is evidence that a chemical substance has displayed carcinogenic potentiality whether in animals or man, its presence in foods must be interdicted. Furthermore, the burden of proof of noncarcinogenicity rests upon the manufacturer or processor, and judgment as to carcinogenicity disregards such questions as amounts in a food, manner of use, possible thresholds of activity, etc. In short, for any substance for which there is any evidence of carcinogenicity there must be a zero tolerance. It was upon these bases of judgment that action was taken in the aminotriazole and stilbestrol episodes of 1959.

Although the laudable intent of the law is beyond dispute, the fact remains that it does raise important questions as, for example, what actually is a carcinogen? In practice it is easy to define a carcinogen as any material which can engender cancer in man or animals under any condition of entrance, or as a material which acts to increase the risk of cancer. Nonetheless, the question remains: How sure can one be that a substance which induces cancer in a highly susceptible inbred strain of mice or rats, and acting under somewhat exaggerated experimental conditions of testing, can do the same in man under natural conditions? By analogous reasoning one could define an infection-producing agent or "pathogen" as one which produces infectious disease in man or animals. In this instance, however, it is known that there are infectious agents which are pathogenic for lower animals but not for man. However, with respect to carcinogenic substances the law assumes that if the material is carcinogenic for a mouse it may also be carcinogenic for man. It will doubtless be a long time before this question of carcinogenicity can be satisfactorily answered. But until it is, the validity of the anticancer clause must continue to rest upon assumption rather than upon demonstrated fact.

Regardless of these and other difficulties inherent in the problem of carcinogenic hazard as it applies to chemical additives, a further question needs to be answered, viz., what should be done with reference to similar problems as they pertain to natural foods? Natural foods also contain chemical substances which may present carcinogenic hazards. Before considering this problem, however, let us look for a moment at its background.

For reasons not easy to explain there has appeared in this country and elsewhere a group of food enthusiasts whose members assume and argue that the addition of chemicals to food products, or even their use as fertilizers, represent unwarranted and even harmful interferences by man with God-given methods for the production of "pure foods." According to this point of view animal manures and rotting composts presumably produce purer and more wholesome foods than do clean solutions of nitrogen, phosphorus and mixtures of trace elements. These enthusiasts in large part ignore the fact that every food constituent is in nature chemical; moreover, they seem to imply an impropriety in man's utilization of chemical analysis and isolation from a food of its chemical constituents, followed by their synthesis and use. Inherent in their thinking seems to be the assumption that for a food to be pure and undefiled it must be natural and untampered with by man. Such an antichemical point of view looks backwards nostalgically to the "good old days" when fruit trees needed no sprays and food technology was of a "meat and potatoes" degree of simplicity.

To such points of view the only question at issue is: What are the facts concerning the purity and wholesomeness of natural foods? Do the peoples in underdeveloped countries who subsist on natural foods enjoy a more enviable nutritive status than do those of us who presumably enjoy the advantages of modern food technology? The answer, of course, is that some of the most powerful poisons known to man occur naturally in plants and that man has had to learn the hard way how to select those foods which are friendly to his taste while avoiding those which have disagreeable consequences. If nature is so kind and beneficent in supplying our daily food, she keeps toxicologists surprisingly busy investigating such naturally-occurring toxic agents as the poisons in certain mushrooms, the atropin in some members of the nightshade family, the digitoxins of foxglove, the ricins of the castor bean, the cyanogenetic glucosides in fruit seeds, the gossypol in cottonseed, the goitrogenic agents in cabbage, kale and rutabagas, the tannins of tea, the alkaloids of the loco weed and of the Senecio species, the ciguatera toxins of certain fish, etc.

Because of these and related problems the Food Protection Committee is looking into the question of toxic substances in natural foods. I shall anticipate this study only to allude briefly to possible relationships of some of these substances to the induction of cancer. I do so mainly because, since the main purpose of the anticancer clauses in the Food, Drug and Cosmetic Act is to minimize the risk of cancer from the ingestion of food containing carcinogenic chemicals, attention should likewise be given to the possible presence of carcinogens in natural foods. Indeed it is surprising that the enthusiasts for natural foods should have been so much less assiduous in their efforts to protect these foods than they have been with reference to agricultural chemicals. At any rate there is now enough suggestive evidence to warrant a more careful scrutiny of such foods as smoked meats, especially smoked fish; oysters and shellfish; wheat and flour from selenium-rich soils; water and other foods containing radioactive materials, either occurring naturally or from fallout; foods containing safrole; plants containing estrogenic materials; plants containing goitrogenic substances; Senecio alkaloids, etc.

With reference to smoked foods, Dungal (1961) has recently reported on the problem of gastric carcinoma in Iceland where this form of cancer is recorded as the cause of death in 35 per cent of all necropsies. In the rural areas, where smoked mutton and fish are kept in smokehouses for long periods and are eaten in large amounts over long periods, the content of 3,4 benzpyrene in the meat and fish has been found to be significantly high. Moreover, by the feeding of smoked mutton and smoked fish to rats, malignant tumors were produced. Geographically, also, cancer of the stomach was most prevalent around rivers and lakes where there was a greater consumption of smoked salmon and trout.

The natural occurrence of inorganic arsenic in oysters, fish and shellfish indicates the impossibility of eliminating this material from a natural food. The only question is whether or not inorganic arsenic is a carcinogen, as is commonly supposed. It is of interest that apparently no one as yet has been able to demonstrate its carcinogenicity in experimental animals. Therefore evidence of its carcinogenicity rests entirely on clinical and epidemiologic findings.

A paradoxical situation exists with respect to selenium. This element in small amounts has been shown to be a dietary essential for cattle and sheep, and in the Pacific Northwest, where the soil is deficient in selenium, so-called white muscle disease develops from the dietary lack of selenium. This deficiency may be prevented by the addition of selenium to fodder. However, in larger amounts selenium is a carcinogen and has produced hepatomas in rats. According to the anticancer clause, therefore, selenium cannot be added to wheat for interstate shipment, although selenium-containing grain can be transported.

The occurrence of radium-containing water in certain areas of the United States represents essentially a theoretical problem, so far as can now be determined. Of more significance, is the problem of radioactive fallout materials in soil, grains, milk, water and fruits. Obviously in both situations the assumption must be made of the probability that there are safe dosage levels or so-called permissible levels of intake. In any case there is no possibility of establishing zero tolerances for such materials.

The recent demonstrations of carcinogenicity of safrole have led to the withdrawal of oil of sassafras in rootbeer. Nevertheless it is still present in cinnamon and nutmeg and represents another example of a carcinogen occurring in a natural food.

The presence of goitrogens in various members of the Brassica species has long been known. This fact assumed greater interest in its relationship to the contamination of cranberries by aminotriazole and to the antithyroid and tumorigenic activity of this herbicide in rats. The question is: If the tumorigenic potentiality of this chemical is applicable to man, as was assumed in the governmental ruling of 1959, what is to be said about the tumorigenic potentialities of somewhat similar compounds derivable from cabbage, turnips, rutabagas, rape and kale?

The above examples are cited merely to indicate that much remains unknown about the possible presence of tumorigenic agents in natural foods. They further emphasize the fact that legislative action with reference to agricultural chemicals does nothing toward the elimination of possible hazards from natural foods.

Finally, consideration should be given to the important problem of dietary excess in its relationship to our mounting surpluses of such foodstuffs as wheat, corn, milk, butter and eggs. Moreover, because the resulting obesity may contribute to diabetes, hypertension, arthritis, and cardiovascular disease, it is obvious that our overabundance of food products is contributing to our unwanted national surpluses of body fat. Granted that these pathologic processes are caused by multiple factors, circumstantial evidence, nevertheless, suggests that excessive intakes of food contribute materially to their worsening.

In considering such pathologic sequences, we must acknowledge the fact that the highest ranking cause of death in this country today is cardiovascular disease. Furthermore, with reference to atherosclerosis and coronary disease, most of the current research, involving the expenditure of vast sums of money, centers around chemical questions, viz., cholesterol, fatty acids, saturated fats, polyunsaturated fatty acids, essential fatty acids, caloric intakes, sodium excesses, sodium-potassium interrelationships, magnesium actions, hormonal imbalances, etc. The crucial question through it all, of course, is: What is the relationship of such diseases to the presence or absence of these and other chemical constituents in food products?

Is it possible that in our current concern in relation to a few molecules of an agricultural chemical with respect to cancer or to some disease of obscure nature we tend to overlook hazards of greater quantitative concern to the population as a whole? Perhaps too often we put ourselves into orbit around the problem and lose sight of events transpiring at the center, events, moreover, which determine disability and death of many persons. If it is the obligation of the law to prevent the presence of small quantities of aminotriazole in cranberries or stilbestrol in caponettes, what should be said about the legal obligation to prevent excessively large intakes of cholesterol or of saturated fats? In fact there is experimental evidence that cholesterol is carcinogenic. It should be added that the fact is still controversial. At any rate one might argue that, since the death rates from atherosclerosis are greater than are those from cancer, the importance of preventing this form of hardening of the arteries might be said to be greater than is the prevention of cancer. Furthermore there is experimental and clinical evidence that hydrogenated fats tend to favor increases in the concentration of cholesterol in the blood plasma, and atheromatous lesions of blood vessels have been repeatedly produced in lower animals, for example, in rabbits, chickens, and rats, by the feeding of cholesterol. Finally, the evidence for the genesis of atherosclerosis in man, based on experiments on lower animals, is as good as is that for the genesis of cancer in man, also based on experiments in animals.

Although these remarks are admittedly overstatements, they may serve to emphasize the problem of atherosclerosis as a type of biological hazard possibly incident to the excessive presence of specific chemicals in food products. However, with specific reference to the actual problem of atherosclerosis, about all that can be reasonably said at present is that there is no general agreement concerning the dominant factors in its pathogenesis. It is generally agreed that several factors are concerned; genetic, hormonal, nutritional, toxic (tobacco), stressful, etc. In any event, no one thus far has been able to furnish convincing proof that either a high-cholesterol diet or a high fatty acid diet functions as a primary cause of atheromatous degeneration of arteries in man. Nevertheless, some of the most experienced students in the field of vascular disease recommend dietary moderation in the consumption of fatty foods, saturated fats, etc. Presumably their advice is being heeded to a considerable degree in view of the increasing popularity of low-calorie diets, anticoronary clubs, and increasing emphasis on unsaturated fats, and the growing apprehensions of dairy farmers and poultrymen over the mounting surpluses of milk, butter and eggs. Whether today's nutritionists are "cholesterol-happy" souls who delight in contributing to the excesses of dietary fear is a matter for the future to decide, but, to those who are accustomed to the sight of coronary arteries occluded by masses of cholesterolesters, calcium, fibrous tissue and a terminal thrombus, the pathological consequences are always impressive. Although the question whether cholesterol deposits are of exogenous or endogenous origin is still to be determined, there are at least hopeful intimations that these vascular accumulations may be to some degree controlled or even prevented by more informed attention to the significance of certain chemicals in so-called atherogenic food products.

In summary it is evident that in a half-century of concern over the problems of chemicals in food products, our scientists, administrators, and legislators have endeavored to protect and to improve our food supply and to better its nutritional potentiality. In these endeavors many of the obstacles have been due to a lack of adequate nutritional and chemical understanding and to an unawareness of the biological hazards incident to the production, processing and consumption of foods. In the last two decades the development of agricultural chemicals has made possible phenomenal increases in food production while also contributing materially to our agricultural and political problems of food surpluses. It is not surprising, therefore, that the increasing complexities of food technology and of nutritional science, coupled with the changing patterns of disease, should have necessitated the holding of symposia such as the one with which we are now concerned. In short, whereas in fifty years the emphases have changed from time to time, the problems of chemicals in food products have continued as before. In the words of the old French saying "the more things change, the more they are the same." And so, today, we still want pure foods; we still need food protection against the hazards of ignorance; we still need legislative aids to guard against error in the use of the array of chemicals available for agricultural use; we still need additional nutritional and technological research.

LITERATURE CITED

Dungal, Niels. 1961. The special problem of stomach cancer in Iceland. J. Am. Med. Assoc. 178: 789-98.

Food Protection Committee, Food and Nutrition Board, National Academy of Sciences — National Research Council. 1960. Problems in the evaluation of carcinogenic hazard from use of food additives. Publication 749.

Joint FAO/WHO Expert Committee on Food Additives. 1961. Fifth report: Evaluation of the carcinogenic hazards of food additives. WHO Technical Report Series No. 29.

PART TWO

Intentional Additives

3

Specific Nutrients

NEVIN S. SCRIMSHAW
MASSACHUSETTS INSTITUTE
OF TECHNOLOGY

THE ADDITION of specific nutrients to foods already has made a number of major contributions to the solution of nutrition problems in the world. Furthermore, as more and more of the essential nutrients become available in quantity and at low cost, the possibilities increase for making greater use of foods or food mixtures which, with simple supplementation, can satisfactorily replace more costly or less available foods.

The use of a specific nutrient as a food additive should depend upon evidence that it will, in fact, improve the food product to which it is to be added and that its use will not result in a total dietary intake of the nutrient in excess of established tolerances. The majority of specific nutrients which can be added to foods are generally recognized as safe for their intended use and are exempt from the requirement for a tolerance limit.

For foods sold in interstate commerce within the United States, the Food and Drug Administration has established a list of such substances — the so-called "GRAS" list (Code of Federal Regulations). The specific nutrients which appear on the current list are presented in Table 3.1. The use of any substance which appears on this list is still subject to the conditions that, (1) the quantity added does not exceed the amount reasonably required to accomplish its intended nutritional effect, and (2) that the substance is of appropriately good grade and is prepared and handled as a food ingredient.

If a substance does not appear on the GRAS list and no prior regulation has been issued, there are two alternatives. One is to petition the Food and Drug Administration for the issuance of a regulation prescribing conditions under which the additive may be safely used. Issuance of such a regulation, exemption from tolerance or denial of petition involves evaluation by qualified

Table 3.1. Nutrient Supplements Appearing on GRAS List

Amino Acids*	Vitamins	Minerals
Alanine	Vitamin A (alcohol, acetate, palmitate)	Ca (6 salts)
Arginine	Vitamin D_2, D_3	Cu (gluconate .005%)
Cysteine	Tocopherols; tocopherol acetate	Fe^{+++} - 3 salts
Cystine		
Histidine	Carotene	Fe^{++} - 3 salts
Isoleucine	Ascorbic acid	Fe - reduced
Leucine	Thiamine (HCl)	Mg^{++} - 10 salts
Lysine	Riboflavin	
Proline	Niacin; niacinamide	KCl
Serine	Pyridoxine	KI (0.01% in salt)
Threonine	Pantothenyl alcohol; Pantothenates	Zn^{++} - 5 salts
Tryptophan		
	B_{12}	
	Inositol	

Source: Code of Federal Regulations.

*May be L or DL form; free, HCl, hydrated or anhydrous.

experts of data regarding chemical nature, conditions of use, analytical methodology for quality control and safety. The other possibility is for a would-be user to employ similar criteria to satisfy himself that the product he intends to use is generally recognized as safe even though it does not appear on the GRAS list. In this case, however, he is subject to challenge by FDA authorities at any time, and this procedure is seldom used.

When preliminary information indicates no undue risk to public health, a temporary permit may be issued for use of additives under specified conditions and for limited periods of time during which investigations proceed to determine final action. It should be noted that when tolerances for specific nutrients are established, the amount which may be used in any given food is determined by the estimated total daily intake from all sources. Thus, as a matter of policy, the sum of additions permitted for all foods should not lead to an intake which will exceed the established tolerance.

Nutrients for which tolerances are currently in effect are listed in Table 3.2. This list includes not only specific nutrients, but also yeast, since yeast is sometimes added to foods as a

Table 3.2. Dietary Supplements: Restricted
Pending Regulation Issue

Product	For Use as a Diet Supplement
Boron	0.1 mg B/day
Cobalt	1.0 mg Co/day
Copper	2.0 mg Cu/day
Fluorine	0.5 mg F/day
Iodine	0.15 mg I/day
Molybdenum	2.0 mg Mo/day
Nickel	1.0 mg Ni/day
Iodine (from kelp)	0.7 mg I/day
Methionine	200 mg/day
Folic acid	0.4 mg/day
Menadione	1.0 mg/day
Torula yeast, dried	0.4 mg folic acid/day
Yeast, dried	0.4 mg folic acid/day

concentrated source of B-complex vitamins. Several other concentrated nutrient sources, including fish flour and leaf meal,
have been added to foods at least experimentally and will be discussed in a later section.

Having established that a specific nutrient may be used under
the conditions which have been described, a further consideration
must determine the manner in which it is used. For many foods,
standards of identity have been established which define in both
qualitative and quantitative terms the required and optional ingredients and additives. If a proposed additive is not included in
the standard of identity definition, the product to which it is added
must be labeled "imitation" or it is subject to the penalties of
regulations against misbranding. The only other alternative is to
secure by petition and submission of arguments to the Food and
Drug Administration an amended standard of identification which
will include the proposed additive. Although applications are required only for foods entering interstate commerce, the majority
of states and a growing number of communities have their own
regulations which are modeled more or less on the federal regulations. A summary of the types of foods for which federal
standards of identity have been established is given in Table 3.3.

ADDITION OF PURE NUTRIENTS TO FOODS

Minerals

Iodine. Endemic goiter occurs as a significant public health
problem in nearly all of the countries of the Americas, much of

Table 3.3. Foods for which Standards of Identity have been Established

Cocoa products (12)*	Frozen desserts (5)
Cereal flours and related products (32)	Dressings (3)
Alimentary pastes (10)	Canned fruits and juices (35)
Bakery products (5)	Fruit preserves and jellies (5)
Milk and cream (10)	Shell fish (12)
Non-fat dry milk solids	Fish (1)
Cheese; processed cheeses;	Oleomargarine
cheese foods; cheese spreads;	Canned vegetables (7)
related foods (66)	Tomato products (6)

*Number of products in class.

Africa, as well as parts of India and Southeast Asia (Kelly and Snedden, 1958). A number of carefully evaluated field trials have shown it to be effectively prevented by the addition of iodine to salt. This measure will not greatly influence the size of the large fibrosed and nodular goiter of adults, but it will rapidly reduce the size of the diffuse goiters of children and can eliminate goiter in a geographical area within a single generation (Matovinovic and Ramalingaswami, 1958).

The iodization of salt was initiated on a voluntary experimental scale in Michigan and northern Ohio in 1921. A rapid decrease in endemic goiter ensued although the increased use of food produced in non-goitrous areas undoubtedly also influenced the reduction. In 1951, 1.4 percent of the school children in Michigan had goiter compared with 38.6 percent in 1924 (Brush and Altland, 1952), and in Ohio the figures were 4 percent in 1954 compared with 32.3 percent in 1925 (Hamwi et al., 1955). The use of iodized salt beginning in the years 1919-1924 is also credited with the virtual disappearance of endemic goiter and even more important, endemic cretinism, from Switzerland (Uehlinger, 1958). More recently, Gongora and Mejia (1952) reported a drop in the prevalence of endemic goiter among school children in the state of Caldas, Colombia, from 83 percent to 34 percent in two years as the result of the introduction of salt iodized at a level of one part of iodine in 20,000 parts of salt with potassium iodide.

Fortunately, the salt available in this Colombian highland province was a relatively pure, dry mine salt and could be iodized with potassium iodide as in the United States. In tropical areas the salt consumed is frequently relatively crude, moist sea salt which rapidly loses the activity of any added iodide. To refine, dry and package such salt to protect it from moisture would not only increase its price, but also change its physical characteristics. Potassium iodate, unlike potassium iodide, is stable when added to crude, moist salt. Studies of the Institute of Nutrition of

Central America and Panama (INCAP) have shown that the effects of potassium iodate and potassium iodide are identical if the calculated amount of iodine supplied is the same (Scrimshaw et al., 1953).

In trials carried out in school children in El Salvador, INCAP workers showed that either form of iodine given to simulate that in iodized salt resulted in a decrease of endemic goiter of 40 percent in 15 weeks. In Guatemala, a reduction of about 65 percent occurred in 25 weeks. Initial endemic goiter levels were 34 percent in El Salvador and 5 percent in Guatemala (Scrimshaw et al., 1953). Despite evidence that potassium iodate is converted to potassium iodide in the intestinal tract and that it has essentially the same margin of safety as shown by extensive toxicity trials, the use of potassium iodate has not yet been authorized in the United States apparently because no one has petitioned for its use.

With the recommendation of the World Health Organization (WHO) and the Pan American Health Organization, potassium iodate is now being employed for the compulsory iodization of all salt for human consumption in Guatemala and Paraguay and a score of other countries are in the process of implementing legislation for this purpose (Food and Agriculture Organization, 1954; Scrimshaw and Ascoli, 1960).

Fluorine. Extensive publicity has been given to the fact that children living in areas in which the soil, and hence both food crops and potable water are rich in natural fluorides, have less dental caries than those in communities in which natural fluoride levels are low. A number of large-scale longitudinal experiments have proved conclusively that the addition of fluorine to the drinking water in those communities in which natural fluorides are relatively deficient reduces the prevalence of dental caries more or less by half (National Research Council, 1953). Gordonoff and Minder (1960) point out that over two thousand communities in the United States have now voluntarily added fluorine to their drinking water so that it contains approximately one part per million. In every case where controlled observations have been made, a substantial reduction in dental caries has been demonstrated. It is apparent that the more widespread application of this single measure would be of major benefit in improving the teeth of the inhabitants of communities where the soil and water are deficient in natural sources of fluorine and would greatly lower the cost of dental care in such communities.

The American Public Health Association, the American Dental Association, the American Medical Association, the U.S. Public Health Service, the World Health Organization, and other professional groups have all strongly recommended the fluoridation

of the water supplies in such communities. It is unfortunate that
this measure is being blocked in many communities by persons
not fully cognizant of the benefits or who have been misled as to
the possibility of harmful effects from this practice.

Phosphorus. Although the mechanism of action is quite different from that of fluorine, there is growing evidence from experimental studies in rats and hamsters that phosphorus added to
the diet in the form of sodium or potassium orthophosphate will
also greatly reduce the prevalence of dental caries (Harris, 1959;
McClure, 1959; Strålfors, 1956). Sodium or potassium trimeta-
phosphate appears to be even more effective (Nizel, Baker and
Harris, 1962). If the results of these studies apply equally to
man, one of these forms of phosphorus added to wheat flour
should give a striking decrease in the frequency of dental caries.

Since the effects of phosphorus and fluorine are apparently
additive, the result of supplying both in appropriate forms on
dental caries in children should be investigated.

Iron. At the last meeting of the FAO/WHO Joint Expert
Committee on Nutrition in Rome in 1957 (World Health Organization, 1958) and in the discussions of the WHO Study Group on Iron
Deficiency Anemia which met in Geneva in 1958 (World Health
Organization, 1961) it was emphasized that nutritional anemias
are much more prevalent and important as a public health problem than has heretofore been recognized. Anemias of the iron deficiency type seem to predominate. There are no contraindica-
tions to the addition of appropriate amounts of iron to wheat flour
and other cereals as part of an enrichment formula.

In technically underdeveloped areas, the addition of iron to
cereal products could benefit large numbers of individuals and is
required by law in wheat flour sold in the countries of Central
America. Even in a country where cereal enrichment does not
appear practical, the enrichment of some other key foodstuff with
iron might be possible. In India, the possible addition of iron to
salt has been under investigation.

On the other hand, the indiscriminate addition of iron or indeed any of the minerals listed in Table 3.2 could well lead to an
undesirable excess of them in the diet. In some cases we know
enough already to be certain that such excesses would be disad-
vantageous to the maintenance of optimum health. The establishment of tolerance limits provides for an orderly enrichment policy which benefits and safeguards the consumer.

Trace Minerals. There is too little information available at
the present time to justify recommending the specific addition of
any of the trace minerals. A number of clinical research studies
are now underway, however, which may alter this conclusion.

Vitamins

B-Complex Vitamins.   When R. R. Williams (1961) finally
succeeded in synthesizing thiamine in 1936, he hoped that he was
providing a tool which would end once and for all the scourge of
beriberi in the Orient. He anticipated that this vitamin could be
produced synthetically at very low cost and that it would be practical to add it to rice. Extensive field experiments carried out on
the Philippine Islands from 1949 to 1952 demonstrated that this
measure was indeed practical and effective, but politics, apathy
and economic considerations have prevented its general application for the prevention of beriberi.

The enrichment of wheat flour in the United States with thiamine, riboflavin and niacin was authorized in 1941, although adequate supplies of the latter two vitamins did not become available
until 1943. By the end of that year, four states had passed legislation requiring the enrichment of wheat flour with the foregoing
vitamins. Similar laws have now been enacted by 28 of the states
and Puerto Rico, and wheat flour is enriched voluntarily in those
states in which flour enrichment is not mandatory.

The introduction of enriched wheat flour into the southern
United States coincided with the beginning of a drastic drop in the
occurrence of pellagra, as well as other deficiencies of the B-
complex vitamins in that area, although it is difficult to say how
large a role the enrichment of flour played since this was also a
time of rapid economic and social improvement.

The currently approved formula for the enrichment of foods
made from wheat flour is given in Table 3.4. By law, enrichment
of wheat flour is also required in Canada, and through the efforts
of INCAP enrichment is mandatory in Guatemala, El Salvador,
Honduras, Nicaragua, Costa Rica and Panama. Since white bread
is tending to replace whole-grain cereal products in Latin America and other technically underdeveloped areas, enrichment would
seem to be a wise measure and it is strongly recommended as a
matter of policy by both the Pan American Health Organization
and WHO.

For some populations the use of a similar formula for the enrichment of rice appears to be of even greater public health importance than the enrichment of wheat. The levels of thiamine
intake in persons consuming a polished rice diet are lower than
when the diet is based on any other common cereal. For Thailand, Burma, Vietnam, China, and parts of India where clinical
beriberi still occurs, the addition of thiamine to rice is of great
importance. Beriberi has also been reported at least sporadi-
cally from countries in Africa, Latin America and the Middle
East.

Table 3.4. Food and Drug Administration Definitions
and Standards of Identity for Enriched Flour[a]

Additive	Minimum mg/lb	Maximum mg/lb
Enriched Flour		
Thiamine	2.0	2.5
Riboflavin	1.2	1.5
Niacin	16.0	20.0
Iron	13.0	16.5
Calcium[b]	500	625
Vitamin D[c]	250[d]	1000[d]
Enriched Bread		
Thiamine	1.1	1.8
Riboflavin	0.7	1.6
Niacin	10.0	15.0
Iron	8.0	12.5
Calcium[b]	300	800
Vitamin D[c]	150[d]	750[d]

[a] Federal Register 8 (July 3, 1943), 9115-9116 and (August 3, 1943), 10780-10788.
[b] Required in self-rising flours with maximum 1500 mg/lb; optional in other types of flour.
[c] Optional.
[d] Expressed as USP units.

Enriched rice is now on sale in the United States and Canada and is the only kind of polished rice which can be sold in Puerto Rico. Although the measure has been discussed extensively in other countries and has been strongly recommended by both the Pan American Health Organization and WHO, its adoption has not yet become widespread.

Because of the association of niacin deficiency and clinical pellagra with predominantly corn diets, the addition of B-complex vitamins to corn has also been advocated. In the case of corn, niacin is likely to be the most important nutrient in the enrichment formula. Studies carried out by Darby (1956) and co-workers in Yugoslavia have demonstrated that the enrichment of corn flour can be effective in preventing the occurrence of pellagra in a population where it is otherwise common. The amounts of B-complex vitamins which can be added to other cereal products are similar to those shown in Table 3.2.

Calcium pantothenate is accepted for use interchangeably with calcium carbonate, but this is for its calcium content rather than the vitamin portion of the molecule.

The case of pyridoxine is a special one. Although there is no evidence that ordinary diets are deficient in pyridoxine, in late 1951 a new method of processing one of the well known infant formulas resulted in an excessive and unrecognized destruction of pyridoxine. As a result, a number of infants to whom this formula was fed as the sole source of food developed clinical manifestations of pyridoxine deficiency before the cause was discovered (Coursin, 1954; Molony and Parmelee, 1954). In this instance the processing procedure was implicated and ultimately modified so as to lessen the destruction of pyridoxine even though the addition of synthetic pyridoxine would also have been an acceptable means of ensuring that this would not occur again.

Vitamin A. Since vitamin A is a fat-soluble vitamin, it has been feasible to add it to fats and oils. The development of an inexpensive, water-miscible form of vitamin A has made practical the supplementation of non-fat foods with this nutrient.

According to Food and Drug Administration regulations, vitamin A may be added to blue cheese, gorgonzola cheese and bleached milk in such amounts as to compensate for the loss of vitamin A activity during the bleaching process. By far the greatest amount of vitamin A used as a food additive goes into margarine where it constitutes an optional ingredient at not less than 15,000 USP units per pound. Some vitamin A is also used in infant dietary formulas, as well as in "imitation" ice cream employing vegetable fat and in skim milk.

The addition of vitamin A to skim milk powder is now strongly recommended by WHO and FAO whenever skim milk is to be consumed by populations whose vitamin A intakes are known to be inadequate. Advocation of this measure is based on the fact that when protein is the limiting nutrient in a deficient diet and is supplied as skim milk, acute clinical signs of vitamin A deficiency are precipitated if the diet is already marginally deficient in this vitamin (Oomen, 1958).

Vitamin E. At present, vitamin E is used in food mixtures as an antioxidant rather than as a specific nutrient. Food and Drug Administration regulations also permit its use as a dietary supplement although currently there is no evidence that it would benefit any normal human population. Recent observations of Horowitt (1962), however, have indicated that persons fed a diet very high in unsaturated fatty acids may develop increased red cell fragility if the diet does not supply adequate vitamin E or other antioxidant.

Vitamin D. A common practice is to irradiate milk in order to increase vitamin D activity. Whole milk, evaporated milk, skim milk and margarine may contain vitamin D as an optional ingredient and be labeled accordingly in a concentration of not less than 25 USP units per fluid ounce. Both calciferol (vitamin D_2) and activated 7-dehydrocholesterol (vitamin D_3) are used to enrich the various types of flour at a level of not less than 250 nor more than 1,000 USP units per pound. Farina can be enriched with not less than 250 USP units of vitamin D per pound and enriched bread and rolls at not less than 150 nor more than 750 USP units of vitamin D per pound.

Ascorbic Acid. Ascorbic acid is added to such foods as canned mushrooms, frozen fruit, frozen fish dip, soft drinks, milk and even candy primarily for its antioxidant properties. As a nutrient, it is most appropriately added to processed fruit juices and purees in order to bring them up to at least the ascorbic acid level of fresh citrus fruit juices. People do rely on fruit juices to supply the needed ascorbic acid in their diets even though there are very wide species variations in the natural ascorbic acid content of such juices. Because ascorbic acid is inexpensive and free of toxicity, there is no objection to its use in any reasonable quantity.

Folic Acid and Vitamin B_{12}. Pernicious anemia is a relatively rare disease and diagnosis depends upon the development of a characteristic type of anemia. The most serious consequences of the disease process are a slow and irreversible myelin degeneration of the posterior and pyramidal tracts of the spinal cord and other neurological changes involving the peripheral nerves, plexus of the gastrointestinal tract and the subcortical areas in the motor region of the brain. Folic acid has no beneficial effect on the neurological lesions and in fact may aggravate them. Fortunately, the normal levels of folic acid required are well below those which induce a hematological response in pernicious anemia. For the reasons cited, therefore, the folic acid content of foods or multiple vitamin preparations is now legally limited to 400 mcg per day.

While no such problem exists in the case of vitamin B_{12}, the animal products in an ordinary mixed diet provide adequate quantities of this vitamin and there is no need for its addition to foods. A possible exception may be for all-vegetable protein mixtures or supplements for populations consuming only very small amounts of animal protein and also in foods for persons who consume, by choice, a completely vegetarian diet. Since vitamin B_{12} deficiency takes many years to be manifest even among strict vegetarians, it is very difficult to determine whether added B_{12} is justified.

Amino Acids

Studies in experimental animals have demonstrated clearly that the amino acid deficiencies of vegetable proteins or of incomplete protein from animal sources, such as gelatin, can be corrected by the addition of the limiting amino acid. In 1955 the FAO Committee on Protein Requirements (Food and Agriculture Organization, 1957) took into account the studies of amino acid requirements in experimental animals and man and designated an amino acid pattern which seemed at the time optimal for a theoretical reference protein of maximum nutritive value. The proportions of amino acids assigned to this pattern closely resembled those of milk, meat, eggs and other animal proteins known to be of high biological value.

Neither tryptophan alone nor lysine alone have more than a transient effect on the retention of nitrogen from a corn diet. When both are added together, however, in the proportions indicated by the amino acid pattern of the FAO reference protein, a sustained improvement in nitrogen retention is observed in both experimental animals and man. INCAP data on balance studies in children have shown that results obtained with amino acid supplementation closely approach those obtained with isoproteic diets in which the protein was milk (Bressani et al., 1958; Scrimshaw et al., 1958). Similar observations have been made in adults by Truswell and Brock (1961) in South Africa.

Elvejhem (1956), Elvejhem and Harper (1955) and co-workers at the University of Wisconsin (Salmon, 1958; Sauberlich, 1956) have demonstrated and developed the concept of amino acid imbalance and have shown that the addition of certain of the essential amino acids in excess can result in an adverse effect on nitrogen retention. In INCAP studies in children, the addition of methionine to wheat protein at the level called for by the FAO reference protein consistently decreased nitrogen balance even in the presence of adequate lysine and tryptophan instead of bringing about the anticipated improvement (Bressani et al., 1960). This suggested that under these circumstances, at least, the allowance for methionine in the FAO reference pattern was too high and also that methionine addition per se can lower as well as raise protein efficiency depending upon the basic amino acid pattern.

Corn protein is characterized by a considerable excess of leucine in relation to isoleucine although the latter is already present in a greater proportion per gram of nitrogen than called for by the reference pattern. The addition of isoleucine corrects the isoleucine-leucine imbalance and when added in the presence of adequate lysine and tryptophan results in a slight further improvement in protein quality.

When similar studies were carried out with wheat in experimental animals (Bressani and Mertz, 1958; Grau, 1948; Harris and Burress, 1959), in children (Bressani et al., 1960), or adults (Clark et al., 1957) the addition of lysine alone in the proportion called for by the amino acid pattern of the FAO reference protein resulted in a marked increase in nitrogen retention. In children fed adequate protein intakes, the net retention of nitrogen approximated that for an isonitrogenous diet in which the source of the protein is milk. Methionine addition, at the same level as that used in the corn experiments, exhibited neither a beneficial nor adverse effect in this case.

Today in the United States the addition of lysine to cereals and vegetable mixtures manufactured as special "high" protein foods is permitted. However, the routine addition of lysine to ordinary wheat flour is not authorized on the grounds that the improvement involved would not be of practical significance in the American diet.

ADDITION OF NUTRIENT CONCENTRATES TO FOOD

Nutrients can also be effectively and economically added to foods in the form of naturally concentrated sources. A common and effective way of adding B-complex vitamins to various foods is the use of yeast. Both food yeast (Torula or Candida utilis) or debittered Brewers yeast (Saccharomyces cerevisiae) contain substantial amounts of thiamine, riboflavin and niacin, as well as other B-complex vitamins and protein in an acceptable form. The use of yeast as a food additive has been questioned recently because of the possibility that its folic acid content might be sufficient to interfere with the diagnosis of pernicious anemia by correcting the blood abnormality without arresting the irreversible nerve degeneration previously mentioned.

Although the total folic acid content of these yeasts is around 10 mcg per gram of dried yeast when the method described by the Association of Official Agricultural Chemists (1955) is used, there is evidence (Binkely et al., 1944) that a very high proportion of the folic acid in yeast as measured by this procedure is bound in the inactive folic acid conjugate form. The latest method of the Association of Official Agricultural Chemists (1960) determines the free or active folic acid in yeast and other natural products. With this method the active folic acid content of these yeasts ranges from 1 to 3 mcg per gram. Yeast causes no gastrointestinal disturbances when consumed in relatively small amounts, but in larger amounts tends to produce loose stools and flatulence. If the more appropriate 1960 AOAC method for free folic acid is used, ingestion of from 150 to 450 grams of yeast would be required to reach the maximum safe allowance of 400 mg. It is inconceivable that this amount of yeast would be either recommended or consumed by any significant number of persons.

In low-cost, protein-rich vegetable mixtures prepared for the prevention of protein malnutrition in technically underdeveloped areas, yeast has proved a convenient means of adding B-complex vitamins. In tropical and semitropical countries this is a feasible and practical measure because theoretically, at least, the necessary torula yeast can be produced from the excess molasses, and the Brewers yeast which is a by-product of beer production can be debittered.

Acerola juice is a naturally concentrated source of vitamin C. Lueng (1961) found that one species (Malpighia punicifolia) of this tropical fruit has nearly 2,000 mg of ascorbic acid per 100 grams. The juice is now produced commercially in Puerto Rico and shipped to the United States for use as a natural source of ascorbic acid in fruit juice and baby food formulas.

African palm oil may contain as much as 14,000 mcg of vitamin A activity per 100 grams (Lueng, 1961). Important quantities of vitamin A are contributed to the diets of persons in some tropical countries by African palm oil, and it is being recommended for addition to foods in Indonesia for the prevention of the xerophthalmia and keratomalacia due to vitamin A deficiency — a condition common in this area.

Another natural source of vitamin A activity often added to mixed feeds for animals and sometimes proposed as a vitamin A source for vegetable mixtures for human consumption is leaf meal prepared from forages such as alfalfa, kikuyu grass, ramie and pangola. Although the fresh meal may have vitamin A activity in the range of 2,000 mcg per 100 mg, it is too unstable to be a dependable source and imparts a green color not generally appreciated in foods for human consumption.

Bressani (1961) and Bressani and Scrimshaw (1961) have shown 3 percent of fish flour or whole egg protein, 4 percent of meat flour, 5 percent of casein, torula yeast or "Dracket" soybean protein, 7 percent of dried skim milk and 8 percent of cottonseed or soya flour are equally effective in increasing the protein efficiency of lime-treated corn flour for rats. In addition to the improvement in net protein quality, each component makes a small contribution to the total protein content. Since up to an average of 80 percent of the calories and 70 percent of the protein may come from corn tortillas alone in some Indian communities in Guatemala and Mexico, this type of enrichment may be of considerable nutritional significance. Of course, the addition of milk solids to wheat flour is now a common practice. The addition of fish flour to bread has been found acceptable in trials in French West Africa and Chile.

The preparation of concentrates from natural sources for use as food additives must, of course, result in a "food grade" product. In the use of leaf meal supplements as vitamin A sources, the material must be clean and the residues of pesticides in the final product must not exceed established tolerance levels for such compounds in foods. The use of fish flour in the United States is limited at the present time to that manufactured from fish fillets on the basis that the use of whole fish would include substances which can be regarded under the law as "filthy, putrid or decomposed."

POLICY CONSIDERATIONS

It is quite obvious that the indiscriminate addition of specific nutrients to foods is contrary to the public interest. As already mentioned, excesses of some nutrients, such as vitamins A and D, can be harmful; an amino acid addition, such as methionine, may be detrimental to protein quality as well as beneficial to its improvement; the use of folic acid in foods and multiple vitamin preparations is contraindicated. Furthermore, the addition of specific nutrients for promotional reasons, rather than nutritional, would raise the cost of food unnecessarily. Since each food manufacturer is interested in increasing the appeal of his product by enrichment, if it is commercially advantageous and if he is permitted to do so, regulation is required.

The least controversial basis for adding specific nutrients to food is the restoration of those present naturally, but destroyed or lost as a result of processing. This is one of the bases on which the addition of thiamine, riboflavin and niacin to wheat flour and rice was originally authorized since these nutrients are lost in the milling of wheat or the polishing of rice. An equally important justification, however, was the frequency of reports of vitamin deficiencies from both rural and urban clinics, particularly in the southern United States and the slums of the large cities. Evidence was available to show that inadequate intakes of thiamine, riboflavin and niacin were common among low-income population groups and that such inadequacies were traceable in substantial part to the predominance of refined cereals in the diets. The addition of ascorbic acid to fruit juices and fruit drinks has received little criticism since people have been conditioned to depend upon these as natural sources of vitamin C.

As early as March, 1939, the Council on Foods and Nutrition of the American Medical Association went on record as favoring the restorative addition of vitamins to foods and opposing indiscriminant fortification. However, they approved the following fortifications: 1) the addition of 400 units of vitamin D to milk, 2) the addition of vitamin A to substitutes for butter, 3) the addition of iodine to table salt and 4) the addition of calcium and iron salts to cereal products.

An argument in favor of addition of such nutrients is that persons known to be receiving substantial amounts of the above nutrients from other sources still benefit by the additions. For example, persons consuming butter were not likely to be deficient in vitamin A. It was argued that persons consuming an otherwise acceptable substitute should not be deprived unnecessarily of this essential nutrient. Because the soils of some areas contain useful quantities of iodine, fluorine or both, their populations have been found free of endemic goiter and with relatively fewer dental caries. The addition of iodine to salt and fluorine to water in areas where these minerals are deficient is a logical public health measure although it is more a matter of equalization than of true restoration. On the other hand, the addition of either iodine or fluorine to many different foods could lead to excessive intakes and harmful consequences.

Thus, there are good medical reasons for adding nutrients to food and for limiting the amounts to be added. Restrictions on the blanket addition of nutrients to any food, even though no physiological harm would result, are reasonable to protect the public interest. It would be a wholly unjustified financial burden on the consumer to add vitamins to all foods that do not happen to be a good source of them. Although the routine addition of vitamins to candy has often been suggested, this would be doubly undesirable since it would be a useless expense and would interfere with educational efforts to persuade persons to eat a balanced diet. B-complex vitamins should be available in candy form for special dietary and therapeutic purposes, but this is using the candy form to make the therapeutic administration of vitamins palatable. It is quite different from using vitamin fortification for the primary purpose of selling candy. On the other hand, it certainly is appropriate to allow essential nutrients, which may be in short supply, to be added to special formulas for infant feeding whenever these are intended as the principal or sole food for the child.

More difficult is the question of whether or not B-complex

vitamins should be added to the ordinary cow's milk routinely on sale in grocery stores or supermarkets or handled by home delivery. Here, it would seem to be against the interests of the consumer to raise unnecessarily the basic cost of this important food even though there may be justification for the addition of vitamins to special milk intended for therapeutic purposes.

The addition of vitamins A and D to both whole milk and skim milk would also appear justified. The vitamin D is significant particularly in northern temperate areas where children may not be exposed to sufficient sunlight during certain seasons of the year to ensure enough conversion of ergosterol in the skin to vitamin D. The assurance of an adequate amount of vitamin D along with added calcium and protein for skeletal growth which milk supplies may well benefit a large number of children.

Frequently, people depend upon whole milk as an important source of dietary vitamin A and when skim milk is consumed instead for reasons of economy, calorie reduction or possible lowering of serum cholesterol, the only important nutritional disadvantage is a loss of vitamin A. It is important that vitamin A either be restored to the milk or made up in the diet in some other way. Within the United States other dietary sources usually suffice. When U.S. skim milk is shipped abroad through welfare agencies, however, it is frequently destined for children who are already showing borderline deficiency signs of vitamin A. The sudden supplying of protein increases their vitamin A requirements, and if additional vitamin A is not supplied, frank deficiency may ensue. Several years ago a serious increase in xerophthalmia and keratomalacia in Indonesian children was traced to this cause.

The simultaneous administration of vitamin A capsules along with skim milk is clumsy and relatively costly. Fortunately, effective and economical methods now exist for producing a stable, water-miscible form of vitamin A which can be readily added to skim milk. WHO is strongly urging that skim milk sent to technically underdeveloped areas be fortified with vitamin A if it is deficient in the diets of the children for whom the milk is destined.

A similar argument can be made for the addition of lysine to wheat flour intended for shipment to technically underdeveloped areas. Persons in such areas are usually consuming predominantly cereal diets which are deficient in net protein value. In other words, the diets contain protein of relatively low biological value, and even if a large quantity were consumed, the intake would undoubtedly be insufficient for optimum growth or recovery from trauma and infection. It should not be considered even ethical to promote the distribution of surplus foods as a nutritional measure in technically underdeveloped areas unless some nutritional benefit will occur. Under the circumstances presently prevailing in many technically underdeveloped areas, the improvement of the protein quality of wheat flour with lysine and the standard vitamins and minerals is indicated.

On the domestic scene the amino acid enrichment of cereal and other foods has been more controversial. Although the addition of a small amount of lysine to white wheat flour will substantially improve its biological value, the amount of protein per 100 grams is still relatively small compared with most foods of animal origin. The improvement in quality with lysine addition becomes significant only when white bread and other flour-based products, such as spaghetti and macaroni, are consumed in substantial quantities by persons whose diets contain relatively little animal protein. There is no certain data as to the frequency with which this occurs in the United States. However, it is presumed that the routine addition of lysine to wheat flour might benefit some individuals in the older age groups whose economic resources are very limited or who do not take the trouble to prepare complete meals for themselves, as well as those children and teenagers who consume excessive amounts of bread and insufficient milk and other sources of animal protein and for other persons consuming certain types of bizarre diets. A cross-sectional, nationwide survey with emphasis on the amount and type of protein in the diets of individuals would be required to determine accurately how much of a contribution the routine enrichment of wheat flour with lysine might make. Obviously, neither average per capita nor family intake estimates provide an answer to the question. It seems probable, however, that the addition of lysine only to high protein breads and cereals, as is the present practice, fails to benefit those persons who would most gain from the additional utilizable protein because such products sell at a premium and, hence, are avoided by persons having limited funds to spend on food.

The question of enriching cereals, other than wheat cereals, with those amino acids necessary to improve protein quality has not arisen for the purely practical reason of excessive cost. This cost is being lowered and would probably drop more rapidly if a market were ensured. It is likely, therefore, that the question of supplementing other cereals and vegetable protein sources, such as legumes, will ultimately arise.

Any attempt to improve the protein quality of all foods by amino acid addition would be economically impractical as would be an effort to ensure that all foods contained a balanced complement of the other essential nutrients. The optimum nutrition of man is based on the concept of a variety of foods in the diet which are nutritionally synergistic. A policy similar to that followed for the addition of vitamins to food would seem indicated; that is, allowing amino acid supplementation only where good reason for doing so can be demonstrated. Obviously, agreement in principle is easier than application.

On an international scale, the role of synthetic amino acids is likely to become increasingly important in the development of low-cost, protein-rich food mixtures intended to extend existing food supplies as population pressures increase. Use of vegetable mixtures which would otherwise be too poor in protein quality for the successful prevention of protein malnutrition will be possible. On a national scale, synthetic amino acids are likely to aid in the solution of special problems of emergency and space feeding where storage space and weight required must be reduced to a minimum.

The addition of fish flour is another means of improving the protein value of a food. In the United States the limitation on the use of fish flour for human consumption is based on the conclusion that even though cooking has rendered it safe, allowing fecal material in fish meal is unesthetic to the American people, and that allowing sterilized feces in fish flour would make it more difficult to enforce regulations to keep such contaminants out of other food products. Therefore, only the use of fish flour prepared from fillets is condoned. In technically underdeveloped areas, the need for additional high quality protein sources at the lowest possible cost outweighs these considerations. Both WHO and FAO advise countries that there is no objection to the use of flour prepared from whole fish providing it is processed in such a way as to retain relatively high protein quality and to be free from viable pathogenic bacteria or toxic substances.

The FAO International Conference on Fish in Nutrition (Food and Agriculture Organization, 1961) held in Washington, D.C. in September last year reinforced a previous recommendation of an FAO Study Group held in Rome in 1960. It was proposed that a series of three standards of identity for fish flour be established on the basis of the amount of residual fat and, hence, on the degree of deodorization and removal of the fish flavor. The three products would be deemed equally acceptable for human consumption from a nutritional and health point of view. It was felt that quite apart from the question as to whether or not the use of whole fish should be permitted, a single standard of identity requiring the flour to be deodorized and decolorized would be undesirable for international use. The reason is that it would restrict fish flour for human consumption to a more expensive type of product, whereas on a world-wide scale a variety of products could be effectively utilized depending upon the degree to which the population would accept flour with the odor of fish and an unesthetic color.

SUMMARY

The following excerpts from the Statement of General Policy in regard to The Addition of Specific Nutrients to Foods adopted jointly by the Food and Nutrition Board of the National Research Council and the Council on Foods and Nutrition of the American Medical Association in May, 1961 will serve as a useful summary.

1. The principle of the addition of specific nutrients to certain foods is endorsed, with defined limitations, for the purpose of maintaining good nutrition in all segments of the population at all economic levels. The requirements which should be met for the addition of a particular nutrient to a given food include (a) acceptable evidence that the supplemented food would be physiologically or economically advantageous for a significant segment of the consumer population, (b) assurance that the food item concerned would be an effective vehicle of distribution for the nutrient to be added, and (c) evidence that such additions would not be prejudicial to the achievement of a diet good in other respects.

2. The desirability of meeting nutritional needs by the use of an adequate variety of foods as far as practicable is emphasized strongly. To that end, research and education are encouraged to insure the proper choice and preparation of foods and to improve food production, processing, storage, and distribution so as to retain their essential nutrients.

3. Foods suitable as vehicles for the distribution of additional nutrients are those which have a diminished nutritive content as a result of loss in refining or other processing or those which are widely and regularly consumed. The nutrients added to such foods should be the kinds and quantities associated with the class of foods involved. The addition of other than normally-occurring levels of nutrients to these foods may be favored when properly qualified judgment indicates that the addition will be advantageous to public health and when other methods for effecting the desired purpose appear to be less feasible.

4. Scientific evaluation of the desirability of restoring an essential nutrient or nutrients to the diet is necessary whenever technologic or economic changes lead to a nutritionally-significant reduction in the intake of a nutrient or nutrients. Such reduction might result either from a marked decrease in the consumption of an important food or from a considerable increase in the consumption of foods of diminished nutritive quality.

Similar evaluation is desirable, with the limitations defined in section (1) above, whenever advances in nutritional science and in food technology make possible the preparation of nutrient-enriched products which are likely to make important contributions to good nutrition.

5. The endorsement of the following is affirmed: the enrichment of

flour, bread, degerminated corn meal, corn grits, whole grain corn meal and white rice; the retention or restoration of thiamine, riboflavin, niacin, and iron in processed food cereals; the addition of vitamin D to milk, fluid skim milk, and nonfat dry milk, the addition of vitamin A to margarine and to fluid skim milk and nonfat dry milk; and the addition of iodine to table salt. The protective action of fluoride against dental caries is recognized and the standardized addition of fluoride to water is endorsed in areas in which the water supply is low in fluoride.

6. The above statements of policy and of endorsement apply to conditions existing in the United States. Recommendations for additions of nutrients to foods for export should be based on similar physiological or economic advantages expected to accrue to the respective consumers.

To these points a seventh should be added in any international consideration of the problem. The justification and value of adding specific nutrients to foods is sometimes quite different in technically underdeveloped areas. For example, the addition of vitamin A to skim milk or amino acids to cereals may make the difference between adequacy and frank clinical deficiency of vitamin A or protein, respectively. Moreover, standards for concentrated nutrients, such as fish flour and leaf meal need to be adjusted to local conditions of use and acceptability. Food enrichment or fortification with specific nutrients may be more important in technically underdeveloped areas. Thus, special regulations for adding specific nutrients to foods should be adapted to the actual needs of a given country.

LITERATURE CITED

Association of Official Agricultural Chemists. 1955. Official Methods of Analysis. 8th Ed., Washington, D.C.

_____. 1960. Official Methods of Analysis. 9th Ed., Washington, D.C.

Binkely, S. B., Bird, O. D., Bloom, E. S., Brown, R. A., Calkins, D. G., Campbell, C. J., Emmett, A. D., and Pfiffner, J. J. 1944. On the vitamin B_c conjugate in yeast. Science 100: 36-7.

Bressani, R. 1961. Enrichment of lime-treated corn flour with deodorized fish flour. Presented at the FAO International Conference on Fish in Nutrition, Washington, D.C. September 19-27.

_____, and Mertz, E. T. 1958. Relationship of protein level to the minimum lysine requirement of the rat. J. Nutrition 65:481-91.

_____, and Scrimshaw, N. S. 1961. The development of INCAP vegetable mixtures. I. Basic animal studies. In:

Meeting Protein Needs of Infants and Children, Publication 843. Nat. Acad. Sci.-Nat. Res. Council, Washington, D.C.

_____, _____, Behar, M., and Viteri, F. 1958. Supplementation of cereal proteins with amino acids. II. Effect of amino acid supplementation of corn masa at intermediate levels of protein intake on the nitrogen retention of young children. J. Nutrition 66:501-13.

_____, Wilson, D. L., Behar, M., and Scrimshaw, N. S. 1960. Supplementation of cereal proteins. III. Effect of amino acid supplementation of wheat flour as measured by nitrogen retention of young children. J. Nutrition 70:176-86.

Brush, B. E. and Altland, J. K. 1952. Goiter prevention with iodized salt: results of a thirty-year study. J. Clin. Endocrin. 12:1380-88.

Clark, H. E., Mertz, E. T., Kwong, E. H., Howe, J. M., and De-Long, D. C. 1957. Amino acid requirements of men and women. 1. Lysine. J. Nutrition 62:71-82.

Code of Federal Regulations, Title 21. Federal Food, Drug and Cosmetic Act. Part 121 (Food Additives) (Added), Subpart B. Exemption of certain food additives from the requirement of tolerances, Section 121.101, Substances that are generally recognized as safe.

Coursin, D. B. 1954. Convulsive seizures in infants with pyridóxine-deficient diet. J. Am. Med. Assoc. 154:406-8.

Darby, W. J. 1956. Report and recommendations concerning the programmes of control of pellagra and endemic goiter in Yugoslavia. WHO Regional Office for Europe, Eur-Yugoslavia -16.1, July.

Elvejhem, C. A. 1956. The effect of amino acid imbalance on maintenance and growth. In: W. H. Cole, [ed.]. Some Aspects of Amino Acid Supplementation. Rutgers Univ. Press, New Brunswick, N. J.

_____, and Harper, A. E. 1955. Importance of amino acid balance in nutrition. J. Am. Med. Assoc. 158:655.

Food and Agriculture Organization. 1954. Report of the Third Conference on Nutrition Problems in Latin America, Caracas, Venezuela, 19-28 October 1953. FAO Nutrition Meetings Report Series No. 8, Rome.

_____. 1957. Protein Requirements. FAO Nutritional Studies No. 16, Rome.

_____. 1961. Proceedings of the International Conference on Fish in Nutrition. Washington, D.C., Sept. 19-27.

Gongora y Lopez, J. and Mejia, C. F. 1952. Two years of the treatment of goitre with iodized salt in the Caldas Department. Medicina y Cirugia, Bogota 16:357-71.

Gordonoff, T. and Minder, W. 1960. Fluorine. In: G. H. Bourne, [ed.] . World Review of Nutrition and Dietetics, Vol. 2. Hafner Publishing Co. New York.

Grau, C. R. 1948. Effect of protein level on the lysine requirement of the chick. J. Nutrition 36:99-108.

Hamwi, G. J., Van Fossen, A. W., Whetstone, R. E., and Williams, I. 1955. Endemic goiter in Ohio school children. Am. J. Public Health 45:1344-48.

Harris, R. S. 1959. Effects of food ash, phosphate and trace minerals upon hamster caries. J. Dental Res. 38:1142-47.

_____, and Burress, D. A. 1959. Effect of level of protein feeding upon nutritional value of lysine-fortified bread. J. Nutrition 67:549-67.

Horwitt, M. 1962. Discussion of Fourth Session on "Recent Advances in the Estimation of Relative Nutrient Intake and Nutritional Status by Biochemical Methods." Proceedings of Symposium on Recent Advances in the Appraisal of the Nutrient Intake and the Nutritional Status of Man. Massachusetts Institute of Technology, March 6-7.

Kelly, F. C. and Snedden, W. W. 1958. Prevalence and geographical distribution of endemic goitre. Bull. World Health Org. 18:5-173.

Lueng, W. T. W. 1961. Food Composition Table for use in Latin America. Interdepartmental Committee on Nutrition for National Defense - Institute of Nutrition of Central America and Panama.

McClure, F. J. 1959. Further observations on the cariostatic effect of phosphates. J. Dental Res. 38:776-81.

Matovinovic, J. and Ramalingaswami, V. 1958. Therapy and prophylaxis of endemic goitre. Bull. World Health Org. 18:233-53.

Molony, C. J. and Parmelee, A. H. 1954. Convulsions in young infants as result of pyridoxine (vitamin B_6) deficiency. J. Am. Med. Assoc. 154:405-6.

National Research Council. 1953. The problem of providing optimum fluoride intake for prevention of dental caries. A Report of the Food and Nutrition Board, Publication 294, Nat. Res. Council. Washington, D. C.

Nizel, A. E., Baker, N. J., and Harris, R. S. 1962. The effect of phosphate structure upon dental caries development in rats. In: Abstracts of the Fortieth General Meeting, March 15-18, Internat. Assoc. Dental Res.

Oomen, H. A. P. C. 1958. Personal communication.

Salmon, W. D. 1958. The significance of amino acid imbalance in nutrition. Am. J. Clin. Nutrition 6:487-94.

Sauberlich, H. E. 1956. Amino acid imbalance as related to methionine, isoleucine, threonine and tryptophan requirement of the rat or mouse. J. Nutrition 59:353-70.

Scrimshaw, N. S. and Ascoli, W. 1960. Endemic goiter in Latin America. Public Health Reports 75:731-37.

_____, Bressani, R., Behar, M., and Viteri, F. 1958. Supplementation of cereal proteins with amino acids. I. Effect of amino acid supplementation of corn masa at high levels of protein intake on the nitrogen retention of children. J. Nutrition 66:485-99.

_____, Cabezas, A., Castillo, F., and Mendez, J. 1953. Effect of potassium iodate on endemic goitre and protein-bound iodine levels in school children. Lancet 265:166-69.

Strålfors, A. 1956. Karieshämning genom fosfater. Svensk tandläk. Tidskr. 49:108.

Truswell, A. S. and Brock, J. F. 1961. Effects of amino acid supplements on the nutritive value of maize protein for human adults. Am. J. Clin. Nutrition 9:715-28.

Uehlinger, E. A. 1958. The influence of iodine on Swiss goiter. In: T. D. Kinney and R. H. Follis, Jr., [ed.]. Proceedings of a Conference on Beriberi, Endemic Goiter and Hypovitaminosis A. Federation Proc. 17 (Supple. 2, Pt. II):66.

Williams, R. R. 1961. Toward the Conquest of Beriberi. Harvard Univ. Press. Cambridge, Mass.

World Health Organization. 1958. Joint FAO/WHO Expert Committee on Nutrition. Fifth Report. Tech. Report Series No. 149.

_____. 1961. Study Group on Iron Deficiency Anemia. Geneva. Sept./Oct., 1958. In: The Medical Research Programme of WHO. 1958-1961. Report by Director-General. Geneva. Sept. 15.

4

Flavors and Colors

LOREN B. SJÖSTRÖM
CHARLES J. KENSLER
ARTHUR D. LITTLE, INCORPORATED

THERE IS A TENDENCY to think of using flavors and colors for enhancing food products as a fairly recent innovation of civilized man. And yet, in rummaging through history evidence appears that these food additives have been used since antiquity.

After the late neolithic period, shortly before the rise of the first urban civilizations, about 3500 B.C., there is evidence that loaves of bread were baked with added spices and condiments. Poppy seeds and mustard seeds were known to have been used at this time.

The use of salt, too, dates back to pre-classical antiquity, when salt was considered a leading economic necessity as reflected in the salt routes which formed a network over Europe and the Mediterranean basin to the near east into Asia. It was obtained from working rock salt deposits and by evaporating sea water and water from saline springs for the preserving of fish and seasoning of food. Fish were salted or pickled in brine as early as 2300-1200 B.C. There is record of salt being marketed in lumps or bricks in ancient Egypt during the 6th Dynasty — 2200 B.C. Salt for everyday use was obtained from evaporated river or sea water and usually was impure. Desert salt, the purest, was used in service to the gods, "for offerings had always to be seasoned with salt" (Singer, Holmyard, and Hall, 1954).

From the recent book by Chadwick (1958), which dates the use of flavoring materials at about 1400 B.C., the following is quoted: "This bread and porridge could be enlivened by spices; coriander is the most frequent, but a list from Mycenae includes also celery, cumin, cyperus, fennel, mint, pennyroyal, safflower (both flower and seeds) and sesame."

Later in Egypt, around 1200 B.C., condiments such as onions, garlic, and the like were commonly in use.

Other records of Greek civilization tell us that pepper and anise were used to season breads and cakes. Spices such as cumin, coriander, sesame, and silphium were either grown or imported, as were flavoring herbs like fennel, marigold, and mint. Although most spices were imported, cultivation of indigenous aromatic or savory herbs as thyme, garlic, and water cress was recorded in 600-300 B.C. Similarly, garlic, salt, and garden herbs were used in Rome.

Much less is known about the ancient use of coloring agents. However, just before the birth of Christ, Pliny the Elder wrote of at least one use in his commentaries on Greece and Rome. In describing the manufacture of groats, he stated that during the final stage of refinement, "an admixture of chalk is added, which passes into the substance of the grain and contributes colour and fineness . . . " (Pliny, trans. by Loeb, 1950).

IMPORTANCE OF FLAVOR AND COLOR

Such evidence of the use of flavoring and coloring materials can be traced through history from ancient man to modern man, with the degree of use of these materials increasing with each passing century. It has been within very recent years, however, that the use of flavors and colors has become exceedingly important — in fact, so important as to be considered almost a necessity rather than a luxury. This has been a direct result of several changing patterns in our way of living. Probably most important is that we have become predominantly an urban rather than a rural society. This, coupled with a many-fold increase in total numbers of people, has created the necessity of drastic changes in the manner of providing food. The multitude of city dwellers demands at least as many types of food as were known to their rural parents. Geneticists, agriculturists, and many others have found ways to increase the yield of crops. Home economists, food technologists, and engineers have found an increased number of ways to preserve the foods grown — picking foods at the proper stage for shipping, improving methods of drying, pasteurization, pickling, canning and packaging, developing the techniques of refrigerating and freezing. Other ways were discovered to reduce spoilage — the use of antioxidants, antibacterial agents, yeast and mold inhibitors, and other agents. Economists and businessmen evolved complicated systems of transportation, storage, and marketing the food products in order that they could be available to the urban people. As a result, food products are subjected to treatment never before experienced. The day of offering yesterday's kill or this morning's freshly picked fruits and vegetables for sale or exchange has become an anomaly. And few treatments to which our modern food products are exposed — except perhaps by rare, lucky chance — produce an enhancement of their intrinsic characteristics of flavor, color, and appeal.

Emphasis has been placed on getting enough food products to the point of need. But in the changing pattern of society, a lagging but influential desire has colored the picture. The consumer wants more than just food. He wants his food to resemble as closely as possible the freshly plucked or slaughtered product. He wants it to look good, taste good, and be good. He even would like to be tempted with new foods, new taste experiences, and also with convenient foods. And with his increased buying power, he can afford to demand these things.

All of these shifts in need, these alterations in patterns of food gathering, these human desires have created a great number of problems for food research and food technology. It has not been sufficient to know simply that the modern manner of handling food products affects their properties of flavor and color — and adversely for the most part. It has not been enough to be aware of extant desires. It has been essential to develop means of restoring or preserving the natural flavors and colors of altered food products. It has been essential to learn ways of creating new foods. To accomplish these tasks, the researcher has depended heavily on the use of food additives. He has found new ways to use old materials about which man has known for centuries, but he has also synthesized new varieties of matter.

PROBLEMS IN THE PERCEPTION
OF COLORS AND FLAVORS

In view of the enactment of the Food Additives Amendment of 1959, and considering recent trends in design and composition of foods, research workers must emphasize increased consciousness of legal and scientific criteria for safe use of materials, but also of scientific knowledge required for more effective use. The problems associated with optimum use of color and flavor are typical. Because color is perceived by a single sensory organ, and because its intensity and composition can be measured by instruments, one might assume that it can be dealt with as a routine matter. Actually this is not true; the choice of a proper color, like the choice of flavor, requires a thorough understanding of consumer attitude, psychological reaction, and conditioned response. Control of subtle differences in tone and balance may be required in both color and flavor. The language of color is highly developed, and for any particular color such as green, there are many variants which have specific associations and meanings. For example, these may suggest such diverse items as green apples, bile, crême de menthe, algae, pistachio, or ascomycetes. A red color may have many connotations, from blood to rose to wine. Colors may be added to food to produce a definite red or green color, or they may be added simply to modify a natural creamlike color. Although color is a measurable property and is adjustable by simple additions, there are a number of natural products which require more attention to quality control for color than for any other property.

Optimum color and flavor are extremely important in consumer acceptance. It has even been suggested that they may be interlinked. In any event, it would be useless to try to persuade the consumer to buy gray raspberries because they have optimum flavor, or tasteless raspberries because of optimum color. We believe that color and flavor are two of the most important independent variables influencing acceptance. While color is straightforward from the standpoint of chemistry and measurement, it is complex from the standpoint of psychology and acceptance. Flavor, on the other hand, is complex from the viewpoint of chemistry and measurement and from the standpoint of psychology and acceptance. Both can exist in an almost infinite number of variations, but the number in the case of flavor is multiplied by a large factor because at least three different sensory mechanisms may be involved in its perception, whereas color is perceived by only one organ.

Flavor is perceived when a food or beverage is taken into the oral cavity. It is perceived as a total sensation composed of the reactions of the taste and olfactory receptors to chemical stimuli, and in many cases the reactions of the tactile, temperature, and pain receptors as well. In other words, the flavor of a food or beverage is the over-all impression gained through taste, odor, texture, and a group of feeling factors.

COLOR AS A FOOD ADDITIVE

As an index of the need for added color, it is estimated that more than 2,000,000 pounds of synthetic Certified Colors are used annually by the food, drug, and beverage industries, in

Table 4.1. Current Color Additives Provisionally Listed for Food,
Drug and Cosmetic Use

FD and C Green No. 1
FD and C Green No. 2
FD and C Green No. 3
FD and C Yellow No. 5
FD and C Yellow No. 6
FD and C Red No. 2
FD and C Red No. 3
FD and C Red No. 4
FD and C Blue No. 1
FD and C Blue No. 2
FD and C Violet No. 1

Color Additives Provisionally Listed for Food Use
Which Have Not Been Subject to Certification

Alkanet
Annatto
Beet juice, powder
Bixin, norbixin
Calcium carbonate
Caramel
Carbon black (Channel or impingement process)
Carmine
Carminic acid
Carotene (Natural and synthetic)
Carrot oil
Charcoal (N.F. XI)
Chlorophyll
Chlorophyll copper complex and Chlorophyllin copper complex
Cochineal
Ferric chloride, Ferrous gluconate, Ferrous sulfate
  (in processing black olives)
Iron oxides
Paprika and paprika oleoresin
Riboflavin
Titanium dioxide (0.4 percent in bakery and confection)
Turmeric-curcumin
Ultramarine blue (0.5 percent in salt for animal feed)
Xanthophyll

addition to a still larger volume of colors of natural origin or derived from natural products such as annato, carmine, caramel,
carotene, chlorophyllin, tumeric, and xanthophyll.

At the time of the passage of the 1938 Food, Drug and Cosmetic Act, about seventeen approved synthetic colors were certifiable for use. Two more were added later. This list was compiled in part from previous history of safe use in food products
and limited additional toxicological and pharmacological data. It
was not until 1950, when a candy company made a quantity of

Halloween candy adding enough FD and C Orange No. 1 to match
the color of pumpkins (Zuckerman, 1962), that a real issue was
raised as to the inadequacy of the law. A number of children who
ate this candy had severe gastrointestinal upsets. This incident
pointed up the absurdity of the 1938 Act under which the Food and
Drug Administration (FDA) practically was required to certify
that this color was edible, and was not authorized to establish any
upper limit for its concentration in foods. At about this time the
FDA, recognizing the lack of adequate toxicological information
on most of the certifiable colors as well as their inability to regulate the amount to be used, initiated long-term animal tests to
gain more adequate information. In 1955 the committee appointed
by the National Research Council to make recommendations on
the FDA certified coal-tar color program considered that only
one of the FD and C colors had been adequately tested and proven
reasonably safe for use by modern toxicological standards under
all probable conditions of use. This was tartrazine (FD and C
Yellow No. 5). It was estimated by this group that if the certified
color-testing program (which included drug and cosmetic colors
as well) were to be continued at the rate then in operation in the
FDA's laboratories, it would probably take 25 years to accumulate the necessary experimental information. The colors for food
use, however, had priority, and it is of interest that since that
time eight (see Table 4.2) of the nineteen certifiable dyes have
been removed from the list for various reasons. FD and C Red
No. 1 was found to be a hepatotoxic agent, and FD and C Yellows
Nos. 3 and 4 were reported to contain small amounts of beta
naphthylamine, a known bladder carcinogen, and were reported
(Harrow and Jones, 1954) to break down in acid media yielding
beta naphthylamine. At present, of the remaining eleven certifiable food colors, adequate six-year feeding experiments have indicated the safety of nine, and studies in progress on the other
two have not indicated any hazard.

It should be noted that color additives now fall under the
Color Additive Amendments of 1960. It includes authorization for
establishment of conditions and tolerances for safe use. For the

Table 4.2. Delisted FD and C Colors

FD and C Red No. 1	FD and C Yellow No. 3
FD and C Red No. 32	FD and C Yellow No. 4
FD and C Yellow No. 1	FD and C Orange No. 1
FD and C Yellow No. 2	FD and C Orange No. 2
Carbon Black (lamp process)	

first time all coloring agents are included — any dye, pigment, or
other substance however derived. Among the natural coloring
agents, only one form of carbon black (channel or impingement
process) has been accepted because of the content of polynuclear
aromatic hydrocarbons in some forms which have been found
carcinogenic for many species. Wood charcoal may be used as a
pigment.

Although color manufacturers are making every effort to supply the demand for food colors, the present restricted list is seriously handicapped by the lack of oil-soluble colors. An effort is
being made to supply these needs by development of insoluble
lake pigments of approved colors, but some difficulties are being
encountered in their use.

FLAVOR MATERIALS AS FOOD ADDITIVES

The list of natural and synthetic flavoring materials used in
foods exceeds the color list by many hundreds. For example, Dr.
Richard L. Hall, Chairman of the Food Additives Committee of
the Flavoring Extract Manufacturers' Association (FEMA), reported that the original list developed by this committee through
their industry questionnaires provided data on some 1400 materials. This data submitted to FDA for their guidance has been
helpful in clarifying the legal status of most flavor additives.
While there is some overlapping of various lists, Dr. Hall summarized the status as follows:

Materials covered by FDA white lists	191
Materials on extension	343
Materials on FEMA generally recognized as safe (GRAS) lists	662
Materials on extension which may be dropped for lack of supporting data or suspected toxicity	100 to 200
	1,396

It is possible that FEMA may find it necessary to drop or cease
to support approximately 120 materials on the basis of low volume of use and lack of data. There is a good possibility that
about 1200 out of 1400 potential flavor materials and adjuncts will
survive this critical screening procedure.

In the case of safrole, a constituent of oil of sassafras, which
was a prominent ingredient of root beer until 1960 — this material
was voluntarily removed from root beer by the manufacturers
when they were informed that it was toxic and possibly a

carcinogenic material. Later that year a panel of technical experts was convened to consider the current status of safrole, and all experimental data collected to date was made available for study. This panel was not asked whether in its opinion safrole would represent a hazard to the consumer, but only whether the evidence indicated that this material was or was not a carcinogen. After examining all available evidence, the panel summarized its response as follows:

This panel has been asked to evaluate pharmacological data on safrole from the standpoint of whether or not it is to be classed as a substance which induces cancer and to report its findings to the Commissioner of Food and Drugs.

. . . .

Study of these documents reveals that:

1. Safrole is an hepatotoxic agent in rats and dogs.

2. In the single life-span experiment which has been completed, a statistically significant number of malignant liver tumors has been found in the group of rats (male and female) fed the highest dose level (5000 ppm) of safrole. When the data for each sex were examined, only the female rats showed a statistically significant incidence. Groups of rats fed lower levels (1000 ppm and below) did not contain an incidence of malignant tumors significantly different from that found in the control group.

At the 1000 ppm level, female rats contained a significantly higher incidence of benign hepatomas than did the female controls. Lower dose level female groups were not significantly different from controls, and none of the male groups had a significantly elevated incidence of benign hepatomas.

Chronic experiments on dogs are in progress in the Food and Drug Administration laboratories, and also in the Food and Drug Research Laboratories. Both of these experiments have been in progress for 2 to 3 years. The dose levels being studied are 20, 10, and 5 mg/kg/day. These animals are alive and in apparent good health at the present time.

3. In the long-term rat-feeding experiment, the total tumor incidence of all types was not significantly different at any dose level of safrole from the incidence in the control group. The increased incidence of liver tumors in the 5000 ppm group was balanced in numbers by a reduction in mammary tumor incidence; this reduction may be due to the nonspecific factors of inanition and increased mortality in this group.

4. A decision at this time as to whether safrole should be classed as a substance which produces cancer thus rests on the one long-term production of malignant liver tumors in female rats. This experiment was carefully and competently executed and the opinion of the panel is that safrole is a weak hepatic carcinogen; to confirm the scientific validity, further experimentation is suggested.

In these experiments, it should be pointed out that a significant incidence of liver tumors was produced only at the highest level of safrole fed; a level which was barely tolerated and markedly decreased survival time. This type of exposure is hard to envisage in a population having access to other food choices. The probability is that exposure to small amounts of safrole would not represent any hazard, although if the committee had been asked the question, it may have answered in the affirmative. In view of the Delaney Clause, which bans the addition of known carcinogens to foods, and the committee opinion that the available evidence indicates that safrole is a weak hepatic carcinogen for the rat, this material was put on the banned list and the directive published in the Federal Register. It is of interest to note the various food flavorings, such as nutmeg and star anise, which contain safrole are still approved for use, and perhaps constitute official recognition of insignificant levels of a potential carcinogen in terms of its threat to the health and well-being of a consumer.

In the last few months, the Department of Health, Education, and Welfare, acting with the advice of a scientific panel of experts which had been asked whether or not quinine should be allowed for use in foods, specifically beverages, received a favorable decision from the panel, and established a tolerance of 83 parts per million as free base in beverages. In this case, there was no evidence of possible carcinogenic activity, but it is known that quinine in high doses can produce harmful effects, a situation which is true for most chemicals, e.g., sodium chloride. When the committee considering medical data was asked about safety under conditions of specified use, it was possible to set a tolerance level for quinine. Both this flavoring agent and safrole had been employed for generations. It is to be hoped that in the not too distant future, it will be possible to bring to bear the same type of scientific judgment and reasonableness in the consideration of certain types of weak carcinogens as hazards to the consumer as was possible in the case of quinine. This approach is obviously impossible under present legislation.

FUTURE PROSPECTS

The short-term outlook for development of new flavors and colors appears to favor conservatism and a search for products of natural origin. On the other hand, the general effect of the larger investment required to qualify food additives may lead to an upgrading of research efforts and more careful screening of candidates for marketing. This may result in a net benefit to the food industry. New approaches to the evaluation of safety of raw materials and additives must be developed as we progress from simple processed foods to more complex convenience products, and finally to foods based on isolated, purified and modified nutrients. In view of projected population increases, and the anticipated need for utilization of new sources of raw materials for food, it is easy to visualize that we are approaching the time when foods must be constructed from nutrients procured from materials heretofore considered to be nonfood. Further, the formulation of palatable products from these sources will require a substantial increase in the skill of development of appealing color, flavor and texture. Since these new developments will obviously be in the public interest, a means must be found to conduct proper biological evaluations to assure the absence of consumer hazard and the preservation of nutritional values. It is perhaps not unrealistic to consider that some public source of funds — perhaps on a matching basis — can be made available for the conduct of toxicological testing to be sure that the public will be protected.

LITERATURE CITED

Chadwick, J. 1958. The Decipherement of Linear B. Random House, New York.

Harrow, L. S. and Jones, J. H. 1954. The decomposition of azocolors in acid solution. J. Assoc. Offic. Agr. Chemists 37: 1012-20.

Pliny. 1950. Natural History, SVIII vxix 112-16. Loeb Translation, Vol. 5, pp. 260 ff.

Singer, C., Holmyard, E. J., and Hall, A. R. 1954. A History of Technology, Vol. I. From Early Times to Fall of Ancient Empires. Oxford Univ. Press.

Zuckerman, S. 1962. Color additive amendments to the Food, Drug, and Cosmetic Act and its meaning to the cosmetic industry. Presented at meeting of the Society of Cosmetic Chemists, New England Chapter, April 26.

5

Antioxidants

KARL F. MATTIL

SWIFT AND COMPANY

IN THE RESEARCH PROGRAM on food fat products at Swift and Company, two basic objectives are recognized. The first of these objectives is to produce the best possible shortenings, margarines, and salad oils by the most efficient and economical processes. It is obvious that this objective must constantly be pursued by any industry that wants its products to remain competitive. It necessitates the evaluation and re-evaluation of the relative merits of available processes and processing equipment. It also requires the constant development of new and improved products for every purpose. It means that new products must be tailor-made to fit specific requirements and specific conditions, such as found in the production of cake mixes, pie mixes, bread, etc.

The second basic objective is to provide a maximum flexibility in the selection of raw materials to be used in any product. This is of tremendous importance to the total economic structure of the edible fat industry because of the seasonal variations in supply that occur from time to time. It is obvious that any product which absolutely requires a specific raw material ingredient suffers a serious disadvantage when that raw material is in short supply or nonexistent. On the other hand, this flexibility helps make it possible to use raw materials that otherwise might become a serious surplus problem. To attain this objective of maximum flexibility, it had to be assumed that the inherent deficiencies of each of the raw materials could be overcome to such a degree that they could then be used successfully in the finest quality products. Twenty years ago this looked like a huge task. As of today, a significant portion of this objective has been accomplished. Antioxidants have played an important role in attaining the objective.

NEED FOR ANTIOXIDANTS

A half century ago lard was the preferred shortening. Although by present standards it was very unstable, it had sufficient stability for the demands that were then put upon it. Hydrogenated, deodorized, vegetable oil shortenings were introduced in 1911 and set a new standard. At first consumers considered these new products as simply lard substitutes, but their popularity grew rapidly until they largely displaced lard in many of its uses. The reasons for preferring the hydrogenated deodorized vegetable shortenings are several:

1. Although in some uses the natural flavor of lard is still preferred (such as, for example, bread), in other uses the flavor is incompatible. Therefore, the flavorless quality of the shortenings is preferred.
2. The natural crystal structure of lard is such that it is the shortening of preference for flaky, tender pastries, but not satisfactory for certain types of products where creaming is necessary (a deficiency which now has been adequately corrected).
3. Vegetable oils contain substantial amounts of natural antioxidants, principally the tocopherols, so that the hydrogenated, deodorized, vegetable shortenings have a higher degree of stability than had ever been known prior to their development. Thus high stability frying fats, biscuit and cracker shortenings and other shortenings for use in products requiring long shelf life became available. Lard, on the other hand, contains little or no natural antioxidant. After lard has been deodorized to remove its natural flavor, it has practically no shelf stability at all.

It was recognized early, that if we were going to retain our markets for lard, we must learn how to stabilize it. Our laboratory began its search for suitable antioxidants for meat fats about 35 years ago. It was recognized that the ideal antioxidant for fatty food products should possess the following characteristics:

1. Exhibit effective inhibitory action.
2. Be easily soluble in fats.
3. Impart no foreign flavor, odor, or color even on long continued storage.
4. Exhibit no harmful physiological effects.
5. Undergo no changes when heated.
6. Possess the ability to retard rancidity in baked goods prepared from the fat treated with it.
7. Be available in quantity and be economical.

Later, from some unhappy experiences, it was learned that another criterion should be added; namely, that the antioxidant not produce colors in the presence of moisture and iron.

In 1933, Drs. Newton and Grettie of Swift and Company reported that gum guaiac was a suitable antioxidant for meat fats. Extensive tests were carried out by Professor A. J. Carlson at the University of Chicago to establish that this material was entirely innocuous physiologically. A request was submitted to the Meat Inspection Division of the Bureau of Animal Industry for approval of the use of gum guaiac in lard and, in 1940, it became the first approved antioxidant.

During the course of their investigation, Drs. Newton and Grettie reported the observation that there was a relationship between the ability of an antioxidant to carry its effectiveness into baked goods and the ratio of the solubility of the antioxidant in fat to its solubility in water (Richardson, Grettie and Newton, 1936). They discovered that substituted polyphenols and polyphenol derivatives which were soluble in oils and fats, and relatively insoluble in water, were effective not only in stabilizing the oils and fats as such, but also in retarding the oxidation of a fat after it had been used as shortening in bakery products. Phenols which were relatively soluble in water were believed to be extracted from the fat when the shortening was mixed with other ingredients containing moisture. This was especially true if the other ingredients were alkaline in reaction. Subsequent investigations in our laboratory with a number of different antioxidants (Mattil and Black, 1947) have verified this hypothesis and led to the development of a rather simple partition test which can be used to predict the potential ability of an antioxidant to stabilize fat in baked goods.

CURRENT USE OF ANTIOXIDANTS AND SYNERGISTS

Since 1940, a number of other antioxidants and synergists have been approved by the federal agencies for use in meat fat products. The currently approved list is shown in Table 5.1, along with the maximum permitted levels of use. This list has been winnowed from literally dozens, perhaps hundreds, of compounds that have varying degrees of antioxidant potency. Some years ago this group of approved antioxidants would have seemed adequate. However, experience with them has shown that no single compound on the list, nor any combination thereof, fully satisfies the original specifications referred to previously.

Table 5.1. Antioxidants Approved for Use in Meat Fats

	Maximum Permitted (percent)
1. Resin guaiac	0.1
2. Nordihydroguaiaretic acid (NDGA)	0.01
3. Tocopherols	0.03
4. Lecithin	-
5. Butylated hydroxyanisole (BHA)	0.01
6. Butylated hydroxytoluene (BHT)	0.01
7. Propyl gallate (PG)	0.01
Synergists	
1. Citric acid	0.01
2. Monoisopropyl citrate	0.01
3. Phosphoric acid	0.01
4. Glycine	0.01

Antioxidants

Resin guaiac has long since gone into disuse. Many people never did learn how to use it successfully, although in many respects it was a good antioxidant. However, it had two serious handicaps: first, it was not always available when needed; and, second, it had a tendency to produce various types of colors that had to be considered definitely uncomplimentary in shortenings. NDGA has never been used widely, largely due to cost. The primary antioxidants that have found the greatest favor are BHA, BHT, and propyl gallate. Each of these has certain specific characteristics in its favor, and they are often used in combination in order to acquire a summation of their favorable attributes. When a combination is used, the total permissible level is 0.02 percent. However, they also have negative attributes that prevent them from being wholly satisfactory. BHA and BHT have strong phenolic odors and flavors which are readily apparent at the approved level of use in shortenings. Propyl gallate, like resin guaiac, has a propensity for producing unwanted blues and greens in the presence of very small amounts (0.1 percent or less) of moisture and iron.

Table 5.2 shows the relative potencies of the approved antioxidants as measured by Moore and Bickford (1952). These data were obtained by measuring the number of hours required to attain a peroxide value of 100(m.eq. per kg.) during aeration at 97.7°C. with an air flow of 2.33 ml./sec. This is a standard procedure for measuring the relative stability of fats, called the active oxygen method (AOM). From a scientific point of view, it is

Table 5.2. Relative Effects of Antioxidants in Lard[*]

	Hours to attain a peroxide value of 100		
	Level of antioxidant added		
Antioxidant	0.01%	0.05%	0.10%
None (control)	4	4	4
α-Tocopherol	17	11	5
γ-Tocopherol	19	18	11
Lecithin	5	6	7
NDGA	50	42	35
Resin guaiac	3	9	12
Propyl gallate	44	90	88
BHA	19	20	21
BHT	23	50	68

[*] Adapted from Moore and Bickford (1952).

particularly interesting to note the variable effects of increased
levels of added antioxidants. Such data serve to confound those
who have attempted to suggest a mechanism of antioxidant action.

One mechanism proposed was first suggested by Christiansen
(1924). According to his so-called chain reaction theory, a fat in
the process of oxidative rancidity is able to unite with oxygen to
form a peroxide only after it is activated by the absorption of energy. After union has occurred and peroxide formation is completed, the activating energy may be released and made available
for the activation of a new molecule, to form a new peroxide. A
chain of reactions is thus set up, and the initial absorption of a
single unit of energy will result in the formation of a great number of peroxide molecules, unless the chain is broken by absorption of the activating energy in an extraneous reaction.

The antioxidants are presumed to be substances which are
capable of absorbing the activating energy and thus preventing it
from being transmitted to further molecules of the oil, or in
other words, which are capable of breaking the chain. The antioxidant may be expected to be oxidized in the process. This has
been demonstrated in a number of experiments where the rate of
peroxide formation was relatively slow during early stages of
oxidation, while the added antioxidant gradually disappeared;
after the complete disappearance of the antioxidant, the rate of
peroxide formation rapidly increased. However, anomalies such
as seen in Table 5.2 cause some doubt that the pattern of antioxidant action is as simple as would appear from the above. On the
basis of such a simple explanation, one would expect the protection to be in direct proportion to the concentration of added antioxidant, but quite obviously, such is not the case. Nor have these
anomalies been adequately explained by the more recent mechanisms which have been proposed (Uri, 1961).

Table 5.3. Relative Effects of Antioxidants in Cottonseed Oil
and Hydrogenated Cottonseed Oil *

Antioxidant	Cottonseed Oil 0.01%	0.05%	Hydrogenated Cottonseed Oil 0.01%	0.05%
None (control)	9	9	121	121
α-Tocopherol	8	7	107	102
γ-Tocopherol	9	8	135	161
Lecithin	11	11	172	186
NDGA	10	19	120	258
Resin guaiac	7	9	100	124
Propyl gallate	16	55	172	495
BHA	7	7	108	158
BHT	9	15	118	172

Header: Hours to attain a peroxide value of 100

* Adapted from Moore and Bickford (1952).

The use of antioxidants has been limited largely to meat fats, since they have shown relatively little effect in vegetable oil products. As stated previously, the latter contain natural antioxidants. At the approved levels, added antioxidants effect little or no increase in the resistance of the vegetable oil products to peroxide formation (Table 5.3).

On the basis of the data shown in Tables 5.2 and 5.3, one could easily conclude that propyl gallate should be the antioxidant of choice for all uses. This is definitely not the case. BHA and BHT are both much more commonly used, very often in combination. This is due to several reasons:

1. Neither BHA nor BHT produces unwanted colors under any known circumstances.
2. BHA and BHT, being relatively more soluble in fat than in water, effectively stabilize the fat in baked goods, while propyl gallate, being more water soluble, does not.
3. When properly used in deodorized shortenings, BHA and BHT provide essentially as good actual shelf life as does propyl gallate, even though stability by the AOM is less.

This raises a rather important point. The AOM method was initially developed for use in natural fats and was intended to estimate the relative stability of the fats to oxidative rancidity. The end point of the determination was set somewhat arbitrarily at peroxide values ranging between 20 and 100, depending upon the fat substrate. These points were meant to coincide roughly with the point at which the fat or oil tasted or smelled rancid. As a practical matter, deodorized fats are not used to the point where they become rancid by the classical definitions. They develop other flavors and odors that make them unacceptable for use long before they become rancid (i.e., peroxide value 20-100). Generally speaking, the odor and flavor are unacceptable by the time the peroxide value has risen to 4 or 5. Consequently, the AOM, while being useful as a guide, is not a reliable index of the actual use stability of a deodorized fat.

Synergists

It was observed many years ago that certain acidic materials enhanced the apparent activity of the primary antioxidants. These same compounds were also observed to effect considerable improvement in the flavor stability of deodorized vegetable oil products. However, in the absence of primary antioxidants they were ineffective. Consequently, they were referred to as synergists. The accumulation of research evidence over the past 20 years now leads to the rather generally accepted conclusion that these compounds act not as co-stabilizers, but rather as metal scavengers. Metals such as iron and copper salts are strong pro-oxidants. The acids, evidently through chelating effects, remove the pro-oxidants and consequently prepare a more favorable climate for the primary antioxidants. The acidic chelating agents, particularly citric acid, are almost universally used in deodorized fats and oils, whether of vegetable or animal origin.

BENEFITS FROM ANTIOXIDANTS

Customers who use lard as such can do so with confidence that their products will have reasonable shelf life. The antiquated and inefficient method of handling lard in tierces has been largely replaced by tank truck or tank car shipping and bulk handling with their attendant economies. Such handling without antioxidants would be precarious if not impossible. The shortening manufacturers can now use lard as an ingredient, and the growth of this use dramatically demonstrates the need for antioxidants (Figure 5.1). Finally, the consumer is protected, in that the industry can provide less expensive fat-containing foods with far greater shelf stability than possible before antioxidants were permitted. In our modern complicated manufacturing, distributing and retailing channels, this added protection means real dollar and product savings to all parties concerned.

USE OF LARD IN SHORTENINGS (1935-1960)

Fig. 5.1.

MORE PROGRESS NEEDED

There is still need for better antioxidants that will meet all the specifications listed previously. The search for improved antioxidants is being continued in certain university and government laboratories. Unfortunately, progress on new antioxidants by antioxidant manufacturers seems to have slowed down.

It should be noted that antioxidant activity is not found only in polyphenolic compounds. Both sulfur and nitrogen bearing compounds have shown activity. The industry came close to adopting thiodipropionic acid and its esters quite a number of years ago, but could never satisfactorily eliminate the characteristic flavor of these compounds. More recently, amino-hexose-reductones have been reported to be strong inhibitors of peroxide formation (Evans et al., 1958).

A second important area where progress is needed, and in this case the research work has all been done, is in the expanded use of antioxidants when and where they will do the most good. Specifically, the entire status of the meat fat, and particularly the lard, industry would be improved if all meat fats were stabilized during or immediately following rendering. This certainly represents the most effective and efficient basis for stabilizing lard regardless of its ultimate use.

For example, if all lard were stabilized at the rendering plant with BHA and/or BHT, it would immediately have adequate stability for distribution in the normal channels. Instead, most lard is now shipped from the rendering plant unstabilized, to be stabilized at some subsequent point prior to shipment to the final consumer. During the initial handling prior to stabilization it is common for some increase to occur in the peroxide value. Even though this lard may then be stabilized by the addition of antioxidant, it can never again be as good nor as stable as if it had been properly treated at the point of rendering. Further, meat fats that have developed some peroxides are more difficult and expensive to use in shortenings, and produce unsatisfactory results. They are harder to bleach, harder to hydrogenate, and cause reduced stability in the finished shortenings. Contrary to the often heard opinion of some, meat fats that have undergone serious deterioration just cannot be upgraded by the use of antioxidants. Such deterioration is not reversible, and antioxidants can only preserve the status of the fat as it is at the time they are used. Consequently, shortening manufacturers definitely prefer to use lard with a minimum peroxide value. This can best be assured by stabilizing at the rendering plant.

Some lard is stabilized when purchased by a specific customer who is willing to pay the extra cost. However, no provision can be made for paying the premium on stabilized lard handled through the Board of Trade. Further, considerable lard goes to export markets that do not permit antioxidants. These are trade problems that should be resolved, because everyone will ultimately benefit when all lard is stabilized at the rendering plants.

LITERATURE CITED

Christian, J. A. 1924. Note on negative catalysis. J. Phys. Chem., 28:145-48.

Evans, C. D., Moser, H. A., Cooney, P. M., and Hodge, J. E. 1958. Amino-hexose-reductones as antioxidants. I. Vegetable oils. J. Am. Oil Chem. Soc., 35:84-88.

Mattil, K. F. and Black, H. C. 1947. A laboratory method for determining the ability of antioxidants to stabilize fat in baked goods. J. Am. Oil Chem. Soc., 24:325-27.

Moore, R. N. and Bickford, W. G. 1952. A comparative evalua-
     tion of several antioxidants in edible fats. J. Am. Oil Chem.
     Soc., 29:1-4.
Newton, R. C. and Grettie, D. P. 1933. U. S. Pat. 1,903,126.
Richardson, W. D., Grettie, D. P., and Newton, R. C. 1936. U. S.
     Pat. 2,031,069.
Uri, N. 1961. Mechanism of Antioxidation. In W. O. Lundberg
     (ed.) Autoxidation and Antioxidants. Interscience, New York.

6

Intentional Additives:
Commentary and Discussion

A. J. LEHMAN
FOOD AND DRUG ADMINISTRATION

COMMENTARY

D R. SCRIMSHAW'S excellent presentation on specific nutrients leaves little for comment. It should be emphasized that unless it has been demonstrated that a nutrient offers a positive effect on the public health, its addition cannot be justified. The guide lines for justification and the endorsement of the addition of specific vitamins and minerals to certain processed foods in the United States are given in the authoritative statement of general policy in regard to The Addition of Specific Nutrients to Foods adopted jointly by the Food and Nutrition Board, National Academy of Sciences-National Research Council and the Council on Foods and Nutrition of the American Medical Association.

An important principle to consider in proposing an addition is the suitability of the food as a vehicle for the essential nutrient. The food must be acceptable to the population for which it is intended. There is also need for the establishment of upper limits for the additive. This is based to a degree upon considerations of safety; however, this aspect is of minor importance since most nutrients possess such a large margin of safety. A limitation based on known human nutritional requirements is important. The addition of quantities beyond those which would be expected to meet nutritional needs introduces a factor of economic importance to the consumer which is generated by the false promise of superior benefits.

Dr. Mattil discussed quite fully the need for, and the efficacy of the presently approved antioxidants. As with all food additives, the basic question is one of safe use. These antioxidants are all phenolic compounds. In general, 10,000 parts per million in the diets of experimental animals begin to show toxic effects. The no-effect level is 5000 ppm. A 100-fold margin of safety would allow 50 ppm daily in the total human diet for a lifetime. It has been estimated (see page 101 of the brochure entitled "Appraisal of the Safety of Chemicals in Foods, Drugs and Cosmetics," published by The Association of Food and Drug Officials of the United States, 1959) that if all of the fats normally consumed by the human were preserved with an antioxidant at presently approved levels, the added phenolic intake would be approximately 30 milligrams daily, or of the order of 20 ppm in the total daily diet. However, these antioxidants are also permissible in other foods such as potato flakes and granules, dry breakfast cereals, dry mixes for beverages, active dry yeast, dry mixes for desserts, and sweet potato flakes in concentrations ranging from 2 to 1000 ppm. The consumption of these food items contributes approximately 18 ppm of the antioxidants to the total daily diet. The sum total daily intake would be 38 ppm, which still allows 12 ppm for other uses.

It has been said that "one eats with his eyes." An uncolored raspberry gelatin dessert probably would not impart the same gustatory sense as one colored red. There is no need to dwell on the suitability of colors for use in foods. All colors, those derived from vegetable sources and inorganic pigments as well as coal-tar colors, are regulated under the Color Additive Amendments of 1960. As with food additives, a color must be shown to be safe by adequate pharmacological testing. Specifications, methods of manufacture, and needed methods of analysis both for the color itself and for determining it in foods, drugs, and cosmetics must be submitted. The Color Additive Amendments do not become fully effective until January 12, 1963. Congress allowed this transitional or provisional period during which any color used in food must be submitted to sufficient pharmacological investigation to prove that it is safe for its intended use. In the meantime, food manufacturers may continue to use colors that have been certified as safe unless evidence develops that they are not safe.

As Mr. Sjöström points out, there are approximately 1200 flavoring agents. Only a very few of these have been investigated pharmacologically. At this writing 268 natural flavors have been listed as generally recognized as safe. A group of 213 have been extended for use pending further testing, but not beyond June 30, 1964. Of the synthetics, 26 have been listed as generally recognized as safe, with 756 on the extension list. The safety of these is based on long history and details regarding use, chemical characterization, toxicity, etc. It is quite likely that as more pharmacological data become available, some of the presently listed flavors may be delisted as being unsafe.

*   *   *

DISCUSSION

Question: Is any attempt being made to standardize terminology for antioxidants which have diverse functions such as oxygen scavengers or fat preservatives?

Lehman: Yes, the naming of such compounds will be decided, but we need help from industry. If industry doesn't help, the problem will be decided for you.

Oser, Bernard L.: I might comment in answer to the question as to how much of a food additive is being ingested by the public, and what sort of a safety factor should be used. In order to determine safe levels for something such as food flavors, we determine a maximum use level derived from data for average daily consumption. We assume that the maximum daily consumption of a food can be obtained from the average daily consumption multiplied by a certain factor. The factor will change depending on the particular food. We then have a maximum use level and assume further that the maximum amount of each food is eaten every day. The maximum use level is then multiplied by 100 as a safety factor. In actuality we therefore have a safety factor which is more like 100 x 100. This level is used in animal feeding trials.

Lehman: We hear the comment that a part per billion can't be very harmful but one must go back to the toxicology of the compound in question. Ten years ago concentrations of a few parts per million were considered quite low. In the last five years we have been dealing with parts per billion, and now some techniques are available for determining parts per trillion or even parts per quadrillion — concentrations of the order of 10^{-15}.

There's the problem of mold on ground nuts or peanuts from the Southern hemisphere which is found to be fatal at a concentration of one part in a million of the diet.

Weed killers and insecticides are creeping into our estuaries from underground water. One to ten parts per billion of weed killers are destroying microscopic algae which form the food source for many larvae. The larvae are being starved or killed by low levels of insecticides, and consequently larger fish are deprived of food — the food chain in this case is being disrupted. This is a very serious matter, and we must think in terms of parts per billion levels.

Dr. Cannon remarked about no effect levels of carcinogens — levels below which no carcinogenic effects are noted. In this connection there was publication some months ago of work done on feeding acetamide to rats. Groups of rats were fed acetamide (5 percent) for varying periods of time, and then put back on normal diet; if acetamide was fed for 4 months or more they later got carcinoma of the liver just as frequently as animals kept steadily on acetamide. Is there such a thing as a no-effect level for a carcinogen? It will take a lot more work before we have the answer.

A lot of current work and some human studies are being done on arsenicals. Organic arsenicals are used as direct feed additives for chickens and swine, and we might expect a breakthrough in this area. Arsenic is apparently detoxified by being bound to a protein, as in shellfish, and is excreted with no damage.

Oser: The remark on no effect levels of carcinogens suggests that we should be thinking about the validity of animal feeding trials. There are now data which show that citrus oils given orally to mice cause cancer.

Lehman: This takes us back to the thought that has been expressed many times — "The way not to have cancer is not to be born."

PART THREE

Incidental Additives

7

Pesticides

F. A. GUNTHER
UNIVERSITY OF CALIFORNIA

P ESTICIDES are chemicals required in agricultural production to protect plant and animal "crops" — and sometimes their transformed food and feed products — against the otherwise disastrous depredations of insects, mites, certain bacteria and fungi, nematodes, and weeds among growing food and fiber plants. In modern practice, these chemicals are used safely as plant-protection or animal-protection agents during production, transfer, and storage of practically all crops and crop products to meet the consumers' expectations of both quality and quantity production in contemporary agriculture. Modern transportation, refrigeration, and the burgeoning advances in food technology have made gourmets of us all, with the result that quality in foodstuffs has become an inexorable right of the consumer. The standards of "acceptable" quality for a really incredible variety of foodstuffs have risen to unprecedented heights through steady gains from the use of chemicals in agriculture. Chemicals necessary to adequate agricultural production are of the most diverse natures and purposes, as listed below:

Antibiotics to control bacterial and viral diseases of plants and animals and as feed additives
Defoliants to make harvesting easier (e.g. cotton)
Desiccants to speed drying of plant tissues (e.g. hay)
Fungicides to control fungi in almost any environment
Fumigants for soils, commodities, and spaces
Hormones for caponizing and other purposes
Insecticides and acaricides for almost any environment
Nematocides to control nematodes in soil
Plant-growth regulators to alter plant behavior in desirable manners
Plant nutrients and fertilizers for almost any deficiency

Seed disinfectants for almost any infection or infestation

Selective herbicides for control of undesired plants

Senescence inhibitors for certain plant parts

Sex sterilants ("chemosterilants") for control of certain insect species

Soil conditioners for altering physical characteristics of certain types of soils

Sprouting agents to promote sprouting processes, as with seed potatoes

Washing and cleaning agents for post-harvest use

As readily imagined, these chemicals vary in nature from simple inorganic and organic compounds to exceedingly complex ones, with the expected gamuts of chemical and biological reactivities. The judicious use of these chemicals has contributed immeasurably to the excellence and abundance of modern agricultural production in this country as attested by manpower-production relationships (Gunther and Jeppson, 1960):

In 1920, 1 farm worker produced food and fiber for 8.

In 1940, 1 farm worker produced food and fiber for 11.

In 1955, 1 farm worker produced food and fiber for 19.

In 1957, 1 farm worker produced food and fiber for 23.

In 1961, 1 farm worker produced food and fiber for 24 (Aldrich, 1961).

Obviously, mechanization, improved strains and breeds of livestock and plants, better marketing facilities and organizations, and conservation and intelligent use of soil and water resources are inseparably associated with chemicals in this increased efficiency of production (Gunther and Jeppson, 1960).

Among these chemicals the pesticides are perhaps most widely known, for their effects as control agents are often immediately apparent, and their omission from almost any horticultural effort is distressingly obvious to even the most casual gardener. Pesticide chemicals are used around the world in many thousands of tons annually, yet there has not been a single human disaster assignable to residues from these often-toxic chemicals in legally permitted uses. Despite the extensive and intensive use of remarkably efficient pesticides, however, enormous losses in agricultural production still occur. For example, many authorities have stated that in an average year present losses caused by all insects in the United States amount to more than four billion dollars, and on a world-wide basis to as much as twenty-one billion dollars (Gunther and Jeppson, 1960). A recent compilation (Sharp, 1960) of annual losses to California agriculture attributes $283,709,000 to damage by insects; $350,000,000 to damage by weeds; and $124,965,000 to damage by nematodes. Even though we still share our crops in this significant proportion with agricultural pests, our markets are plentifully supplied with the widest selections ever of attractive, sound, and wholesome produce, meats, and dairy products. The once expected "worm-in-every-apple" is now an association of historical interest only, but within the current state of knowledge withdrawal or even curtailment of present chemical pest-control practices would within a single season return this and other equally unappetizing commodities to our tables along with slow starvation of our population from half-rations, and total disappearance of many foodstuffs. Thus, Boyce (1962) has estimated that pesticides are responsible for up to half of the earlier mentioned very large increase in the productiveness of our food and fiber farming operations from the use of all agricultural chemicals. Equally striking numerical significance has been accorded this increase by Clarkson (1960), who reported that in 1959 there were produced 58 percent more farm commodities on fewer acres than in 1939, a span of only 20 years.

In 1940 when one farm worker produced food and fiber for 11, the total number of commercially available agricultural pesticide chemicals was about 50 (Boyce, 1962); today there are more than 200 basic chemicals registered federally for pesticidal use in more than 7,000 formulations (Decker, 1960). Nearly 638 million pounds of pesticide chemicals were produced in 1960 in the United States (Schechter and Westlake, 1962). California, which produces 139 major farm crops and accounts for 9 percent of the total cash receipts from farming in this country (Special Committee Report, 1960), has registered over 14,000 brand-name and other registrations of these pesticide chemicals for in-state use. During the three years 1958-1960 fungicides comprised 22.7 percent of the total production of synthetic organic pesticides in the United States, herbicides 26.4 percent, and insecticides and acaricides together 50.9 percent (Shepard, 1961).

PESTICIDE RESIDUES

Organic pesticides — in contrast to the usually highly polar (ionizable) inorganic pesticides such as the most-used salts of arsenic and lead — as a class are to some extent soluble in plant and animal tissues. In general the non-systemic materials can dissolve in fats, waxes, and oils whereas the systemic materials are usually deliberately designed to pervade aqueous tissues. Thus, both types may exist for short periods external to the cuticle or skin of the treated item, then they migrate in substantial amounts into cuticular and subcuticular regions. Nature and extent of this migration depend upon many factors including structure (including electrokinetic details) and stability of the chemical, the detailed structure and composition of the substrate surface, and natures of any discontinuities in this surface. Crafts and Foy (1962) have recently summarized and discussed the many modes of entry of both polar and nonpolar compounds into and through plant cuticle, and Ebeling (1962) has extensively reviewed the several influences of formulation and physical environment upon these migrations.

Many pesticide chemicals are fugitive on the substrate due to hydrolysis, spectral radiation or thermal instability, physical dislodgment by the actions of wind and rain, etc. Some of those that survive to penetrate into tissues below their original surface placement are also fugitive by virtue of possible enzyme attack, hydrolysis, and other in situ alterations as they migrate from cell to cell or from tissue to tissue. Both externally produced and internally produced alteration products may be toxicologically quite harmless, or they may be more toxic to mammals than were the parent compounds. These products, and the pesticide molecules unaltered by micro-environments, in residence on and in plant and animal tissues, constitute "pesticide residues."

Because it is conceivable that under conditions of gross misuse residues from some pesticides could constitute possible hazards to the public health, they are regulated by law in many countries. In the United States and some other countries they are regulated through the medium of tolerances, or maximum amounts permissible in foodstuffs; since late 1954 the U.S. Food and Drug Administration has established more than 2,000 tolerances for residues of these pesticide chemicals in or on a huge variety of raw agricultural commodities (Gunther, 1962). For crops and crop products these tolerances are for residues greater than zero, except for milk. Several compounds and mixtures of compounds are exempt from the requirement of a tolerance if applied to a growing crop, whereas others are exempt if applied after harvest; a few pesticide chemicals have zero tolerances on any edible crop, and a very few of them do not leave residues from accepted use and accordingly also have zero tolerances.

Tolerance values are based upon the careful interpretation of extensive residue, pharmacological, toxicological, and use data with a conservative built-in safety factor to permit declaration that these permitted levels are safe for human consumption and that they represent the maximum possible levels that could occur in agricultural practice. They are commonly expressed as ratios: parts of pesticide per million parts of commodity, or its equivalent in the metric system as milligrams per kilogram.

LEGISLATIVE CONTROL OF RESIDUES

Around the world increasing moral, scientific, and concomitant legislative attention is being devoted to the useful control of the types and amounts of these chemicals and their alteration products persisting as residues in our food (Gunther, 1962; Dormal and Hurtig, 1962). Control measures variously include regulation of dosage and timing of applications, required evaluations of natures and magnitudes and locales of persisting residues, tolerance assignments, and the enforcing of minimum intervals between applications and harvest.

As discussed by Gunther (1961, 1962), prior to about 1952 a few countries had old legislation to help prevent excessive residues of specific inorganic chemicals on certain foodstuffs, such as those containing arsenic and lead on apples (Canada, England, U.S.A.), boron on citrus fruits (Switzerland), and arsenic and copper in wines (France). As the in situ chemical and biochemical fates of both penetrated and unpenetrated residues of these pesticides have been elaborated, the agricultural chemicals industry around the world has recognized the magnitude of its responsibility in sponsoring and supporting broad investigations of the residue behavior of its products. To formalize continuity of this attitude and these responsibilities the Miller Pesticide Amendment (Public Law 518 of the 83rd Congress) was enacted in the United States.[1] Many other governments have also recognized that without these chemicals their populations could not be adequately fed, yet, at the same time, that pesticide chemicals must be so regulated as to maintain the present extraordinary record (Zavon, 1959; Special Committee Report, 1960) of safety in every valid agricultural use. Existing or pending legislation by these governments necessarily acknowledges that successful regulation of pesticide residues must be based upon often highly accurate determinations of their amounts and natures if they persist to the consumer of the treated commodity; the several types of existing controls are illustrated in Table 7.1 (Gunther, 1961, 1962; Dormal and Hurtig, 1962).

---

[1] This 1954 amendment, effective in 1956, included insecticides, acaricides, and fungicides; the Colley Amendment of 1959 specifically added growth regulators, herbicides, and nematocides to subject pesticides.

Table 7.1. Types of Legislative Control of Pesticide Residues in Foodstuffs [a]

Country	Residue Control Program	Timing Restrictions	Source of Residue Data
Australia	State jurisdiction[b,c]	Recommended[d]	State
Austria	Regulated by law[b]	Regulated	. . . .
Belgium	Regulated by decree[b]	Regulated	State
Canada	Compulsory, comprehensive[c]	Regulated[e]	Applicant, state
Denmark.	State jurisdiction[f]	Regulated[d]	Food institute
France	Restricted by decree[b]	Regulated[d]	State
Great Britain.	Voluntary: new chemicals	Regulated[d]	Public analysts
Greece	Compulsory: olives	Regulated	. . . .
India	State jurisdiction	Optional	State
Indonesia	. . . .	Regulated	State
Israel	. . . .	Regulated[d]	State and institutes
Italy.	Compulsory: olives, grain	Regulated[e]	Provinces
Luxemburg	Regulated by decree	Recommended	. . . .
Mexico	None	None	None
New Zealand	Regulated by law[c]	Regulated	State
Norway.	None[b]	Probable[d]	Food institute
Spain	None[b]	None	Institutes
Sweden	State jurisdiction[b]	Regulated[d]	State
Switzerland.	Regulated[b,c]	Regulated	Cantons
The Netherlands.	Regulated by law[b,g]	Regulated[d]	State
United Arab Republic .	Comprehensive	Regulated	Ministry of Agriculture
United States	Compulsory, comprehensive[b,c]	Regulated	Many
U.S.S.R.	Compulsory, comprehensive[c]	Regulated	Ministry of Public Health
W. Germany	Comprehensive[d,f]	Probable	Institutes and Industry

[a] Largely from Gunther (1961 and 1962) with additions from Dormal and Hurtig (1962).
[b] Certain materials prohibited.
[c] With official tolerances for permitted residues.
[d] Extensive revision anticipated or in progress.
[e] Not by statute, but minimum interval is often recommended on label.
[f] Complete revision of old legislation.
[g] Unpublished tolerances.

In Table 7.2 (from Dormal and Hurtig, 1962) are reproduced some numerical examples of pesticide residue tolerances in several countries to illustrate divergencies among these "constants" as derived by somewhat different ways of extrapolating raw animal-toxicity data to levels safe for human consumption. Except for DDT, these numerical differences are analytically slight yet there is growing apprehension that they may hinder the free international movement of foodstuffs. Also, this list of so-called "multiple" tolerances is illustrative only, for a complete listing shows even greater lack of uniformity in that in some countries certain pesticide chemicals are forbidden by law yet permitted in others. Dormal and Hurtig (1962) have recently summarized the several pros and many cons of possible international regulation of pesticide residues, but the problem is a complicated one (Gunther and Jeppson, 1960; Gunther, 1961, 1962).

Table 7.2. Some Maximum Tolerances for the Same Insecticide Residues
in Various Fresh Foodstuffs in Different Countries [a]

Insecticide Chemical (Common Name)	Tolerance (p.p.m.) [b]						
	Australia	Canada	New Zealand	Switzerland	The Netherlands	U.S.A.	U.S.S.R.
Aldrin	0.1	0.25	0.25	0.1	0.1	0.25	0.0
Arsenic - containing. . . .	...	2.0[c]	1.0[c]	1.0[c]	0.7[d]	2.3[c]	0.0
DDT.	7.0	7.0	5.0	4.0	5.0	7	1.0
Diazinon	0.15[e]	0.75	...	0.75	1	0.75[f]	...
Methyl demeton .	0.15[e]	...	...	0.5	0.5	1.25[g]	...
Methyl parathion	0.15[e]	...	...	0.75	0.5	1	...
Parathion	0.15[e]	1.0	1.0	0.75	0.5	1	5.0[h]

[a] Abridged from Dormal and Hurtig (1962).
[b] Some of these tolerances are less for certain crops: the values reproduced here are maximum values for any crop.
[c] Calculated as arsenic; 3.5 p.p.m. of combined arsenic as AS_2O_3 in the U.S.A.
[d] For Tuzet only.
[e] Calculated as organically bound phosphorus. It is not clear if this refers to phosphorus content or to parent compound; if the latter, the tolerance is about 1.5 p.p.m. as parent compound.
[f] For most crops; 1 p.p.m. on olives.
[g] As demeton, not methyl demeton; mostly 0.75 p.p.m.
[h] The purified compound only; the "impure" compound is restricted to 0.0 p.p.m.

RESIDUE CHEMISTRY

Exigencies of pending legislations and regulations, moral responsibilities, fascinating research possibilities in the ultra-micro chemistry and biochemistry of residues, and intense interest in developing a sound rationale for the realistic control of essential pesticide chemicals in good agricultural practice have resulted in the new field of residue chemistry (Gunther and Jeppson, 1960; Gunther, 1961, 1962; Blinn and Gunther, 1962; Schechter and Westlake, 1962).

This highly specialized field of chemistry achieved awakening recognition about twenty years ago under the impetus of the lipophyllic DDT and its many successors, and has matured under the pressure associated with the essential survival of practicable chemical pest control newly threatened with the novel and incredibly complicated requirements of proving safety in every agricultural use. Residue chemistry is a composite of analytical, biological, organic, physical, and physiological chemistries, all at the microgram level, and confounded by the existence of micro-residues dynamically existing in macro-amounts of great varieties of plant and animal tissues. Most chemists are not trained to work with microgram amounts of materials nor are they trained to work with unusually rigorous isolative and determinative techniques. During this gestation period it has been necessary, therefore, for chemists thrust into this virgin field to acquire the necessary background and skills almost empirically, and there is still much to learn.[2] To illustrate, Schechter and Westlake (1962) have protested that a legally set "zero tolerance ...automatically established for such [carcinogenic pesticidal][3] compounds presents the analytical [residue][3] chemist with the problem of developing a sensitive and specific method to detect residues at levels approaching zero." To preserve sanity in mathematics, "zero" must remain a constant less than any assignable quantity, or "none." A pesticide residue at a level of one part per million (p.p.m.) in a foodstuff exists as about 10^{15} molecules of that residue per gram of that foodstuff; if the compound is present at 0.01 p.p.m. there are still about 10^{13} molecules of pesticide per gram; at 10^{-6} p.p.m. there are still about 10^9 molecules present, and so on. Clearly as analytical methodology has improved to allow us to detect and demonstrate some residues successively at 0.1 p.p.m., at 0.01 p.p.m., at 0.001 p.p.m., and recently (Moore, 1962) at less than 0.0001 (10^{-4}) p.p.m., we are indeed "approaching zero," but not very rapidly. The prompt and almost frantic adoption of almost every type of analytical instrumentation in this field — often in striving for closer approaches to zero residues — has sometimes been associated with inadequate allowance for proper training of the "analyst." Consequent perpetrations in the literature of misobservations are detrimental to the literature searching novice who has to rely heavily upon published material for his basic training and acquisition of perspective in any area of science. An example to illustrate this point, and chosen because it was encountered in one of our literature searches on analytical methodology, concerns a recent report (Rosenfield et al, 1960) on an infrared method for the flavor evaluation of insecticide-treated vegetables. The authors attributed probable flavor-correlation significance to IR fingerprint spectra from isolates of insecticide-treated samples of certain vegetables, whereas their "significant" spectrum is actually and unfortunately that of a silicone oil or grease contaminant from their laboratory equipment.

Dust, granular, and spray applications of pesticides to plant surfaces afford extracuticular "deposits" of material, portions of which enter into cuticular and subcuticular tissues to become residues. Pesticides are also acquired by plant parts through soil applications through sorption processes at root hairs and other subterranean organs and tissues; so acquired, these

---

[2] Formal graduate-level training in pesticide residue chemistry is now available in at least one institution.

[3] Bracketed words inserted by present author.

chemicals are also residues. Depending largely upon formulation (Ebeling, 1962), deposits may dissipate or attenuate by various means, but some portion of each non-ionic pesticide[4] will penetrate into and perhaps beyond the treated plant surface. If penetration is partial and localized, the pesticide is said to be non-systemic; if penetration is essentially complete, rapid, and "systemic" the pesticide is a systemic chemical. Regardless of locale, these chemicals are exposed to many opportunities for chemical alteration (Gunther and Blinn, 1955; O'Brien, 1960; and many others):

Exterior deposits:	Weathering, air oxidation, ultraviolet-induced decompositions, thermal alterations
Cuticular deposits and residues:	Ultraviolet, thermal, and enzymatic alterations
Subcuticular residues:	Enzymatic alterations, hydrolysis, ligand formation
Systemically distributed residues:	Diverse enzymatic alterations, hydrolysis, ligand formation

These alteration products may be more toxic or less toxic than the parent molecule but if they persist in or on the treated commodity, they are assumed guilty until proven innocent. The residue chemist must therefore note and elaborate any in situ alterations that take place so that the necessary and sufficiently incisive information is available for eventual establishment and assessment of possible hazard from these metabolic products. With only microgram quantities of materials in immense dilution in gross amounts of exceedingly complex substrate mixtures, and with multitudinous opportunities for chemical and biochemical transformations, the residue chemist must consider the use and the propriety of every tool and technique to acquire the necessary information, as illustrated in Table 7.3. Fortunately, pesticide chemicals belong to a relatively few broad reaction classes of compounds and à priori can often be categorized as to the precept-established behavior expected, as dehydrohalogenation, hydrolysis, epoxidation, thiol-thiono interconversion, sulfoxidation, ligand retention, and a few others. Ever-present interfering substances always accompany extraction and other isolative procedures employed, and attempts to circumvent or minimize

---

[4] Some ionic pesticide chemicals will penetrate plant tissues by several pathways available in most leaf surfaces (Crafts and Foy, 1962; Ebeling, 1962).

Table 7.3. Instrumental Analytical Methods Applied to Pesticide Residue Evaluations
(Gunther, 1962)

Instrumental Technique	Evaluative Character Qual.[a]	Quant.	Approx. No. Applications	Approx. Min. Detectable Unit [b]
Completely automated	?	+	Many [c]	Micrograms
Spectrophotometric				
Infrared				
Regular	+	+	30	Micrograms
Attenuated total reflection	+	–	– –	?
Visible	?	+	>100	Micrograms
Ultraviolet				
Regular	+	+	ca. 40	Micrograms
Far (to ca. 180 mu.)	+	+	ca. 5	Micrograms
Fluorescent	+	+	ca. 10	Microgram
X-ray [d]	+	+	1 [e]	Micrograms
Scintillation	–	+	Many [f]	<Microgram
Neutron-activation	?	+	Many [f]	<Microgram
Nuclear magnetic resonance	+	?	Several	Milligrams
Mass spectrometry	+	?	Several	Milligrams
Electrochemical				
Potentiometric	–	+	Many [f]	Microgram
Polarographic				
Regular	?	+	25	Micrograms
Oscillographic	?	+	7	<Microgram
Coulometric	?	+	Many [f,g]	<Microgram
Thermal conductivity	?	+	Many	Microgram
Electron capture (transfer)	?	+	Many [f]	<Microgram
Gas chromatography	+	+	>100	<Microgram

[a] Question mark means qualitative in special situations.
[b] Under ideal conditions.
[c] Enzyme assay of organophosphorus insecticides.
[d] Diffraction, emission, absorption, and fluorescence.
[e] Organobromine pesticides.
[f] Primarily organobromine and organochlorine compounds.
[g] Organosulfur compounds.

these often can alter both amount and nature of sought isolate, and these operations, too, must be proven for both reliability and efficiency.

CONCLUSIONS

Residue chemistry is largely empirical in outlook and necessarily pragmatic in philosophy so that theory is lagging far behind enthusiastic practice. For most purposes it is an economic activity required in the interests of the public health. In this field, therefore, the use of precision and other analytical instrumentation is increasing unusually rapidly because the desired results cannot be obtained in any other way, and often because more reliable results can be achieved more quickly. This combination of reasons has enormous persuasive power (Gunther, 1962).

As hinted above, however, excessive enthusiasm for the

remarkable possibilities inherent in analytical instrumentation without simultaneous and elaborate acquisition of the capability to evaluate properly the derived data is often encountered in this field. It is appropriate here to recall the conversation quoted by H. N. Wilson (1960): "Twenty years ago if I were given a can of paint on a Monday morning, by Wednesday morning my boss expected to know the mineral constituents of the pigment, what oil had been used, and the nature of the thinners. Nowadays if I give a man a can of paint on Monday, by Wednesday morning he has not even decided what wavelength to use."

Technically speaking, the pesticide residue situation is competently and confidently under continuing control; confusions and disturbances still occur, largely because of stigmas that will accrue against a particular food or food product. Unfortunately, the public does not seem to know that all our foods contain pesticide residues, but in amounts of no concern under present watchful legalized use. Harmful amounts are not allowed, and the enforcing agencies are adequate to their task.

LITERATURE CITED

Aldrich, D. G., Jr. 1961. In Proceedings Tenth Annual Meeting Agricultural Research Institute, p. 4, Washington, D. C.

Blinn, R. C. and Gunther, F. A. 1962. The utilization of infrared and ultraviolet spectrophotometric procedures for assay of pesticide residues. Residue Rev. 2: in press.

Boyce, A. M. 1962. Chemicals in agriculture and crop production. Presented 131st National Meeting, Am. Chem. Soc., Washington, D. C.

Clarkson, M. R. 1960. Chemicals in agricultural practices. Nat. Agr. Chem. Assoc. News 18 (No. 4):3.

Crafts, A. S. and Foy, C. L. 1962. The chemical and physical nature of plant surfaces in relation to the use of pesticides and to their residues. Residue Rev. 1:112.

Decker, G. C. 1960. Agricultural chemicals. Federation Proc., Part II, Vol. 19, suppl. No. 4, p. 17.

Dormal, S. and Hurtig, H. 1962. Principles for the establishment of pesticide residue tolerances. Residue Rev. 1:140.

Ebeling, W. 1962. Analysis of the basic processes involved in the deposition, degradation, persistence, and effectiveness of pesticides. Residue Rev. 2: in press.

Gunther, F. A. 1961. Pesticide residues in foods: analytical philosophy, technology, and capabilities in other countries. J. Assoc. Offic. Agr. Chemists 44:620.

_____. 1962. Instrumentation in pesticide residue determination. Adv. Pest. Control Research 5:191.

_____, and Blinn, R. C. 1955. Analysis of Insecticides and Acaricides. Interscience, New York-London.

_____, and Jeppson, L. R. 1960. Modern Insecticides and World Food Production. Chapman & Hall, London. Wiley & Sons, New York.

Moore, A. D. 1962. Electron capture with an argon ionization detector in gas chromatographic analysis of insecticides. J. Econ. Entomol. 55:271.

O'Brien, R. D. 1960. Ionic Phosphorus Esters: Chemistry, Metabolism, and Biological Effects. Academic Press, New York-London.

Rosenfield, D., Epstein, E. I., Stier, E. F., and MacLinn, W. A. 1960. A study of a spectrophotometric method for the flavor evaluation of insecticide-treated vegetables. Food Research 25:513.

Schechter, M. S. and Westlake, W. E. 1962. Chemical residues and the analytical chemist. Anal. Chem. 34:25A.

Sharp, P. F. 1962. Unpublished data.

Shepard, H. H. 1961. The Pesticide Situation for 1960-1961. Agricultural Stabilization and Conservation Service, USDA, Washington, D. C.

Special Committee Report. 1960. Agricultural Chemicals and Recommendations for Public Policy. California State Printing Office, Sacramento.

Wilson, H. N. 1960. The changing aspect of chemical analysis. The Analyst 85:549.

Zavon, M. R. 1959. Residues — a medical appraisal. Nat. Agr. Chem. Assoc. News 18(2):3.

8

Plant Growth Regulators

A. J. VLITOS

*TATE and LYLE CENTRAL AGRICULTURAL
RESEARCH STATION Trinidad, W. I.*

I T IS OFTEN difficult to describe a <u>plant growth regulator.</u>
Superficially it would seem that any chemical which regulates
or modifies growth of plants should, <u>ipso facto,</u> be considered
a growth regulator. However most plant physiologists accept a
much narrower definition. Therefore in this chapter a plant
growth regulator is considered as any organic chemical which,
in relatively small amounts, either stimulates, inhibits, or in
some other way alters the growth or development of plants.

Many types of plant growth regulators are being used in agriculture for a variety of purposes. Some function to control
weeds, others stimulate flowering or fruiting, still others are
used to increase fruit size, to induce rooting of cuttings, to produce parthenocarpic fruits, to modify the chemical composition
or nutritive qualities of leaves, fruits, or stalks, to control pre-
harvest fruit drop, and to regulate dormancy in stored crops.
Other uses for plant growth regulators are rapidly being developed. Within a few years it is quite likely that most fruits and
vegetables shall have been exposed to some type of growth regulator at one stage or another during production. It is appropriate
therefore that some thought and discussion be given to these "incidental food additives," why they are used, and their ultimate
fate within plant tissues. The presentation has been divided into
three parts: I. Historical Aspects of the Plant Growth-Regulator
Concept. II. Agricultural Applications. III. Plant Growth Regulators and their role in Human Physiology.

HISTORICAL ASPECTS OF THE PLANT
GROWTH-REGULATOR CONCEPT —
NATURALLY-OCCURRING PHYTOHORMONES

Darwin to Went

It comes as a surprise to most people to learn that the modern history of the plant growth-regulator concept began with Charles Darwin in 1881. In that year Darwin published a volume called "The Power of Movement in Plants," describing some novel and ingenious experiments which he carried out with a few grass seedlings, a razor blade, a bit of tin foil, and light. Darwin noted that when coleoptiles[1] of canary-grass seedlings (Phalaris canariensis) were illuminated from one side the seedlings curved towards the source of light. The curvature took place at a considerable distance below the tip of the seedlings. If, however, the tip of the coleoptile was removed or if it was covered with a light-proof material, such as tin foil, bending did not occur, even though the rest of the plant had been exposed to light. Darwin concluded that light was "perceived" by the tip of plants and that some unknown "influence" must have been transmitted from the stimulated tip to the basal areas of the coleoptile, where curvature occurred.

Then, in 1911, Boysen-Jensen proved that the unknown "influence" or "stimulus" produced in the tip of coleoptiles could pass through nonliving materials such as gelatin. Eight years later Paál (1919) confirmed Boysen-Jensen's work and also showed that replacement of the severed tip on one side of the coleoptile stump would produce curvatures away from the treated side. This suggested to Paál that there was a replacement of the effect of lateral light by an asymmetrical distribution of the "stimulus" produced by the tip. It became apparent that the "tip" of plants is the seat of a "growth-regulating center." Paál concluded that 1) a substance or mixture of substances is manufactured in the tip and 2) this substance(s) is secreted internally, moves downwards, and diffuses to all sides of the plant. This concept, with minor modifications, is still widely accepted today (Audus, 1959).

Probably the single most important impetus to the plant growth-regulator concept came as a result of the research of Went (1928), who managed to collect in agar blocks the substances postulated by Darwin, Boysen-Jensen, and Paál. Went developed the now classical Avena coleoptile curvature assay. This technique was important for three reasons: 1) it stimulated

---

[1] A coleoptile is a tubelike first leaf produced by grasses upon germination.

research, 2) it permitted the convenient concentration of active substance, and 3) the bioassay was sensitive enough to detect as little as 10^{-2} micrograms of auxin, based on the standard, indole-3-acetic acid. Thus, twenty-nine years after Darwin's simple experiments, plant physiologists realized that phototropism, responses to gravity, and the growth of plants were in some mysterious way regulated by minute quantities of chemicals. At present not all of these substances have yet been characterized nor are their modes of action fully understood.

Kögl to the Present

The determination of the chemical structure of naturally-occurring plant growth regulators has lagged behind similar studies in human and animal endocrinology. It should be realized however that the history of plant growth regulators has been studded with most unusual circumstances. First of all, the initial isolation of a crystalline substance with plant hormone activity was not from plants but from human urine! Kögl, Haagen-Smit, and Erxleben (1933, 1934) turned to plant sources and managed to isolate two auxins from corn germ oil. The two substances were termed auxin a and auxin b. Kögl and Erxleben (1934) suggested the following structural formulas for the two compounds:

Auxin a

Auxin b

Unfortunately these structures have never been verified by synthesis nor have other workers been able to repeat the isolations (Wieland, Ropp, and Avener, 1954). Few, if any, plant physiologists accept auxin a and b as authentic naturally-occurring plant growth regulators. However if it had not been for Kögl's interest in these substances, it is doubtful that "heteroauxin" (the other auxin) would have been identified at that time. After Kögl and his co-workers found that the plant growth regulating activity of urine varied as much as five-fold they attempted to improve their isolation procedures. In so doing they obtained an entirely different product which they named "heteroauxin." Analysis of heteroauxin proved it to be indole-3-acetic acid (IAA). Kögl considered IAA to be an "excretory product," but history has confirmed that IAA is probably the major naturally-occurring phytohormone. Actually the compound had been synthesized much earlier by Ellinger (1905). Thimann, in 1935, showed that IAA was identical to "rhizopin" a plant growth regulator that had been isolated previously from the fungus Rhizopes simnus by Nielsen (1930). Finally, after this tortuous course of scientific history, the plant growth regulators, though not ready for agricultural applications, began to interest practical people.

Several early investigators had expressed the then unpopular view that a single substance such as IAA could not possibly "catalyze" all of the diverse and complicated processes involved in plant growth. Zimmerman and Hitchcock (1937, 1942, 1951), Zimmerman and Wilcoxon (1935) and all three of these workers (1939) at the Boyce Thompson Institute for Plant Research, Inc. demonstrated that a variety of snythetic compounds (i.e. indole-3-butyric acid, 2,4-dichlorophenoxyacetic acid, ethylene, naphthalene acetic acid, substituted benzoic acids, etc.) could act as plant growth regulators and could produce a variety of physiological responses when applied to plants or to parts of plants. They were able to show that many of the responses of plants attributable to naturally-occurring IAA (such as induction of rooting, curvature of stems, suppression of flowering, etc.) could be induced with synthetic compounds, some of which were unrelated to IAA. Based on their prolific data Zimmerman and his associates maintained that the single hormone concept was oversimplified and at best represented an idealized scheme. In recent years it has become quite clear that plants contain a number of non-indolic plant growth regulators (Crosby and Vlitos, 1961; Macmillan and Suter, 1958). The most important of these, and by far the most dramatic (Phinney et al, 1956; Vlitos, 1960), are the gibberellins.

An excellent historical review of the gibberellins is already available (Stowe and Yamaki, 1957), therefore the present discussion only deals briefly with the main points. Through a series of fascinating researches Japanese workers were able to isolate the chemical component responsible for the stimulation of growth in rice seedlings infested with Gibberella fujikuroi. Following

World War II, when the Japanese literature was available, British and American teams confirmed the earlier studies. More recently the gibberellins have been isolated from higher plants and have been shown to occur in seeds and fruits of several genera (Phinney et al, 1956). Although the chemical structure of each of the gibberellins occurring in higher plants or in fungi has not been established, some similarities in structure have been noted. Cross et al (1956) proposed the following structure for gibberellic acid, isolated from the imperfect stage of Giberella fujikuroi.

Among some of the other types of plant growth regulators occurring naturally are the kinins (Jablonski and Skoog, 1954; Miller, Von Saltza and Strong, 1955) and the endosperm factors described by Steward and Shantz (1956).

Jablonski and Skoog reported in 1954 that various natural products such as extracts of malt and vascular tissues contained compounds which stimulated the division of excised pith cells of tobacco in culture. In 1955 Miller, Von Saltza and Strong isolated 6-furfurylaminopurine from an autoclaved sample of desoxyribonucleic acid. They proved that this compound, which they named kinetin, would induce cell division in tobacco pith cultures. Skoog and Miller found that many other adenine derivatives are also effective as cell division factors or "kinins." A recent report from Australia indicates that a kinin occurs in apple fruitlets (Goldacre and Bottomley, 1959).

Steward and Shantz (1956) have stressed for many years that coconut milk and other endosperm tissues contain growth factors active in stimulating cell division and that these substances are apparently unrelated structurally to other of the naturally-occurring plant growth regulators.

New plant growth regulators shall undoubtedly be isolated in the future. Many may prove to be unrelated to those known today. From the discussion which follows it will be readily appreciated that the synthetic growth regulators being used in agriculture at present represent a vast array of chemical configurations.

AGRICULTURAL APPLICATIONS

Herbicides

The Chlorophenoxy Compounds. Zimmerman and Hitchcock
demonstrated in 1942 that 2,4-dichlorophenoxyacetic acid (2,4-D)
was a synthetic plant growth regulator. Two years later Mitchell
and Hamner (1944) suggested that the compound might be useful
as a selective weed killer. In Great Britain, 2-methyl-4-chloro-
phenoxyacetic acid (MCPA) was discovered as a herbicide in 1942
by Slade, Templeman, and Sexton but, owing to security regulations during World War II, their discovery was not made public
until 1945. By 1952 the annual production of 2,4-D in the United
States alone exceeded 28 million pounds (Van Overbeek, 1952).
Almost all of this was used to control weeds.

Uses. Several factors were responsible for the ready acceptance of chlorophenoxy chemicals as herbicides. First, they
are selective; that is they destroy most dicotyledenous (broad-
leaved) weeds without injuring certain monocotyledenous (grass)
crops. Prior to the discovery of the plant growth regulants,
weeds and crops were subject to an equivalent risk of injury from
the so-called "contact" type herbicide. A second reason for the
widespread popularity of 2,4-D and its relatives was its low
price, considering the ultimate benefits resulting from its use.
Thirdly, the chlorophenoxy derivatives are of relatively low
mammalian toxicity. Perhaps the most important reason for the
successful entry of 2,4-D into agriculture was that the chemical
fulfilled a practical need which, prior to its discovery, was not
being met effectively by other means.

From an economic standpoint the three most important chlorophenoxy herbicides are:

2,4-dichlorophenoxyacetic acid (2,4-D)

2,4-D

2,4,5-trichlorophenoxyacetic acid (2,4,5-T)

2,4,5-T

and,

2-methyl-4-chlorophenoxyacetic acid (MCPA)

O-CH$_2$COOH

CH$_3$

Cl   MCPA

The 2-methyl-4-chloro analog (MCPA) has been used mainly in Great Britain, primarily because it can be synthesized economically, using cresol as one of the starting materials. MCPA is also less damaging than 2,4-D to certain types of cereals (Audus, 1959).

2,4,5-T is used either alone or in combination with 2,4-D for controlling undesirable woody vegetation in forests and on ranges.

In the temperate zones more 2,4-D has been used for the control of weeds in cereals than for any other purpose. In the tropics, 2,4-D has been, and is still being, used to control weeds in sugar cane, in rice, and to some extent in coffee. Combinations of 2,4-D with substituted ureas or with symmetrical triazines have proved useful in sugar cane (Annual Report, 1960-1961). Although more effective herbicides than the chlorophenoxy compounds may be available at present, this group of growth regulators is still attractive to growers because of economy and versatility.

2,4-D is an extremely versatile weed killer. Originally suggested as a post-emergence herbicide, the compound has been used as a soil sterilant, a selective pre-emergence herbicide in corn, sugar cane, flax, wheat, barley and oats, and in combination with 2,4,5-T and other compounds for woody plant control (Crafts, 1961). In general, one to two pounds of 2,4-D in 100 gallons of water per acre will control dicotyledenous weeds in sugar cane, wheat, oats, barley, corn, and other monocotyledenous crops. The quantities of chemical required for best results depends upon the type of soil to be treated, susceptibility of the crop to damage, total rainfall, soil moisture, and economic considerations.

Many cereals are themselves sensitive to chlorophenoxy herbicides. Different varieties of winter wheat, for example, show varying degrees of susceptibility. Maximum sensitivity to 2,4-D in wheat is at the time of heading (Longchamp, Roy and Gautheret, 1952). Barley is also susceptible to 2,4-D damage, especially as a seedling up to the 5-leaf stage and then later at heading (Derscheid, 1952). Staniforth (1952) has shown that flower parts of field corn are susceptible to 2,4-D. If the chemical is applied when flower parts are being formed, reductions in yields may be expected. It may be concluded from these and other observations that applications of plant growth regulators require critical timing if maximum benefits are to be realized.

In some areas of the world the qualitative composition of the weed flora in certain sugar-cane fields in Trinidad where previously dicotyledenous species were the most abundant. This ecocotyledenous weeds now comprise the greater part of the total weed flora in certain sugarcane fields in Trinidad where previously dicotyledenous species were the most abundant. This ecological shift has coincided with the widespread use of 2,4-D in such fields, but it does not necessarily follow that the change would not have occurred as a result of other factors.

Metabolism. What is the fate of a chlorophenoxy plant growth regulator once it enters the plant? The answer to this question is by no means simple, and depends a great deal on the amount of the growth regulator which eventually reaches active metabolic sites. When an ester of 2,4-D is applied to a leaf at a concentration normally used for herbicidal effects, the ester is hydrolyzed rapidly to the free acid and the acid form of 2,4-D is then transported to other portions of the plant (Crafts, 1960). Within the cell the growth regulator complexes with either proteins or with glucose (Jaworski and Butts, 1952), but the exact nature of the binding is not clearly defined. It is known that certain enzymes, for example emulsin and takadiastase, release 2,4-D slowly from the protein complex. This may explain why the morphological aberrations, resulting from treatment of a plant with herbicidal doses of 2,4-D, may persist for several weeks (Audus, 1959). On the other hand, nonherbicidal doses of 2,4-D applied to bean leaves are rapidly degraded with release of CO_2 (Weintraub et al, 1952). Employing radioactive 2,4-D with the C^{14} label in the side chain, Weintraub et al (1952) showed that most of the C^{14} could be accounted for as $C^{14}O_2$, released from bean leaves a few hours after initial treatment. Similarly small amounts of 2,4-D may be readily inactivated by riboflavin in strong light (Carroll, 1949), but it is doubtful that at herbicidal concentrations such inactivation would be of much consequence. Where 2,4-D is used as a pre-emergence herbicide and comes into contact with soil, it is rapidly attacked by microbial enzymes (Audus, 1951; Boysen-Jensen, 1911).

Wain and Wightman (1954) have shown that certain species of plants can degrade the lengthy side chain of chloro-substituted phenoxy compounds and, by beta-oxidation, produce within their tissues the acetic acid analogs (i.e. 2,4-D, 2,4,5-T, etc.). One may therefore predict selectivity once it is known whether a particular species of weed can beta-oxidize the side chain and whether the crop lacks this capability.

Leafe (1962) has recently reported on the metabolism of MCPA in tissues of the weed Galium aparine. This species is resistant to MCPA. By introducing a single alpha-alkyl substituent into the MCPA molecule one obtains DL-alpha-(4-chloro-2-methylphenoxy) propionic acid (CMPP), a chemical which is toxic to Galium.

$$Cl \overset{H}{\underset{CH_3}{\bigcirc}} -O-\overset{H}{\underset{H}{C}}-COOH \longrightarrow Cl \overset{H}{\underset{CH_3}{\bigcirc}} -O-\overset{H}{\underset{CH_3}{C}}-COOH$$

MCPA   CMPP

(inactive)   (active)

MCPA-2-^{14}C and CMPP-2-^{14}C were applied to leaves of G. aparine and their fate followed over a ten day period. Very little MCPA could be recovered as the intact molecule, but most of the radioactivity was located in water-soluble fractions, in proteins and nucleic acids, in starch, and in tissue debris. Nearly all of the CMPP was recovered unaltered after ten days within the tissue. According to Leafe, two possibilities may account for the metabolism of MCPA in G. aparine: 1) either the side chain is degraded with the liberated carbon being incorporated as one- or two-carbon fragments, some released as CO_2, or 2) MCPA is conjugated to other cellular components. MCPA-[36]Cl was metabolized as rapidly as MCPA-1-^{14}C or MCPA-2-[14]C. Interestingly, the radioactivity in protein-nucleic acid and residue fractions, isolated from tissues treated with MCPA-1-^{14}C or 2-^{14}C, is derived from the carbon of the side chain. The residue fractions do not contain the phenoxy moiety. Resistance of G. aparine to MCPA appears to be due to a detoxication of the molecule involving the loss of both carbon atoms of the side chain. The introduction of an alpha-alkyl group into MCPA blocks the degradation and presumably accounts for the effectiveness of CMPP in controlling G. aparine.

In view of the many mechanisms by which chlorophenoxy growth regulators may be degraded, either prior to or after entrance into the plant, it is not surprising that this group of chemicals has rarely been encountered as a serious toxicological hazard in foods. According to Woodford (1960) the acute oral LD_{50} for rats of the important chlorophenoxy herbicides is as follows:

a) 2,4-D = 375 mg./kg.
b) 2,4,5-T = 300 to 500 mg./kg.
c) MCPA = 700 mg./kg.

For comparison, Woodford has presented the acute oral LD_{50} for rats of common aspirin as 1200 mg./kg.

Cows which were fed $5\frac{1}{2}$ gms. of 2,4-D daily for 106 days did not suffer in performance nor did they show pathological symptoms (Mitchell, Hodgson and Gaetjens, 1946). On the other hand, a human male can ingest as much as 0.5 gm. of 2,4-D daily for three months without showing ill effects.

Halliday and Templeman (1951) fed cows 18 gm. of 2,4-D per day without affecting quality of the milk or disturbing the health of the animals.

The accumulation of nitrates in forage grasses following treatment with 2,4-D has been reported to account for death of cattle which had ingested such plants. The nitrate is converted to nitrite by microorganisms in the rumen. The nitrite in the blood stream converts hemoglobin to methemoglobin. Methemoglobin is ineffective as an oxygen-carrier and anoxia may result. Weeds which are resistant to 2,4-D are generally the ones which accumulate nitrates. Certain poisonous weeds (i.e. hemlock) after treatment with 2,4-D become more palatable to cattle (Audus, 1959).

Substituted Ureas

<u>Uses</u>.   1,3-Bis(2,2,2-trichloro-1-hydroxyethyl) urea, better known

```
        Cl  H   H   O   H   H   Cl
        |   |   |   ||  |   |   |
  Cl -  C - C - N - C - N - C - C - Cl
        |   |               |   |
        Cl  OH              OH  Cl
```

as dichloral urea, was the first substituted urea to be used as a herbicide. The compound was synthesized by chemists at Union Carbide and was discovered to be a herbicide by L. J. King at the Boyce Thompson Institute for Plant Research.

Dichloral urea was employed for several years to control monocotyledenous weeds in sugar beets. The compound was unique for its time because it was toxic to almost all germinating grasses but innocuous to sugar beets, especially once the beet root had become established (Crafts, 1961). Owing to its rather narrow spectrum of selectivity and also because more potent substituted ureas were discovered shortly thereafter, dichloral urea has never found a large commercial market.

Monuron, 3-(p-chlorophenyl)-1,1-dimethyl urea

rapidly replaced dichloral urea and the chlorophenoxy derivatives, especially where long residual herbicidal activity was required. By virtue of a relatively low solubility in water (i.e. 230 ppm. at 25°C), monuron persists in upper layers of soil for several weeks longer than comparable quantities of chlorophenoxy derivatives (du Pont, Monuron, 1958). The chemical is active against both monocotyledenous and dicotyledenous weeds and has been used successfully in sugar cane, pineapples, citrus, asparagus, avocados and a number of other crops. More recently the closely-related diuron 3-(3,4 dichlorophenyl)-1,1-dimethylurea

has replaced monuron in certain areas (du Pont, Diuron, 1958). Diuron is less water-soluble than monuron (i.e. 42 ppm. vs. 230 ppm.). This is advantageous especially in areas with more than 40 inches of rain per year. An important use for diuron is in sugar cane where at 4 lbs. per acre (of the 80 percent wettable powder) a single pre-emergence application controls weeds for periods exceeding six weeks (Annual Report, 1960-61; Lawrie and Vlitos, 1962). In contrast, single pre-emergence applications of 2,4-D rarely control weeds under similar conditions for more than four weeks. Mixtures of diuron and 2,4-D have also proved effective in sugar cane. The combinations are less expensive and are as effective as the higher rates of diruon used singly. Some varieites of cane show less tolerance to the substituted ureas than others. It has been reported that sugar cane grown in soils deficient in phosphorus is more susceptible to damage by diuron than cane grown on fertile soils (Lawrie and Vlitos, 1962). In addition to monuron and diuron, several other substituted ureas have also shown considerable promise as herbicides (Crafts, 1961).

3-Phenyl-1,1-dimethylurea or fenuron

appears to be effective for controlling deep-rooted perennial
weeds in regions with moderate rainfall. Under dry conditions
fenuron is effective against shallow-rooted annuals. The compound is soluble in water at the extent of 2900 ppm., leaches from
soil more rapidly than monuron or diuron, and consequently is of
more interest to growers in temperate zones than in the tropics.
In contrast to fenuron, a substituted urea with extremely low
water-solubility is 1-n-butyl-3-(3,4-dichloro phenyl)-1-methyl-
urea or neburon.

Neburon by virtue of its poor solubility in water is active against
shallow-rooted weeds in crops with a deep root system.

Some of the substituted ureas have proved to be selective
when used as post-emergence herbicides. Friesen and Forsberg
(1959) have shown that neburon at 3 lbs. per acre (active ingredient) destroys wild buckwheat without injuring cereals, but the
crop must not have advanced beyond the 6-leaf stage. Diuron at
3 to 4 lbs. per acre has been used successfully as a post-
emergence treatment in sugar cane (Annual Report, 1960-61). The
herbicide was effective against grass and broad-leaved weeds but
did not injure sugar-cane ratoons. It may readily be appreciated
that a herbicide which destroys established weeds and then acts
to prevent the growth of newly-germinated weeds, without injuring the crop, is indeed versatile.

Versatility is one of the most attractive practical features of
a plant growth regulator. Each year new uses are developed for
"old" compounds. For example, 2,4-D is still a popular herbicide and is being utilized in combinations with newer materials.
Similarly, the chloro-substituted benzoic acids (discussed in detail below) which have been known as growth regulators since
1942 are only now being fully exploited as herbicides. The potential uses for the substituted ureas have probably yet to be fully
developed, even though newer types of growth regulators are con-
tinually being uncovered.

Metabolism. The substituted ureas are applied chiefly to the soil as pre-emergence herbicides. Therefore it is appropriate to consider first their stability in soil. According to Hill and Mc-Gahen (1955) monuron and diuron are rapidly oxidized in a number of different types of soils as a result of microbial activity. In fact, a Pseudomonas sp. was isolated which could utilize monuron as a sole source of carbon.

If a substituted urea herbicide is not immediately attacked and degraded by soil microbes it may enter the roots of higher plants within 30 minutes after application (Crafts and Yamaguchi, 1960). Within two hours the chemical moves into stems and leaves. Accumulation of monuron in upper plant parts continues for up to 8 days after the initial application to roots.

Fang and his co-workers (1955) treated bean plants with carbonyl-C^{14}-labeled monuron and found that, within four days, two radioactive compounds could be detected in treated tissues. One of these metabolites was unreacted monuron while the second was an unidentified compound which released monuron upon acid hydrolysis. After four days very little residual radioactivity remained in treated tissues. It was concluded that monuron is degraded quite rapidly within plant tissues. Crafts (1961) has presented an interesting account of the unpublished work of Todd and his associates at Du Pont who have studied the metabolism of substituted ureas for many years. These workers have found that one molecule of monuron inhibits the photosynthetic activity of 125 or more molecules of chlorophyll. The reaction is reversible. Upon removal of monuron the chloroplasts resume evolution of oxygen. Competitive inhibition by monuron (and diuron) of flavin mononucleotide is now considered to be a likely explanation for its mode of action.

It is of considerable interest that compounds as phytocidal as the substituted ureas nevertheless possess a very low order of toxicity to mammals. The acute oral LD_{50} of monuron for rats is 3700 mg./kg. while that of dichloral urea is greater than 31,600 mg./kg. (Woodford, 1960).

The Symmetrical Triazines

Uses. The symmetrical triazines are a relatively new group of plant growth regulators which are of interest chiefly as herbicides. They are characterized by a narrow spectrum of selectivity. For example 2-chloro-4,6-bis (ethylamino)-s-triazine (or simazine)

simazine

and 2-chloro-4-ethylamino-6-isopropylamino-s-triazine (atrazine)

atrazine

are innocuous to corn and to sugar cane but are highly active against most germinating monocotyledenous and dicotyledenous weeds. Corn tolerates up to 8 lbs. (active) per acre of simazine while sugar cane tolerates over 16 lbs. per acre of the herbicide. Both compounds are effective chiefly as pre-emergence treatments. Cox (1962) has reported also that 20 lbs. per acre of various triazine derivatives is an effective dosage for non-selective uses. Methoxy- and methylmercapto-substituted triazines were found to destroy established weeds more quickly than chloro-substituted derivatives, although atrazine resembled the methoxy substituted triazines in this respect. Granular formulations of simazine and atrazine have performed more satisfactorily than the wettable powders for nonselective uses.

Residual weed control with pre-emergence applications of simazine and atrazine is roughly equivalent to that obtained with the substituted ureas. Variations in soil type and rainfall may modify the relative residual persistence. The degree of water solubility of the various triazines determines, to some extent, their specific use. Simazine which is soluble at the rate of 5 ppm. in water is considered more desirable for use in areas of heavy rainfall, while atrazine which is more soluble (i.e. 70 ppm.)

is better fitted for areas with moderate rainfall. There are notable exceptions to this general rule. Atrazine for example has
given much better residual control of weeds than simazine in
Trinidad sugar-cane fields (Annual Report, 1960-61). This is difficult to understand in terms of water-solubility but it is quite
likely that differences in adsorptive capacities of soils, stability
to elevated soil temperatures, and other undetermined environmental factors influence relative performance.

Some features which have attracted growers to the triazines
are 1) marked selectivity and "safety" to the crop; this is particularly so in sugar cane, 2) relatively low mammalian toxicity,
and 3) long residual control of weeds with single pre-emergence
applications.

Metabolism. According to Gysin and Knüsli (1960) simazine
may be degraded in corn tissues as follows:

$$C_2H_5HN-C \quad C-NHC_2H_5 \qquad \xrightarrow{corn\ juice} \qquad C_2H_5HN-C \quad C-NHC_2H_5$$

simazine   hydroxysimazine

(R = some natural plant
constituent, possibly a protein)

$$H + CO_2$$

Montgomery and Freed (1959, 1960) found that corn plants
treated with C^{14} ring-labeled simazine or atrazine released $C^{14}O_2$
shortly after treatment. At harvest only traces of unreacted triazine could be recovered from treated tissues. Zweig and Ashton
(1961) have reported that atrazine greatly inhibits $C^{14}O_2$ fixation
in excised bean leaves. They found that sucrose synthesis was
inhibited in atrazine-treated bean tissues. There was also
marked modification of organic acid synthesis. Considerable
evidence exists that the selectivity of symmetrical triazines is
due to differential degradation rates within tissues of different

species (Roth, 1957). If simazine is added to fresh juice of corn (a tolerant species) the compound is degraded within 100 hours, but if the chemical is added to wheat (a susceptible species) juice over 90 percent of unreacted simazine can be recovered unaltered after 100 hours. Davis, Funderburk and Sansing (1959) have reported that considerable degradation of simazine occurs in the roots of intact plants. Sheets (1960) has shown that the triazines are susceptible to breakdown by soil microorganisms, particularly in soils of high colloidal content. The acute oral LD of simazine for rats is greater than 5,000 mg./kg.

The Carbamates

Uses. Of the carbamate growth regulators there are six compounds which have been used as herbicides. Their structures are as follows:

isopropyl N-phenylcarbamate (IPC)

isopropyl N-(3-chlorophenyl) carbamate (CIPC)

4-chloro-2-butynyl N-(3-chlorophenyl) carbamate

ethyl N,N-di-n-propylthiolcarbamate (EPTC, Eptam)

sodium N-methyldithiocarbamate

$$CH_3-NH-\overset{\overset{S}{\|}}{C}-S-Na\cdot 2H_2O \text{ (Vapam)}$$

and

2-chlorallyl diethyl dithiocarbamate (CDEC, Vegedex)

$$\underset{C_2H_5}{\overset{C_2H_5}{\diagdown}}N-\overset{\overset{S}{\|}}{C}-S-CH_2-\overset{\overset{Cl}{|}}{C}=CH_2$$

Isopropyl N-phenylcarbamate (IPC) is effective against annual grass weeds in spinach, mustards, and sugar beets (Crafts, 1961). The chloro-substituted analog, CIPC, provides greater residual control of weeds than IPC and is more selective to certain crops. CIPC is less volatile than IPC and is therefore less damaging to crops sensitive to carbamates. Weeds susceptible to CIPC include crabgrass, chickweed, watergrass, and Johnson grass.

Carbyne (4-chloro-2-butynyl N-(3-chlorophenyl) carbamate) is used to control wild oats in wheat, barley, flax, and to some extent in sugar beets.

Eptam (N,N-di-n̲-propylthiolcarbamate) is effective against a broad range of grasses and sedges but is tolerated by beans, beets, broccoli, and a number of other vegetable crops. The compound when incorporated in moist soil, produces volatile vapors which are effective in preventing establishment of germinating weed seeds. Best results are obtained when the compound is incorporated into soil to depths at which contact may be made with germinating weeds.

Vapam (sodium N-methyldithiocarbamate) is a water-soluble herbicide which decomposes in soil to release methylisothiocyanate. Methylisothiocyanate is a fumigant with phytocidal activity against many deep-rooted perennial grasses.

Vegedex (2-chloroallyldiethyldithiocarbamate) has been used in field corn, soybeans, sugar beets, peanuts, and in a number of vegetable crops (Crafts, 1961). The compound is effective against most annual grasses.

Metabolism.   A thorough investigation of the fate of N-phenyl-carbamates in crops which had received heavy doses of IPC and CIPC revealed that residues at harvest did not exceed 0.05 ppm. (Gard and co-workers, 1954, 1957, 1959). Although not proved, it has been suggested that IPC is oxidized to N-hydroxy IPC within plant tissues (Shaw et al, 1960). Studies with

S^{35}-labeled Eptam indicated that at several weeks after treatment, less than 3 percent of the quantity of herbicides initially applied to crops persisted in tissues (Fang and Theisen, 1959). Fang and Yu (1959) have suggested that the sulfur atom is oxidized to the sulfate and that the sulfate is then incorporated into several amino acids and into two unidentified, S-containing, compounds.

The acute oral LD_{50} for rats of IPC is 1,000 mg./kg., of CIPC 3800 mg./kg., of Eptam 3160 mg./kg., and of Vapam 820 mg./kg. (Woodford, 1960). Ivens and Blackman (1949) have presented strong evidence that the carbamate esters may associate with lipophilic side chains of proteins within the nuclear spindle of meristematic tissues. The herbicidal activity of the carbamates may be linked to the precipitation of proteins and disruption of the nuclear spindle in susceptible cells.

Substituted Benzoic Acids

Uses.   Zimmerman and Hitchcock demonstrated in 1942 that halogenated benzoic acids were active growth regulators. Several years elapsed before it was realized that these substances could be used as herbicides. One of the essential features of chloro-substituted benzoic acids is their residual persistence in soils. Another characteristic is their narrow spectrum of selectivity. The extensive trials of Furtick (1958) proved that 2,3,6-trichlorobenzoic acid

is the most active of the chloro-substituted analogs, followed by 2,3,5,6-tetrachlorobenzoic acid

More recently 2,4-dichloro-3-nitrobenzoic acid (Dinoben)

and 3-amino-2,5-dichlorobenzoic acid (Amiben)

have shown promise for weed control in tomatoes, beans, pumpkins, corn, soybeans, carrots, peppers, and cole crops. Preliminary experiments indicate that the substituted benzoic acids may be of value for control of grasses and broad-leaved weeds in sugar cane (Annual Report, 1960-61). The prolonged persistence of halogenated benzoic acids in clay soils and under heavy rainfall favors their use in the tropics. Their rather narrow range of selectivity limits their use to either pre-emergence herbicides or soil sterilants.

Metabolism. Very little is known about the fate of substituted benzoic acid within plant tissues. Tritiated 2,3,6-trichlorobenzoic acid is translocated out of bean leaves within 3 hours after application (Mason, 1961). The compound is accumulated in the trifoliate leaf buds and more than 50 percent of the radioactivity is located in the bud two days after treatment. 2,3,6-TBA is translocated much more slowly in corn, a species relatively resistant to the chemical. There is no evidence, however, that corn tissues can degrade 2,3,6-TBA more rapidly than do bean tissues. In general the substituted benzoic acids are relatively resistant to degradation within the plant (Crafts, 1961) and accumulate readily in meristematic regions. The LD_{50} of 2,3,6-TBA (acute oral for rats) is 705 to 1500 mg./kg., while that of 3-amino-2,5-dichlorobenzoic acid is 3500 mg./kg. (Woodford, 1960).

OTHER TYPES OF PLANT GROWTH REGULATORS

Employed as Herbicides

Dalapon (2,2-Dichloropropionic Acid). Dalapon is used chiefly to control monocotyledenous weeds. Usually employed as the sodium salt, the compound is translocated readily from leaves or from roots and either inhibits the growth of the weed or destroys it. Dosages of 4 to 8 lbs. per acre are effective in controlling a broad spectrum of grass species under tree and bush fruit, in ornamental crops, and in asparagus. The compound has also proved useful in controlling most grass species in sugar cane or along edges of field and in drains. Best results have been obtained when dalapon is applied to actively growing grass weeds at an early stage in their growth (Crafts, 1961).

At temperatures between 25° C. and 50° C. dalapon hydrolyzes to yield pyruvic acid, sodium chloride, and hydrochloric acid according to the following equation:

$$C - H_3 - \underset{\underset{Cl}{|}}{\overset{\overset{Cl}{|}}{C}} - \underset{\overset{|}{O \ Na}}{\overset{\overset{O}{\parallel}}{C}} \quad + H_2O = CH_3 - \overset{\overset{O}{\parallel}}{C} - \underset{OH}{\overset{\overset{O}{\parallel}}{C}} \quad + NaCl + HCL$$

Foy (1961) using Cl^{36}-labeled and C^{14}-labeled dalapon noted that the compound was not metabolized very quickly in cotton, corn, or in sorghum plants. Eight days after treatment of plants with Cl^{36}-labeled compound only unreacted dalapon could be recovered. In an interesting series of experiments, Foy was able to show that seeds of wheat and barley that had received pre-emergence treatments of dalapon (i.e. 4 lbs. per acre) carried unreacted dalapon through the second and into a third generation! Fortunately the acute oral LD_{50} for rats of dalapon is 6590 to 8120 mg./kg., a figure which places it among the least toxic herbicides, despite its slow rate of metabolism in plant tissues (Woodford, 1960).

Maleic Hydrazide (1,2-Dihydro-pyridazine-2,6-dione)

Although maleic hydrazide

has not been used as extensively as a herbicide as much as for other uses (see sections below), the compound has stimulated a great deal of research. There is probably more published data dealing with the growth-regulating activity of maleic hydrazide than for any other single herbicide with the possible exception of 2,4-D. Schoene and Hoffman (1949) of the U. S. Rubber Co. reported in 1949 that maleic hydrazide would inhibit the growth of tomato plants. Later Currier and Crafts (1950) showed that the compound would inhibit the growth of grasses selectively.

Grasses which are affected by maleic hydrazide exhibit a dark-green foliage initially, followed by desiccation and death. The compound is effective against a number of important weeds including the following: Johnson grass, quackgrass, wild onions, wild oats, white cockle, and curled dock. By far the most extensive use of maleic hydrazide has been either in nonselective weed control to inhibit rapidly growing grasses and for a number of other nonherbicidal uses which require a potent plant growth regulator.

Maleic hydrazide is readily decomposed by soil microbes (Lembeck and Colmer, 1957). Alcaligenes faecalis and Flavobacterium diffusum can multiply on a medium consisting of the growth regulator, agar, and mineral salts. The persistence of maleic hydrazide in plant tissues has been discussed by Smith et al (1959). The compound forms physical complexes with a number of plant constituents, including gentiobiose, proteins, and ribonucleic acid. Towers, Hutchinson and Andreae (1958) obtained evidence that MHC[14] or MH + glucose[14] when fed to leaves of apples, tobacco, and wheat were rapidly bound to a sugar and formed a stable glycoside. On the other hand it has also been reported that carbonyl labeled C[14]-MH binds strongly in vitro with egg albumin, zein, hexokinase, and with RNA (Shaw et al, 1960). The acute oral LD$_{50}$ of maleic hydrazide for rats is 4000 to 5800 mg./kg. (Woodford, 1960).

Amino Triazole (3-Amino-1,2,4-triazole)

Amino triazole (or amitrol)

$$H - N - N$$
$$H - C - C - NH_2$$
$$C$$

is a growth regulator with rather interesting properties. The compound stimulates growth of plants at low concentrations but at higher concentrations results in severe chlorosis and death. Since the compound is absorbed and translocated rapidly it is useful in controlling perennial weeds and woody species such as poison ivy, poison oak, red and white oaks, and prickly ash. Certain crops, for example, strawberries, cranberries, gladiolus, and citrus are resistant to the compound.

In more sensitive crops, amino triazole has been recommended as a pre-planting treatment. Mixtures of amino triazole and other growth regulators, such as dalapon, have been used for control of perennial grasses.

The rate of metabolism of amino triazole within plant tissues appears to be dependent upon the species. In corn C^{14}-labeled amino triazole is not accumulated, but in soybeans the unaltered molecule accumulates, but only for a short period of time. Over a three-week period there is a steady degradation both in soybeans and in corn (Yost and Williams, 1958). Pinto beans degrade amino triazole within 5 days, forming an unidentified amino-acid-like compound (Racusen, 1958). Convolvulus arvensis degrades amino triazole within 90 hours after application (Herrett and Linck, 1961). French dwarf beans on the other hand form 3-amino-1,2,4-triazolyl alanine from which alpha-alanine can be obtained upon further degradation (Massini, 1959). According to the Federal Register of November 28th, 1959, lots of cranberries or cranberry products which show more than 0.15 ppm. of amino triazole are to be destroyed.

In view of the residue problem the herbicide is now recommended for use in cranberry fields only after the crop has been removed. The acute oral LD_{50} for rats of amino triazole is 1100 mg./kg. (Woodford, 1960).

Chloroacetamides. The chloroacetamides, as exemplified by 2-chloro,N,N-diallylacetamide (CDAA) and 2-chloro,N,N-diethylacetamide (CDEA) are pre-emergence herbicides with activity against germinating grass weeds.

(CDAA)   (CDEA)

At rates of 2 to 8 lbs. per acre CDAA and CDEA have been used successfully for weed control in celery, onions, spinach, carrots, beans, table beets, and other vegetables. Soybeans and corn, as well as other crops resistant to CDAA, can convert the compound to glyoxylic and glycolic acid within 4 to 5 days after the seedlings emerge from the soil. According to Wangerin (1955) susceptible crops cannot degrade CDAA. The acute oral LD_{50} of CDAA to rats is 700 mg./kg. (Woodford, 1960).

Control of Pre-Harvest Fruit Drop

In addition to their widespread use as weed killers, plant growth regulators have also been employed, on a smaller scale, to prevent the abscission of mature fruit prior to harvest. This is particularly important in apple orchards, where the loss of fruit before harvest may result in serious economic losses. Growth regulators are also used to prevent fruit drop in pears, citrus, and apricots.

Abscission of mature apple fruits occurs as a result of weakening of the middle lamella by enzymatic activity. Van Overbeek (1952) has suggested that plant growth regulators may inhibit those enzymes responsible for solubilizing pectins. Dissolution of the middle lamella is thereby prevented and abscission delayed.

Alpha-napthalene acetic acid (NAA) was one of the most popular growth regulators for preventing pre-harvest fruit drop of apples, until it was discovered that the substituted phenoxy acids were more effective (Batjer and Marth, 1945). NAA is not active in the McIntosh, a variety which is prone to losing its fruit prior to harvest. 2,4-D has been used effectively in Winesap apples. Hoffman and Edgerton, in 1952, found that 2,4,5-trichlorophenoxyacetic acid (2,4,5-T) would control fruit drop of McIntosh apples for 4 to 6 weeks after application. 2,4,5-Trichlorophenoxypropionic acid (2,4,5-TP) has a broader spectrum of activity than 2,4-D and has proved effective in McIntosh, Winesap, and other varieties (Edgerton and Hoffman, 1953). In general, NAA acts more quickly than other growth regulators to prevent fruit drop of apples but the phenoxy acids act over a longer period of time (Erickson, Brannaman and Hield, 1952). The phenoxy acids also cause an intensification of red coloration of apples at harvest, a side effect which is advantageous (Audus, 1959).

Control of fruit drop in oranges is achieved with aqueous sprays containing 8 to 10 ppm. of the isopropyl ester of 2,4-D (Erickson, 1951). Stewart, Klotz and Hield (1951) reported that the growth regulator also stimulated the growth of individual fruit. 2,4,5-T has proved to be effective in preventing abscission of lemons and of apricots (Crane, 1953). Crane found that 2,4,5-T would prevent abscission of apricots for 50 to 60 days with a single treatment.

The fate and metabolism within plant tissues of the chlorophenoxy growth regulators has been discussed under Herbicides. Much lower dosages of chemical are used to prevent fruit drop than are normally employed to control weeds. However the growth regulators used to prevent pre-harvest fruit drop are applied much closer to harvest. NAA is known to persist in apple tissues for up to fourteen days after application, but not thereafter (Erickson, Brannaman and Hield, 1952).

Rooting

It has been known since 1934 (Went, 1934) that indole compounds stimulate the growth of root primordia. A practical use of this property is to hasten the propagation of plants from cuttings. Zimmerman and Wilcoxon (1935) found that for the induction of rooting, indole-3-butyric acid (IBA) and alpha-napthalene-acetic acid were more effective than the naturally-occurring indole-3-acetic acid (IAA). Hitchcock and Zimmerman (1930) have reported also that a number of substituted phenoxyacetic acids are equal to or more effective than IBA and NAA. Mono-halogen-substituted derivatives are less active than IBA or NAA, but 2,4-dichloro and 2,4-dibromo derivatives are more effective than the standards. In some species IBA and NAA are ineffective. For example Rhododendron cuttings will root if treated with various phenoxy acids but will not root if exposed to the indoles or to naphthalene derivatives (Wells and Marth, 1954). A high degree of specificity is shown by the Rhododendron variety, E.S. Rand, which responds only to the triethanolamine salt of alpha 2,4,5-trichlorophenoxypropionic acid.

Rooting hormones are applied to cuttings in a number of ways. A practical method is to immerse the base of the cutting into a dilute solution of the growth regulator for two to three hours. Generally 0.0005 to 0.01 percent of compound is sufficient, but the optimum concentration depends upon the species. The rooting compounds are also easily applied in more concentrated solutions but for shorter periods of exposure. Dusts and sprays of the growth regulators also have proved effective.

Several factors influence the ability of cuttings to respond to the growth regulator. Hitchcock and Zimmerman (1930) had found as early as 1930 that the new growth of fruit trees in the spring rooted better than cuttings taken in late autumn or winter. On the other hand some woody tissues root more readily during the winter before buds are formed. Cuttings exposed to the light do not form roots as readily as those kept in darkness (Went, 1935). However if the leaves of a cutting are exposed to light while the basal portion is placed in a rooting medium, rooting may be stimulated (Stoutemeyer and Close, 1946). Similarly, the length of day to which cuttings are exposed influences rooting behavior. Some species root profusely under short photoperiods while others prefer long day lengths (Moskov and Koschezhenko, 1939; Stoutemeyer and Close. 1946).

The metabolism within plant tissues of the growth regulants which are used to induce rooting has not been studied in detail, with the exception of the chlorophenoxy acids. IBA is known to be more persistent than IAA in most tissues, but at the concentrations employed to induce rooting (i.e. 0.0005 to 0.01 ppm.) it is doubtful that IBA presents a residue problem at harvest.

Flowering

The only extensive use of plant growth regulators to induce flowering has been in pineapples. Rodriguez in 1932 noted that brush fires adjacent to pineapple fields in Puerto Rico stimulated the crop to flower. Later, Lewcock (1937) found that ethylene and acetylene would induce a similar response. Clark and Kerns in 1942 reported that 10 to 50 ppm. of NAA sprayed onto foliage of pineapples forced the formation of flower buds. Higher concentrations of NAA inhibit flowering especially if applied late in the crop's development.

In addition to NAA other growth regulators are also effective in promoting floral induction of pineapples. IBA, 2,4-D, ethylene, and acetylene have all been used successfully. Van Overbeek (1946) found that 2,4-D at 5 to 10 ppm. was an effective floral initiator. He calculated that approximately 0.25 mg. of 2,4-D was required, per plant, to give the desired effect.

One of the important practical benefits of forcing pineapples to flower artificially is that the fruits are produced, and are ready to harvest, simultaneously.

The physiological basis for flowering in the pineapple may be related to an accumulation of the natural auxin at the stem apex. Van Overbeek and Cruzado (1948) placed pineapple plants in a horizontal position to reorient the distribution of natural auxin. Such plants flowered precociously, suggesting that an uneven distribution of auxin, with greater quantities concentrated at the apex, induced the formation of flower buds. The stimulation of flowering with applications of synthetic growth regulators would suggest that flowering in the pineapple is favored by high levels of natural auxin. However authentic IAA applied to pineapple foliage has little or no effect on flowering. Perhaps the fact that pineapple tissues are rich in IAA-destroying enzymes (Gordon and Nieva, 1949) accounts for the ineffectiveness of exogenously-applied IAA in this species.

Parthenocarpy

Another important practical use for plant growth regulators is in supplementing natural pollination and in promoting the set of

seedless fruit. Lack of fruit set is a serious problem in certain
crops. In the tomato, for example, low night temperatures and
cool, wet weather may often limit the extent of pollination and result in reduced yields. Applications of growth regulators not only
induce parthenocarpy, but may delay abscission of flowers, may
increase the rate of ripening, and may in some cases increase
fruit size.

Among the growth regulators which have been used to overcome poor fruit set in the tomato, beta-napththoxyacetic acid
(NOA) and p-chlorophenoxyacetic acid (DCA) are two of the most
effective. These two compounds are "weak" growth regulators
and therefore do not produce undesirable side effects on other
portions of the plant. Audus (1959) has presented an extensive
list of growth regulators recommended for inducing parthenocarpy in tomatoes and other crops. The reader is referred to
that work for horticultural details.

Derivatives of phthalamic acid, for example N-meta-tolyl-
phthalamic acid, have shown considerable activity in setting fruit
of the tomato (Hoffman and Smith, 1949). Renewed interest in the
substituted benzoic acids has also taken place, but this group
produces a number of leaf deformations in the tomato and it is
doubtful that it may be used successfully in that crop.

Benzotriazole-2-oxyacetic acid is effective in inducing parthenocarpy in the Calimyrna fig (Crane, 1952). The chemical
does not eliminate the development of the seed coat which is considered desirable from a consumer standpoint. Previously phenoxyacetic acids had been used to produce seedless figs but since
most people are accustomed to eating figs with seeds the practice
was discontinued.

Most plant physiologists accept the theory that fruit set and
parthenocarpy are under the control of naturally-occurring
auxins. Pollen extracts have been known for many years to be
active in inducing parthenocarpy (Yasuda, 1934). Gustafson demonstrated in 1936 that synthetic growth regulators would produce
a response similar to that induced by the natural products. To
date, only tomatoes and figs have benefited substantially from the
use of synthetic growth regulators to induce parthenocarpy.

Wittwer et al in 1957 showed that the gibberellins are extremely active in inducing parthenocarpy in the tomato. At concentrations as low as 0.001 percent in lanolin, gibberellic acid
was as effective as 0.1 percent of IAA. Undoubtedly the gibberellins will eventually rival the older types of plant growth regulators for such uses.

The Use of Plant Growth Regulators To Promote
Abscission of Leaves, Flowers, and Fruits

It may seem paradoxical that, on the one hand, growth regulators are used to prevent pre-harvest drop of fruit and, on the other hand, they also stimulate abscission of a variety of plant organs. However the difference in effect may be due to differences in the capability of various species to set parthenocarpic fruit. For example, Bartlett pears set fruit in response to sprays of NAA but other pear varieties (i.e. Anjou) respond by shedding their fruit.

Some of the most effective growth regulators for inducing abscission are "antagonists" of IAA and of other auxins. It has been proposed that the onset of abscission is under the control of naturally-occurring auxins (i.e. IAA), and as the supply of the natural substance to a given organ is interrupted, abscission follows. However certain types of defoliants act directly on the abscission layer without showing any obvious relationship to natural auxin systems (Audus, 1959).

The use of synthetic growth regulators to promote abscission of leaves has facilitated mechanized harvesting of cotton, potatoes, and beans. Desiccants have been used more extensively than "true" growth regulators to defoliate potatoes and beans, although sprays containing triodobenzoic acid (TIBA) are effective in promoting the abscission of bean leaves (Audus, 1959). Growth regulators have found more of a market for regulating the abscission of flowers and fruits.

Some orchard fruit trees tend to bear fruit in a biennial cycle. If excessive numbers of flowers are produced on a tree in any single year, there is always the danger that the tree will not produce an economical crop the following year. Therefore it is good orchard practice to "thin" the tree of excess flowers or fruits. Hand-thinning is laborious and expensive. A spray of 10 to 30 ppm. of NAA or napthalene acetamide (NAA $_c$) applied just prior to petal fall will cause apple fruits to absciss. 2,4,5-T is more effective than NAA for thinning certain varieties of pears. Olives may also be thinned with applications of about 125 ppm. of NAA in a light oil carrier (Hartmann, 1952), but the sprays must be applied only after the fruit have reached a diameter of 3 to 5 mm. The metabolism within tissues of the growth regulators used to induce abscission has rarely been followed closely.

Other Important Uses

In addition to the foregoing, there are a number of other practical uses for plant growth regulators. They have been used to prevent sprouting in stored crops, to increase fruit size, and to modify the chemical composition of crops.

Dormancy of stored tubers, such as the potato and onion, may be prolonged with treatments of NAA or maleic hydrazide (Guthrie, 1938; Paterson et al, 1952). Apple fruits remain firmer and store for longer periods if, prior to harvest, the trees are sprayed with maleic hydrazide (Smock, Edgerton and Hoffman, 1951). Citrus fruits pre-treated with sprays containing 500 ppm. of 2,4-D or 2,4,5-T resist attacks in storage by the Black Button organism (Alternaria spp.) (Stewart, Palmer, and Hield, 1952).

Increases in sizes of fruits as a result of treatments with growth regulators have been recorded by many workers (Clark and Kerns, 1943; Crane and Blondeau, 1949; Crane and Brooks, 1952; Erickson and Brannaman, 1950; Kraus et al, 1948; Weaver and Mc Cune, 1958; Weaver and Williams, 1950; and Zielinski and Garren, 1952). Although Clark and Kerns have shown that the pineapple is a classical example of a fruit whose size may be regulated by applications of NAA and other substances, the startling effect of gibberellic acid on the size of Thompson seedless grapes is likewise a notable achievement (Weaver and McCune, 1958). Other fruits which have responded to treatments with various growth regulators include apricots, strawberries, and figs.

One of the most neglected potential uses for plant growth regulators is for the modification of chemical composition in food crops. Preliminary experiments in sugar cane indicate that a number of growth regulators, applied as pre-harvest sprays to foliage, may accelerate ripening to permit harvesting when the maximum quantity of sucrose is in the stalk (Annual Report, 1960-61). Beauchamp (1950a, 1950b) obtained gains in sucrose following sprays with acetylene and 2,4-D. His results were confirmed by Mathur (1955) in India and by Brazilian workers (Anon., 1951) but not by others (Loustalot, Cruzado and Muzik, 1950; Lugo-López and Grant, 1952).

An interesting application of growth regulators is their use to modify the synthesis of peppermint oil by Mentha piperita. Dilute solutions of NAA applied to peppermint seedlings increase oil content of mature plants by over 40 percent. The menthol content of the oil of treated plants is increased by 9 percent (Audus, 1959).

Of considerable economic importance is the use of various growth regulators to stimulate the latex flow of rubber trees. It has been suggested that the effect of the growth regulator is not merely to increase water flow but is directly related to increasing rubber production (Audus, 1959).

In the future it would not be too surprising to find that the nutritive value of many crops, including those rich in proteins, may be augmented by specific growth regulators applied at the same stage during production.

Plant Growth Regulators and Human Physiology

Any discussion of the possible role of plant growth regulators in human physiology must, at present, be speculative. It is known that several groups of compounds with plant growth-regulating activity occur in foods (see Table 8.1).

Table 8.1. Food Plants Known To Contain Indole-3-Acetic Acid, Indole-3-Acetonitrile, or Indole-3-Pyruvic Acid (Vlitos, 1960)

Plant	Compound (Isolated or Detected)
Bamboo Shoots	Indole-3-acetic acid
Beans	Indole-3-acetic acid
Black Currant	Indole-3-acetic acid
Brussels Sprouts and Cabbage	Indole-3-acetaldehyde Indole-3-acetic acid Indole-3-acetonitrile
Cauliflower	Indole-3-acetic acid Indole-3-acetonitrile
Citrus (flowers)	Indole-3-acetic acid
Coconut Milk	Indole-3-acetic acid
Corn	Indole-3-acetic acid Indole-3-pyruvic acid
Grapes	Indole-3-acetic acid Indole-3-acetonitrile
Gooseberry	Indole-3-acetic acid
Oat (Coleoptiles)	Indole-3-acetic acid
Peas	Indole-3-acetic acid
Soybean (leaves and stems)	Indole-3-pyruvic acid
Strawberry	Indole-3-acetic acid

Upon ingestion it is quite likely that these compounds undergo
a sequence of metabolic changes. Most of the reaction sequences
for indole growth regulators have been worked out in plants but
some information is available from studies in humans (Jepson,
1958). It is known that an enzyme system catalyzes the synthesis
of IAA from tryptophan in spinach leaves; possibly a similar system occurs in the human. Zinc deficiencies are linked with the
conversion of tryptophan to IAA (Skoog, 1940; Tsui, 1948). Unless
zinc is present in sufficient quantities in plant tissues the subsequent synthesis of IAA is suppressed. It has been accepted for
many years that the chemical pathway leading from tryptophan to
IAA is as outlined in Figure 8.1 (Gordon, 1954). The reaction sequence is based on a number of studies with crude enzyme preparations devised mainly from plant sources.

The fate of plant growth regulators in human metabolism has
been studied by Jepson (1958). He has reported that when IAA is

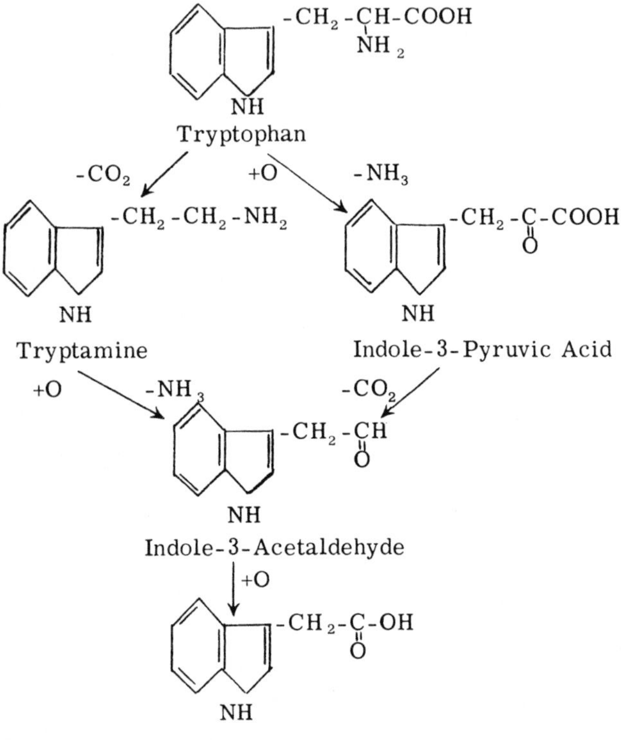

Fig. 8.1. Indole-3-Acetic Acid.

fed to humans, the growth regulator is conjugated to form indole-3-acetylglucosiduronic acid. A 500 mg. oral dose of IAA was accounted for in the urine excreted over a 24 hour period; 200 mg. were present as the glucuronide and the rest as free IAA. Normal urines were also found to contain traces of the glucuronide, which is rapidly converted to IAA or to indole-3-acetamide in the presence of ammonia. Jepson has also reported that indole-3-acetylglucosiduronic acid will yield glucuronic acid and IAA after alkaline hydrolysis or after exposure to the enzyme, glucuronidase. It is of interest to record that when IAA is "fed" to plant tissues it is also conjugated to form indole-3-acetylaspartic acid (Good, Andreae and Van Ysselstein, 1956). It seems reasonable to assume that when an indole growth regulator is ingested by animals the compound is detoxified as a result of conjugation with other cellular metabolites.

LITERATURE CITED

Annual Report. 1960-1961. Central Agr. Res. Sta. Carapichaima, Trinidad.

Anonymous. 1951. Brasil acucareiro 37:328.

Audus, L. J. 1951. The biological detoxication of 2,4-dichlorophenoxyacetic acid in soil. Plant and Soil 3:170-92.

_____. 1959. Plant Growth Substances. London. Leonard Hill, Ltd.

Batjer, L. P. and Marth, D. C. 1945. New materials for delaying fruit abscission of apples. Science 101:363-64.

Beauchamp, C. E. 1950a. Effects of 2,4-D on sugar content of sugar cane. Sugar J. 13:57-70.

_____. 1950b. Experiments to increase the sugar content of sugar cane by means of hormones. Proc. 24th Ann. Conf. Cuban Sugar Technol. Assoc. 147-64.

Boysen-Jensen, P. 1911. La transmission de l'irritation phytotropique dans l'Avena. K. Danske Vidensk. Selsk. 3:1-24.

Brown, J. W. and Mitchell, J. W. 1948. Inactivation of 2,4-D in soil as affected by soil moisture, temperature, the addition of manure and autoclaving. Bot. Gaz. 109:314-23.

Carroll, R. B. 1949. Apparent inactivation of 2,4-D by riboflavin in light. Am. J. Bot. 36:supp. 1. 821.

Clark, H. E. and Kerns, K. R. 1942. Control of flowering with phytohormones. Science 95:536.

_____. 1943. Effects of growth-regulating substances on parthenocarpic fruits. Bot. Gaz. 104:639-44.

Cox, J. R. 1962. Triazine derivatives as nonselective herbicides. J. Sci. Food and Agr. 13:99-103.

Crafts, A. S. 1960. Evidence for hydrolysis of esters of 2,4-D during absorption by plants. Weeds 8:19-25.

_____. 1961. The Chemistry and Mode of Action of Herbicides. New York. Interscience Publ.

_____, and Yamaguchi, S. 1960. Absorption of herbicides by roots. Am. J. Bot. 47:248-55.

Crane, J. C. 1952. Growth-regulator specificity in relation to ovary wall development in the fig. Science 115:238-39.

_____. 1955. Further responses of the apricot to 2,4,5-trichlorophenoxyacetic acid application. Proc. Am. Soc. Hort. Sci. 61:163-74.

_____ and Blondeau, R. 1949. Controlled growth of fig fruits by synthetic hormone application. Ibid. 54:102-8.

_____ and Brooks, R. M. 1952. Growth of apricot figs as influenced by 2,4,5-trichlorophenoxyacetic acid application. Ibid. 59:218-24.

Crosby, D. and Vlitos, A. J. 1961. New auxins from Maryland Mammoth tobacco, in Plant Growth Regulation. Ames, Iowa. Iowa State Univ. Press.

Cross, B., Grove, E., Macmillan, J. F., and Mulholland, T. P. C. 1956. Gibberellic acid IV. The structures of gibberic and allogibberic acids and possible structures for gibberellic acid. Chem. and Ind. Rev. 954-55.

Currier, H. B. and Crafts, A. S. 1950. Maleic hydrazide, a selective herbicide. Science 111:152-53.

Darwin, C. 1881. The Power of Movement in Plants. New York. D. Appleton and Co.

Davis, D. E., Funderburk, H. H., Jr., and Sansing, N. G. 1959. The absorption and translocation of C^{14}-labeled simazine by corn, cotton, and cucumber. Weeds 7:300-309.

Derscheid, L. A. 1952. Physiological and morphological response of barley to 2,4-dichlorophenoxyacetic acid. Plant Physiol. 27:121-34.

E.I. du Pont de Nemours & Co. 1958. Monuron. Mimeo.

_____. 1958. Diuron. Ibid.

Edgerton, L. J. and Hoffman, M. B. 1953. The effect of some growth substances on leaf petiole abscission and preharvest fruit drop of several apple varieties. Proc. Am. Soc. Hort. Sci. 62:159-66.

Ellinger, A. 1905. Uber die Konstitution der Indolegruppe im Eiweiss II. Mittheilung: Synthese der Indole-Pr-3-propionsäure. Ber. deut. Chem. Ges. 38:2884.

Erickson, L. C. 1951. Effects of 2,4-D on drop of navel oranges. Proc. Am. Soc. Hort. Sci. 58:46-52.

_____ and Brannaman, B. L. 1950. Effects on fruit growth of a 2,4-D spray applied to lime trees. Ibid. 56:79-82.

_____, _____, and Hield, H. Z. 1952. Response of Delicious and Rome Beauty apples to a preharvest spray of 2,4,5-T in southern California. Ibid. 60:160-64.

Fang, S. C. and Theisen, P. 1959. An isotopic study of ethyl N,N-di-n-propylthiolcarbamate (EPTC-S^{35}) residue in various crops. J. Agr. Food Chem. 7:770-71.

_____ and Yu, T. C. 1959. Absorption of EPTC-S^{35} by seeds and its metabolic rate during early stages of germination. Res. Prog. Rept. Western Weed Control Conf. pp. 91-92.

_____, Freed, V. H., Johnson, R. H., and Coffee, D. R. 1955. Absorption, translocation and metabolism of radioactive 3-(p-chlorophenyl)-1,1-dimethylurea (CMU) by bean plants. J. Agr. Food Chem. 3:400-402.

Foy, C. L. 1958. Ph.D. dissertation, Univ. of California, Davis. (Cited by Crafts 1961.)

Friesen, H. A. and Forsberg, D. E. 1959. Neburon as a selective herbicide for Tartary buckwheat and wild buckwheat. Weeds 7:47-54.

Furtick, W. R. 1958. Hormolog 2 (1):7.

Gard, L. N. and Reynolds, J. R. 1957. Residues in crops treated with isopropyl N-(3-chlorophenyl) carbamate and isopropyl N-phenylcarbamate. J. Agr. Food Chem. 5:39-41.

_____, Ferguson, C. E., Jr., and Reynolds, J. L. 1959. Effect of higher application rates on crop residues of isopropyl N-phenylcarbamate and isopropyl N-(3-chlorophenyl) carbamate. Ibid. 7:335-38.

_____, Pray, B. O., and Rudd, N. G. 1954. Residues in crops receiving pre-emergence treatment with isopropyl N-(3-chlorophenyl) carbamate. Ibid. 2:1174-76.

Goldacre, P. L. and Bottomley, W. 1959. A kinin in apple fruitlets. Nature 184:555-56.

Good, N. E., Andrea, W. A., and Van Ysselstein, M. H. 1956. Studies on 3-indoleacetic acid metabolism II. Some products of the metabolism of exogenous indoleacetic acid in plant tissues. Plant Physiol. 31:231-35.

Gordon, S. A. 1954. Occurrence, formation, and inactivation of auxin. Ann. Rev. Plant Physiol. 5:341-78.

_____ and Nieva, F. S. 1949. The biosynthesis of auxin in the vegetative pineapple. I. The nature of active auxins. II. The precursors of indoleacetic acid. Arch. Biochem. and Biophys. 20:256-66 and 267-85.

Gustafson, F. G. 1936. Inducement of fruit development by growth-promoting chemicals. Proc. Nat. Acad. Sci. (Wash., D. C.) 22:628-36.

Guthrie, J. D. 1938. Effect of ethylene, thiocyanohydrin, ethyl carbylamine, and indoleacetic acid on the sprouting of potato tubers. Contrib. Boyce Thompson Inst. 9:265-72.

Gysin, H. and Knüsli, E. 1960. Chemistry and herbicidal properties of triazine derivatives, in Advances in Pest Control Research. Interscience Publ. 3:289-358.

Halliday, D. J. and Templeman, W. G. 1951. Field experiments in selective weed control by plant growth regulators. IV. The effect of plant-growth regulators upon the productivity of grassland. Emp. J. Exp. Agr. 19:104-12.

Hartmann, H. T. 1952. Spray thinning of olives with naphthaleneacetic acid. Proc. Am. Soc. Hort. Sci. 59:187-95.

Herrett, R. A. and Linck, A. J. 1961. The metabolism of 3-amino-1,2,4-triazole by Canada thistle and field bindweed and the possible relation to its herbicidal action. Physiol. Plant. 14:767-76.

Hill, G. D. and McGahen, J. W. 1955. Southern Weed Control Conf. Proc. 8:284.

Hitchcock, A. E. and Zimmerman, P. W. 1930. Rooting of greenwood cuttings as influenced by the age of tissue. Proc. Am. Soc. Hort. Sci. 27:136-38.

_____. 1942. Root-inducing activity of phenoxy compounds in relation to their structure. Contrib. Boyce Thompson Inst. 12:497-507.

Hoffman, M. B. and Edgerton, L. J. 1952. Comparison of NAA, 2,4,5-TP, and 2,4,5-T for controlling harvest drop of McIntosh apples. Proc. Am. Soc. Hort. Sci. 59:225-30.

Hoffman, O. L. and Smith, A. E. 1949. A new group of plant growth regulators. Science 109. 588.

Ivens, G. W. and Blackman, G. E. 1949. The effects of phenylcarbamates on the growth of higher plants. Soc. Exp. Biol. Symp. 3:266-82.

Jablonski, J. R. and Skoog, F. 1954. Cell enlargement and cell division in excised tobacco pith tissue. Physiol. Plant. 7: 16-24.

Jaworski, E. G. and Butts, J. S. 1952. Studies in plant metabolism. II. The metabolism of C^{14}-labeled 2,4-dichlorophenoxyacetic acid in bean plants. Arch. Biochem. Biophys. 38: 207-18.

Jepson, J. B. 1958. Indolylacetamide: a chromatographic artifact from the natural indoles, indolylacetylglucosiduronic acid and indolepyruvic acid. Bioch. J. 69:22P.

Kögl, F., Haagen-Smit, A. J., and Erxleben, H. 1933. Uber ein Phytohormon der Zellstreckung Reindarstellung des Auxins aus Menschlichen Harn. Hoppe-Seyler's Z. Physiol. Chem. 214:241-62.

_____. 1934. Uber ein neues auxin (heteroauxin) aus Harn. Ibid. 228:90-103.

_____ and Erxleben, H. 1934. Uber die Konstitution der auxin a und b. Ibid. 227:51-73.

Kraus, B., Fo, J., Clark, H. E., and Nightingale, G. T. 1948. Use of BNA sprays to delay ripening, improve pineapple yields and fruit shape. Pineapple Res. Inst. Special Rept. 10:1-14.

Lawrie, I. D. and Vlitos, A. J. 1962. Chemical weed control in Trinidad sugar cane. Trop. Agr. 39:33-48.

Leafe, E. L. 1962. Metabolism and selectivity of plant growth regulator herbicides. Nature 193:485-86.

Lembeck, W. J. and Colmer, A. R. 1957. Aspects of the decomposition and utilization of maleic hydrazide by bacteria. Weeds 5:34-39.

Lewcock, A. K. 1937. Acetylene to induce flowering in pineapple. Queensland Agr. J. 48:532-43.

Longchamp, R., Roy, M., and Gautheret, R. J. 1952. Récherches sur les modifications du rendement des céréales par les heteroauxins desherbantes. C. R. Acad. Sci. Paris 4:669.

Loustalot, A. J., Cruzado, H. J., and Muzik, T. J. 1950. The effect of 2,4-D on sugar content of sugar cane. Sugar J. 13:78.

Lugo-López, M. A. and Grant, R. 1952. J. Agr. Univ. Puerto Rico 36:187.

Macmillan, J. and Suter, P. J. 1958. The occurrence of gibberellin A in higher plants. Isolation from the seed of runner bean (Phaseolus multiflorus). Naturwissen. 45:46.

Mason, G. W. 1959. Ph.D. dissertation, Univ. of California, Davis. (Cited by Crafts, 1961.)

Massini, P. 1959. Synthesis of 3-amino-1,2,4-triazolyl alanine from 3-amino-1,2,4-triazole in plants. Biochem. et Biophys. Acta 36:548-49.

Mathur, P. S. 1955. Proc. 2nd Bien. Conf. Sugar cane Res. and Dev. Workers in Indian Union 637.

Miller, C. O., Von Saltza, M. H., and Strong, F. M. 1955. Structure and synthesis of kinetin. J. Am. Chem. Soc. 77:2662.

Mitchell, J. W. and Hamner, C. L. 1944. Polyethylene glycols as carriers for growth-regulating substances. Bot. Gaz. 105:474-83.

_____, Hodgson, R. E., and Gaetjens, G. F. 1946. Tolerance of farm animals to feed containing 2,4-D. J. Anim. Sci. 5:194-96.

Montgomery, M. and Freed, V. H. 1959. The uptake and metabolism of simazine and atrazine by corn plants. Western Weed Control Conf. Prog. Rept. 93-94.

_____. 1960. The absorption, translocation, and metabolism of triazine herbicides by corn. Weed Soc. Am. Proc. 41.

Moskov, B. S. and Koschezhenko, I. E. 1939. The rooting of woody cuttings as dependent upon photoperiodic conditions. Compt. Rend. (Doklady) Acad. Sci. (Moscow) 24:489.

Nielsen, N. 1930. Untersuchungen uber einen neuen Wachstumregulierenden Stoff: Rhizopin. Jahr. Wis. Bot. 73:125-91.

Paál, A. 1919. Uber phototropische Reizleitung. Ibid. 58:406-58.

Paterson, D. R., Wittwer, S. H., Weller, L. E., and Sell, H. M. 1952. Effect of preharvest foliar sprays of maleic hydrazide on storage of potatoes. Plant Physiol. 27:135-42.

Phinney, B. O., West, C. A., Ritzel, M., and Neely, P. M. 1956. Evidence for gibberellin-like substances from flowering plants. Proc. Nat. Acad. Sci. (Wash., D. C.) 42:185-89.

Racusen, D. 1958. The metabolism and translocation of 3-aminotriazole in plants. Arch. Biochem. Biophys. 74:106-13.

Rodriguez, A. K. 1932. Smoke and ethylene in fruiting of pineapple. J. Dept. Agr. Puerto Rico 26:5-18.

Roth, W. 1957. Étude comparée de la reaction mais et de blé la simazin, a substance herbicide. Compt. rend. 245:942-44.

Schoene, D. L. and Hoffman, O. L. 1949. Maleic hydrazide, a unique growth regulant. Science 109:588-90.

Shaw, W. C., Hilton, J. L., Moreland, D. E., and Jansen, L. L. 1960. The nature and fate of chemicals applied to soils, plants, and animals. Agr. Res. Serv. 20-9:119-33.

Sheets, T. J. 1960. The uptake and distribution of 2-chloro-4,6-bis (ethylamino)-s-triazine in oat and cotton seedlings. Weed Soc. Am. Proc. pp. 44-45.

Skoog, F. 1940. Relationships between zinc and auxin in the growth of higher plants. Am. J. Bot. 27:939-51.

Slade, R. E., Templeman, W. G., and Sexton, W. A. 1945. Plant-growth substances as selective weed killers. Differential effects of plant-growth substances on plant species. Nature 155:497.

Smith, A. E., Zukel, J. W., Stone, G. M., and Riddell, J. A. 1959. Factors affecting the performance of maleic hydrazide. J. Agr. Food Chem. 7:341-44.

Smock, R. M., Edgerton, L. J., and Hoffman, M. B. 1951. Effects of maleic hydrazide on softening and respiration of apples. Proc. Am. Soc. Hort. Sci. 58:69-72.

Staniforth, D. W. 1952. Effects of 2,4-dichlorophenoxyacetic acid on meristematic tissues of corn. Plant Physiol. 27:803-11.

Steward, F. C. and Shantz, E. M. 1956. The chemical induction of growth in plant tissue culture, in Chemistry and Mode of Action of Plant Growth Substances. London. Butterworths.

Stewart, W. S., Klotz, L. J., and Hield, H. Z. 1951. Effects of 2,4-D on fruit drop and quality of navel oranges. Hilgardia 21:161-93.

_____, Palmer, J. E., and Hield, H. Z. 1952. Use of 2,4-D and 2,4,5-T to increase storage life of lemons. Proc. Am. Soc. Hort. Sci. 59:327-34.

Stoutemeyer, V. T. and Close, A. W. 1946. Rooting cuttings and germinating seeds under fluorescent and cold cathode lighting. Ibid. 48:309-25.

Stowe, B. B. and Yamaki, T. 1957. The history and physiological action of the gibberellins. Ann. Rev. Plant Physiol. 8:181-216.

Thimann, K. V. 1935. On the plant growth hormone produced by Rhizospus suinus. J. Biol. Chem. 109:279-91.

Towers, G. H. N., Hutchinson, A., and Andreae, W. A. 1958. Formation of a glycoside of maleic hydrazide in plants. Nature 181:1535-36.

Tsui, C. 1948. The role of zinc in auxin synthesis in the tomato plant. Am. J. Bot. 35:172-78.

Van Overbeek, J. 1952. Agricultural applications of growth regulators and their physiological basis. Ann. Rev. Plant Physiol. 3:87-108.

_____. 1946. Control of flower formation and fruit size in pineapple. Bot. Gaz. 108:64-73.

_____ and Cruzado, H. J. 1948. Flower formation in the pineapple plant by geotropic stimulation. Am. J. Bot. 35:410-12.

Vlitos, A. J. 1960. The plant hormones - their occurrence in foods and possible role in human physiology. Rev. of Nutrition Research (Borden's) 21:13-26.

Wain, R. L. and Wightman, F. 1954. The growth-regulating activity of certain omega-substituted arylcarboxylic acids in relation to their beta-oxidation within the plant. Roy. Soc. London Proc. (B) 142:525-36.

Wangerin, R. R. 1955. CDAA - Controls weed grasses in grass family crops. Farm Chem. 118:47-49.

Weaver, R. J. and McCune, S. B. 1958. Gibberellin tested on grapes. Calif. Agr. 12.

_____ and Williams, B. O. 1950. Response of flowers to Black Corinth and fruit of Thompson Seedless grapes to applications of plant growth regulators. Bot. Gaz. 111:477-85.

Weintraub, R. L., Brown, J. W., Fields, M., and Rohan, J. 1952.

Metabolism of 2,4-dichlorophenoxyacetic acid. I. $C^{14}O_2$ production by bean plants treated with labeled 2,4-dichlorophenoxyacetic acid. Plant Physiol. 27:293-301.

Wells, J. S. and Marth, P. C. 1954. Evaluation of halogen-substituted phenoxyacetic acids and other growth regulators in rooting Rhododendron and Ilex. Proc. Am. Soc. Hort. Sci. 63:765-68.

Went, F. W. 1928. Wuchsstoff und Wachstum. Rec. Trav. Bot. Neerl. 25:1-116.

_____. 1934. A test method for rhizocaline, the root-forming substance. Proc. Kon. Nederl. Akad. Wetenshap. Amsterdam 37:445-55.

_____. 1935. Hormones involved in root formation. Proc. 6th International Bot. Cong. 2:267-69.

Wieland, O. P., de Ropp, R. S., and Avener, J. 1954. The identity of urine in normal urine. Nature 173:776.

Wittwer, S. H., Bukovac, M. J., Sell, H. M., and Weller, L. E. 1957. Some effects of gibberellin on flowering and fruit setting. Plant Physiol. 32:39-41.

Woodford, E. K. 1960. Weed Control Handbook. Oxford. Blackwell.

Yasuda, S. 1934. The second report on the behavior of the pollen tubes in the production of seedless fruits caused by interspecific pollination. Japan. J. Genet. 9:118-24.

Yost, J. F. and Williams, E. P. 1958. Some residue and metabolism findings in the development of 3-amino-1,2,4-triazole. Northeast. Weed Control Conf. Proc. 12:9-15.

Zielinski, Q. B. and Garren, R. 1952. Effects of beta-naphthoxyacetic acid on fruit size in the Marshall strawberry. Bot. Gaz. 114:134-39.

Zimmerman, P. W. and Hitchcock, A. E. 1937. The combined effect of light and gravity on the response of plants to growth substances. Contrib. Boyce Thompson Inst. 9:455-61.

_____. 1942. Flowering habit and correlation of organs modified by triiodobenzoic acid. Ibid. 12:491-96.

_____. 1951. Growth-regulating effects of chlorosubstituted derivatives of benzoic acid. Ibid. 16:209-13.

_____, and Wilcoxon, F. 1939. Responses of plants to growth substances applied as solutions and as vapors. Ibid. 10:363-76.

_____ and Wilcoxon, F. 1935. Several chemical growth substances which cause initiation of roots and other responses in plants. Ibid. 7:209-29.

Zweig, G. and Ashton, F. M. 1961. (Cited by Crafts.)

9

Animal Growth Promoters

J. KASTELIC
UNIVERSITY OF ILLINOIS

A MONG THE REMARKABLE ACHIEVEMENTS in biological research within the past twenty-five or thirty years are the diverse discoveries which have given us a clearer understanding of the basic biochemical and physiological processes occurring in animal and plant cells. As a result of these achievements, it is now possible to describe a rather large number of molecular reactions associated with the anabolism and catabolism of vital cellular constituents and to specify in a rather precise way, many of the basic functions served in the animal body by the dietary essentials, the proteins, fats, minerals, vitamins and water. Indeed, owing to these remarkable results, experiments such as those so elegantly summarized by Fraenkel-Conrat, Stanley and Calvin (1961), we appear to be on the very threshold of developing a detailed understanding of the mechanism by which cells synthesize proteins, many of which serve as major components of enzymes and hormones, or are associated with genetic processes per se. Thus, we have at hand an enormous amount of detailed information about the complicated but beautifully ordered metabolic pathways serving as biochemical mainstreams by which nutrients absorbed from the gastrointestinal tract are altered, degraded and/or synthesized into a vast variety of cellular constituents. The principal mechanisms through which energy is produced to serve the needs for endergonic cellular processes and for transformations of dietary materials into cellular constituents have also been well established. The successful rearing of at least some species of laboratory animals, through several generations, fed diets that are chemically well defined, attests to the fact that the essential nutritional elements for animals can be rather definitely stated.

Certainly the means to control malnutrition caused by dietary deficiencies are available and reasonably well understood. The commonly occurring deficiencies in the nutrient content of a large variety of practical animal rations have been quite well pinpointed and it is usually possible to correct these difficulties by appropriate nutrient supplementation. Human nutrition and the feeding practices used in meat-animal agriculture are based upon these considerations.

Deficiencies in Knowledge of Growth Regulation

These facts would suggest that we are in a position to specify the physiologic and metabolic details of the roles served by growth-stimulating compounds when these are administered to animals receiving nutritionally adequate diets. Unfortunately it is not possible to do this satisfactorily. As a consequence the traditional view which implied that once all the nutritional elements became known, maximum growth performance in animals could be attained, has had to be modified many times in recent years. The frequent observation that many compounds which stimulate more rapid growth and affect the animal's ability to cope with disease and stresses of environment or influence the balance of endocrine secretions in its body is only now beginning to be fully appreciated.

Such a large and diverse number of animal growth-stimulating compounds have been so well described in the literature that it is quite unnecessary in this discussion to attempt to review the studies that have been reported about them. It is only necessary to note that a comprehensive consideration of this subject is given in the following representative publications: Casida et al. (1959), Luckey (1959), American Cyanamid Co. (1955), Braude, Kon and Porter (1953), Jukes (1955), Andrews, Beeson and Johnson (1954), Burroughs et al. (1954), Clegg and Cole (1954), and Taylor, Hale and Burroughs (1957).

But despite the voluminous amount of information currently available about animal growth-stimulating substances and the great activity in experimental studies being directed to them, it is somewhat embarrassing that it is not yet possible to categorize these compounds into groups that clearly differentiate their unique physiological activities or biological capabilities.

The primary difficulty is that we do not as yet know enough about the neural and humoral agents which govern most metabolic processes. Each endocrine gland exerts a variety of influences upon the animal and there is much direct evidence that such glands affect the functions of one another. We need to remind ourselves occasionally that we are yet without a clear under-standing of the mechanism whereby thyroxine regulates cellular reactions, though there must now be thousands of published re-search reports which deal with the thyroid gland, the function of thyroxine and other iodinated derivatives. The uncoupling effect of thyroxine on oxidative phosphorylation, which may account for its effect on the basal metabolic rate, still fails to provide all the necessary explanations for the varied experimental and clinical symptoms seen in animals surgically deprived of their thyroid glands.

Such questions raise serious problems about the role of growth-promoting substances when administered to animals and greatly complicate any attempt to define the limits of safety which attend their use in the production of animals used for hu-man food.

Dosage and Safety

Adequate references have been cited concerning the difficul-ties in classification of potentially toxic drugs used in food pro-duction and processing. The simplest definition is to regard ani-mal growth-promoting substances as drugs since their uses in meat-animal production involve legal regulatory considerations. The animal growth-stimulating compounds in current use include many antibiotics, synthetically produced chemicals possessing hormonal activity, and some specific metabolic inhibitors as, for example, thyrogenic substances. Moreover, there is the consid-eration that many of these substances play multiple roles in af-fecting the animal's overall response to them. Finally, there is the vexing phenomena that the dosage level may be quite critical. Many of these substances are stimulatory only when administered at a specific level but are inhibitory and unquestionably harmful to the physiological well-being of animals at other dosage levels. Combinations of one or more of some of these substances have been shown in many experiments to produce effects that can only be explained on the grounds that they act synergistically.

There is the obvious need to carefully explore and to under-stand the detoxification mechanisms which play such an important role in protecting animals against the toxic effects of drugs. Certainly we must also clearly recognize the ever present possi-bility that there may be significant differences in the response of the various animal species to administrations of a specific drug. Further, the lack of immediate and apparent toxicity symptoms, does not guarantee that subsequent toxic manifestations will not occur.

The foregoing considerations, though deliberately brief and admittedly inadequate, provide the setting for the subsequent discussion. The judgment about the safety for human consumption of foods produced from animals fed rations containing deliberate additions of growth-stimulating compounds, which cannot be included with the traditionally accepted forms of essential dietary nutritional elements, must come from an understanding of the physiological effects they may produce.

On one hand, one could argue that any substance which improves the rate of growth or weight gain and general health of animals ought to be considered safe and that residues remaining in the tissues of the animal used as human food ought not to present any hazard to the health of the human. On the other hand, the unique nature of some physiological phenomena which they produce in animals may be easily regarded as being "abnormal" and hence their use ought to be viewed with suspicion and concern. Beyond these simplistic positions there remains the ever-present and vexing question of the criteria upon which decisions about safety ought to be made. No one can possibly doubt the limitations of data obtained from short-term acute toxicity experiments but who will want to decide when enough toxicity testing has been done and to determine at what point potential hazards can be safely disregarded?

The long-term view and the soundest approach to this problem is to seek information which explains the basic mechanisms responsible for the physiological, endocrinological and biochemical changes wrought in the tissue and organ systems of animals treated with growth-promoting compounds. Once this kind of information becomes available, arguments about the merits of feeding only conventional diets as opposed to the deliberate use of exotic compounds to induce improved growth performance and increased disease resistance in the animal can be much more easily settled. Two obvious, and perhaps trite, questions about the use of growth-stimulating compounds in meat-animal production can be raised. The first relates to the economic advantages to the primary producer of meat animals and to the meat consumer, which attend their use; the second relates to the human health hazard of residues that may remain in the tissues of meat from treated animals.

Antibiotics

The most widely used animal growth-promoting substances in practical use today are the antibiotics. There is no need for a complete description of their effects upon animals since this is a well documented matter and it is certain they do result, at least under many conditions, in improved animal performance. Of the thousands of antibiotics isolated to date not all of them have proved to be useful in practical animal agriculture for a variety of reasons. Among those in common use, chlortetracycline, penicillin, bacitracin and streptomycin or combinations of them are frequently added to common animal feeds. The amounts added to feeds rarely exceed 50 grams per ton of feed.

Mode of Action

Proposals made for the mode of action of antibiotic growth stimulation have been enumerated by Luckey (1959). Among the many proposals cited, the following are representative:

1. Effects on microflora in the intestinal tract of animals
   (a) Increase microbial synthesis of vitamins
   (b) Reduce the populations of organisms which use critical nutrients, produce toxins, decrease animal's resistance to pathogens, etc.
   (c) Increase rate of nutrient absorption by causing intestinal wall to be more "permeable"
   (d) Enhance rate of liberation of certain nutrients from food

2. Effects on cells and organs
   (a) Enhance apparent utilization of nutrients, efficiency of nutrient metabolism, "spare" vitamins, etc.
   (b) Affect rate of secretions of endocrine glands

It is obvious that there is no conclusive agreement among scientists about the mode of action of antibiotics as growth-stimulating factors.

Qualification must be made that effects of antibiotics vary with the species of animals, environment, age, nutritional status, and exposure to infections. Furthermore, their growth-stimulating effects cannot be easily correlated with level fed or kind of antibiotic used. It is generally conceded that antibiotics appear to reduce the requirements for at least some vitamins, minerals and protein (Jukes, 1955). Whether this phenomenon alone leads to increased food efficiency observed in some antibiotic experiments on animals is equivocal. The most impressive evidence to date is that all antibiotics even when given to animals at low levels protect animals against so-called, "subclinical" microbial infections. Chicks have been observed to grow better in clean quarters than chicks in previously used quarters. Under these circumstances only the chicks reared in previously used quarters respond well to antibiotic administrations. There are some exceptions to this observation, particularly when good animal management, sanitation and feeding practices are employed.

Residues

There is general agreement that gross carcass composition is not affected by feeding low levels of antibiotics to animals. Very few and often questionable data have been obtained which indicate that anitbiotics bring about deleterious changes in the components of blood, vitamin content of tissue, the constituents of muscle, liver, kidney or other vital organs. Nor has there been much evidence that antibiotics produce disturbances in metabolism, reproduction and growth of animals. Tissue storage of antibiotic residues has been found to be nonexistent or negligible when use is carefully controlled and limited to feeding levels which are well below dosages normally used to control serious disease infections (Broquist and Kohler, 1953). Nonetheless, clinical evidence and reports of serious toxic reactions and often fatal allergic reactions to therapeutic dosages of the commonly used antibiotics in human medical practice must be taken into account. The question is whether the eating of meat and other food products obtained from animals fed low levels of antibiotics is creating a reservoir of sensitized individuals who would subsequently react on exposure to antibiotics administered to them for medical reasons. To my knowledge this has not been shown to have occurred. The possibility remains that antibiotic residues, which may be in milk obtained from dairy cattle carelessly treated for mastitis, or in meat from animals fed therapeutic doses of these drugs immediately prior to slaughter, may sensitize individuals to subsequent injections. One highly significant aspect of this problem is the fact that antibiotic residues in meats during cooking are degraded as has been shown by Escanilla, Carlin and Ayres (1959) and Shirk, Whitehill and Hines (1956).

From these and other considerations it may be concluded that health hazards of antibiotic residues in meat have not been shown to exist. Exposure to minute amounts of antibiotics may encourage development of resistant pathogenic organisms in both animals and man but there is little evidence to support this proposition. This problem has been well explored but it has not

produced a public health hazard nor resulted in data which show
that low antibiotic residues in meat produce adverse reactions in
the human.

MISCELLANEOUS GROWTH PROMOTERS

There are many compounds other than the antibiotics which
give growth response in animals when they are added in small
quantities. Among these are included a variety of surfactants,
tranquilizing drugs, arsanilic acid and its derivatives, some sulfa
drugs and copper. The arsenicals, in poultry and swine particularly, appear to parallel most of the effects of antibiotics in
growth and food efficiency except that they do not possess bacteriostatic activity when fed at low levels, Frost and Spruth (1956).
It has long been known that arsenic is constantly present in all
animal tissues. Estimates of the total amount in the adult human
body vary greatly but about 20 mg or 0.3 ppm appears to be a
reasonable figure. Sea fish, oysters and clams may contain as
much as 10 ppm or more. It is interesting to note that dietary
arsenic has been found to be effective in alleviating the symptoms
of selenium poisoning, Moxon (1938). Sharpless and Metzger (1941)
have shown that arsenic when fed to rats is goitrogenic. It is believed that residues of arsenic in the tissues of animals can be reduced to extremely low levels provided arsenic-containing feeds
are not fed to animals several days before slaughter, Bucy et al.
(1955). However, in these particular experiments and in another
reported by Bucy et al. (1954), it could not be demonstrated that
3-nitro-4-hydroxyphenylarsonic acid improved weight gains in
lambs and sheep. Levels greater than .02 percent of the ration
produced pathological changes in the liver, kidney, decreased red
blood cell count and other tissue changes. But it has been conceded that if appropriate care is employed in the use of arsenicals, and they are removed from the diets of animals 6 days or
more before the animals are slaughtered, the amounts which remain in tissue reach very low levels similar to levels found in
nontreated animals. Despite the lively discussion currently being directed towards the safety hazards of feeding arsenicals to
animals, there seems to be no general agreement about their potential toxicity. Perhaps this is the primary reason why the use
of arsenicals is being restricted and subjected to continuous
scrutiny by the U.S. Food and Drug Administration.

HORMONES

The use of chemicals possessing estrogenic activity and natural estrogenic and androgenic hormones for the purpose of increasing rate of gain in meat-producing animals presents problems that are not so easily defined. There can be little doubt about the economic advantage which attends the practice of administering estrogens to cattle and lambs since the literature on this subject is extensive, well documented and reasonably conclusive. Hundreds of thousands of beef cattle and lambs are routinely treated orally or by subcutaneous implantation with the synthetic estrogenically active compound, diethylstilbestrol. A rather extensive and carefully assembled summary of the researches done on this subject has been presented in a Ph.D. thesis by Hinds (1959).

Mode of Action

Among the many doubts about the effect of hormones on the physiological well-being of animal life, none are more vexing than those which are related to their mode of action. A few of the more important effects of diethylstilbestrol on animals are listed in Table 9.1 (Hinds, 1959).

Table 9.1. Primary "Hormonal" Effects Produced
in Animals by Diethylstilbestrol

Effect	Active Compound	Animal
Protein anabolism	DES	Primarily ruminants
Enhanced fat deposition	DES	Avian
Epiphyseal closure	DES	Prepuberal mammals, avians
Reduced fat deposition	DES	Ruminants
Enhanced Ca, P retention	DES	Young ruminants

The Effects on Organs Produced by Diethylstilbestrol

Tissue or Gland Affected	Nature of Effect
Anterior pituitary	Increased weight. Content of gonadotrophic, adrenocorticotrophic and thyroid stimulating hormone may increase.
Adrenal, ovaries, thyroid	Effect on weight or size variable.
External genitalia, females	Hypertrophy.

Summary of Effects Produced by Diethylstilbestrol

Long-term, low-level administrations of these compounds to humans has not been carried out, and there are few data which provide for an adequate understanding of their influence upon the endocrine system of animals. Virtually all the data currently available on the effects of estrogenic or androgenic hormones on the human are limited to the treatment of metabolic or physiological disorders. The dosages involved in clinical practice cannot be compared with those routinely given to meat-producing animals. There is also the further complication that effects of hormones on animals vary widely depending upon species, sex, physiological status, level of dosage and type of tissue or organs which are affected. If these facts can be laid aside for the sake of expediency, it can be stated that the economic advantage accruing from use of hormones in production of meat animals lies in their capacity to induce anabolic effects upon tissue production, bone and muscle particularly. Increase in feed efficiency; improved utilization of nutrients; retention of some nutrients (nitrogen especially); improved digestibility of the ration; increased synthesis of collagen or less specifically, connective tissue; hyperplasia of the adrenal, thyroid and anterior pituitary tissues have been reported by many investigators, but there are also some reports to the contrary. The most intriguing phenomenon is the wide divergence in response to estrogen treatment exhibited by the various animal species, Gaarenstroom and Levie (1939) and Preston, Cheng and Burroughs (1956).

It is now generally conceded that there is a sharp contrast between the growth response effects of estrogens administered to the ruminant species and that which is observed in the monogastric animals including avians. Precise reasons for this phenomenon have not been established. The concensus, reduced to the simplest possible explanation, is that diethylstilbestrol produces cytological changes in the anterior pituitary possibly resulting in an altered production of growth hormone, gonadotrophic hormone, and thyroid-stimulating hormone, Clegg, Cole and Guilbert (1951), Clegg and Cole (1954), Hinds (1959) and Meyer and Clifton (1956). Since the anterior pituitary, the adrenals and thyroid glands of monogastric animals are also affected by diethylstilbestrol administrations without a concomitant improvement in rate of gain or increased food intake, the basic mechanisms involved are not very well understood, Walker and Stanley (1941) and Hartsook and Magruder (1956).

Residues

Considerations of hazards to human health have resulted in a number of investigations to determine whether significant residues of physiologically active residues are present in the tissues of diethylstilbestrol treated animals. Biological assays as well as chemical procedures have been employed, Stob et al. (1954), Andrews et al. (1956), Burroughs et al. (1954), Preston et al. (1956), Turner (1956), Umberger, Gass and Curtis (1958) and Mitchell, Neumann and Draper (1959). The recoveries of diethylstilbestrol reported by these investigators show that kidney and liver tissue, if residues are present, contain the greater amounts as compared to the levels found in fat and muscle. Representative data are not easily given since it has been observed that the period for elimination of diethylstilbestrol from most tissues is about 72 hours or less. Mitchell, Neumann and Draper (1959) administered 100 mg of tritium labeled diethylstilbestrol to an 800 pound steer over a period of 11 days and observed that phenolic material identified as diethylstilbestrol could be detected in various tissues. The following table shows the quantity of diethylstilbestrol found in 6 different tissues.

Tissue	Parts per Billion Diethylstilbestrol
Lean meat	0.30
Internal fat	0.35
Liver	9.12
Kidney	4.15
Heart	-
Blood	-

It may be noted that the steer was slaughtered after feed containing diethylstilbestrol was withheld for 27 hours. While measurable amounts of estrogenic activity have been found in extracts of tissue from lambs and cattle given diethylstilbestrol, in no instance has anyone reported values which exceed 10 micrograms of residue of diethylstilbestrol equivalent per kilogram of tissue when the material was removed from the ration 48 hours before slaughter of animals. These data are quite similar to those reported by investigators who have used biological assay procedures involving the use of young virgin or ovariectomized female mice as assay animals. An increase in uterine weight resulting from small amounts of diethylstilbestrol added to a standard diet has been observed to be proportional to the logarithm of the dose and is often used as a basis for estimating the amount of estrogenic activity present in various animal tissues. While this procedure is extremely useful for obtaining quantitative data about the estrogenic activity present in tissues of treated animals, it provides very little direct information about the metabolism of diethylstilbestrol. Data currently available show that diethylstilbestrol is not rapidly inactivated in animal tissue. It is excreted in the urine as a conjugate, (monoglucuronide) and in the free form. Much of the remaining portion of the absorbed material is excreted via the bile into the gastrointestinal tract. Traces of it appear to be oxidized as indicated by the appearance of expired $C^{14}O_2$ when C^{14}-labeled diethylstilbestrol is injected into mice, rats or dogs (Twombly and Schoenewaldt, 1951). Apparently very little diethylstilbestrol is excreted as the etheral sulfate.

Since it has been shown that the liver and kidney tissues represent the only significant localization centers for residue retention in the animal body, the rate of release of active material from them becomes exceedingly important. It now seems well accepted that if the active substance is removed from the feed between 48 and 72 hours before the animal is slaughtered, the amounts of residues present in the tissues of the body are reduced to such low levels that they often cannot be detected even with the most sensitive procedures. On the basis of data obtained for the meat of a steer treated with labeled diethylstilbestrol, Mitchell, Neumann and Draper (1959) concluded that approximately 7610 pounds would have to be consumed to provide 1 mg of diethylstilbestrol.

The foregoing discussion makes it appear that estrogenic residues in the meat of diethylstilbestrol treated lambs and beef cattle need present no real safety hazard to the human, providing the hormone is used carefully, and the treated animals are given drug-free rations for several days before slaughter. The use of implants involve more difficult problems particularly if they are placed in parts of the bodies of meat animals which may be subsequently used for human food. Unabsorbed residues in poultry carelessly implanted with diethylstilbestrol resulted in disapproval of this practice by the U.S. Food and Drug Administration.

CONCLUSIONS

These few references to the problem of drug residues in tissues of meat animals merely illustrate the need for careful application and adequate understanding of the hazards involved. Yet despite the potential health problems involved, it would be exceedingly difficult to select enough data to sustain an argument that their use ought to be prohibited. But one cannot conclude at this point.

At the present time, despite all the research that has been done, little is known about the effects upon the health of humans chronically exposed to traces of antibiotics, estrogenic or androgenic hormones (either of synthetic or of natural origin), arsenicals and numerous other growth-affecting compounds including those which affect the function of the thyroid gland.

One frequently encounters statements that the rate of tumor growth in hormone or antibiotic treated animals ought to be studied. There is concern about the possibility of long-term, cumulative, antibiotic sensitivity in humans exposed to antibiotics employed in nonmedical practices. Mutagenic activity of some drugs may lead to induced drug resistance of pathogenic microorganisms. But it is equally obvious that natural foods are not entirely without their hazards. Lathyrism, mushroom poisoning, the goitrogens that are present in cabbage, parasitic animal organisms which may infect those eating improperly cooked meat constitute some well-known examples of possible hazards we sometimes overlook in foods eaten by the human.

The conclusion one ought to draw from this discussion is that the only satisfactory answers to these questions lie in continued long-term animal experimentation directed particularly to the biochemical and physiological effects growth-promoting substances may induce in the tissue systems of the bodies of several different species of animals. Ideally these experiments ought to be carried out in human volunteers before one can decide whether or not a real hazard exists from exposure of the public to foods obtained from meat animals treated with growth-promoting substances.

A conservative attitude must prevail about the safety of these practices until definite information is gained about the actual hazards involved. In the meantime it can be certain that their widespread, carefully controlled use provides the only practical means by which their toxicity or lack of it, can adequately be tested. Strict prohibition of their use under practical conditions can only delay the emergence of the needed answers. The toxicity risks involved must therefore remain a question for careful scientific and value judgment appraisals by the legal agencies and scientific organizations charged with the responsibility of protecting the interests and health of the general public.

LITERATURE CITED

Andrews, F. N., Beeson, W. M., and Johnson, F. D. 1954. The effects of stilbestrol, dienestrol, testosterone and progesterone on the growth and fattening of beef steers. J. Anim. Sci. 13:99-107.

_____, Stob, M., Perry, T. W., and Beeson, W. M. 1956. The oral administration of diethylstilbestrol, dienestrol, and hexestrol for fattening calves. J. Anim. Sci. 15:685-88.

Braude, R., Kon, S. K., and Porter, J. W. G. 1953. Antibiotics in nutrition. Nutrition Abst. and Rev. 23:473-95.

Broquist, H. P. and Kohler, A. R. 1953-54. Studies of the antibiotic potency in the meat of animals fed chlortetracycline. Antibiotics Annual. Pp. 409-15.

Burroughs, W., Culbertson, C. C., Kastelic, J., Cheng, E., and Hale, W. H. 1954. The effects of trace amounts of diethylstilbestrol in rations of fattening steers. Science 120:66-67.

Bucy, L. L., Garrigus, U. S., Forbes, R. M., Norton, H. W., and James, M. F. 1954. Arsenical supplements in lamb fattening rations. J. Anim. Sci. 13:668-76.

_____, _____, _____, _____, and Moore, W. W. 1955. Toxicity of some arsenicals fed to growing-fattening lambs. J. Anim. Sci. 14:435-45.

Casida, L. E., Andrews, F. N., Bogart, R., Clegg, M. T., and Nalbandov, A. V. 1959. Hormonal relationships and applications in the production of meats, milk, and eggs. (Supplement.) A report of the Committee on Animal Nutrition, Agricultural Board. P. 53.

Clegg, M. T. and Cole, H. H. 1954. The action of stilbestrol on the growth response in ruminants. J. Anim. Sci. 13:108-30.

_____, _____, and Guilbert, H. R. 1951. Effects of stilbestrol on beef heifers and steers. J. Anim. Sci. 10: 1074-75.

Escanilla, O. I., Carlin, A. F., and Ayres, J. C. 1959. Effect of storage and cooking on chlortetracycline in meat. Food Technol. 13:520-24.

Fraenkel-Conrat, H., Stanley, W. M., and Calvin, M. 1961. The chemistry of life. Chem. Eng. News 39:80-89, 136-44, 96-104.

Frost, D. V. and Spruth, H. C. 1956. Arsenicals in feeds. In "Proceedings of the Symposium on Medicated Feeds." H. Welch and F. Marti-Ibañez, Ed. Pp. 136-49.

Gaarenstroom, J. H. and Levie, L. H. 1939. Disturbance of growth by diethylstilbestrol and oestrone. J. Endocrinol. 1: 420-29.

Hartsook, E. W. and Magruder, N. D. 1956. A study of the mode of action of stilbestrol. J. Anim. Sci. 15:1298-99.

Hinds, F. C. 1959. Studies on the metabolism and physiological action of diethylstilbestrol in ruminants. Ph.D. Thesiᵤ, University of Illinois, Urbana. P. 122.

International Conference on the Use of Antibiotics in Agriculture, Proceedings of American Cyanamid Co., Scientific Session, October 28, 1955. P. 87.

Jukes, T. H. 1955. Antibiotics in nutrition. New York. Medical Encyclopedia, Inc. P. 128.

Luckey, T. D. 1959. Antibiotics in nutrition. In "Antibiotics, their Chemistry and Non-Medical Uses." H. S. Goldberg, Ed. Chap. III. New York. D. Van Nostrand Co. Pp. 174-321.

Meyer, R. K. and Clifton, K. H. 1956. Effect of diethylstilbestrol-induced tumorigenesis on the secretory activity of the rat anterior pituitary gland. Endocrinol. 58:686-93.

Mitchell, G. E., Neumann, A. L., and Draper, H. H. 1959. Animal growth promoters. Metabolism of tritium-labeled diethylstilbestrol by steers. J. Agr. Food Chem. 7:509-12.

Moxon, A. L. 1938. The effect of arsenic on the toxicity of seleniferous grains. Science 88:81.

Preston, R., Cheng, E., and Burroughs, W. 1956. Growth and other physiological responses to diethylstilbestrol in diet of rats and guinea pigs. Proc. Iowa Acad. Sci. 63:423-27.

_____, _____, Story, C. D., Homeyer, P., Pauls, J., and Burroughs, W. 1956. The influence of oral administration of diethylstilbestrol upon estrogenic residues in the tissues of beef cattle. J. Anim. Sci. 15:3-12

Sharpless, G. R. and Metzger, M. 1941. Arsenic and goiter. J. Nutrition 21:341-46.

Shirk, R. J., Whitehill, A. R., and Hines, L. R. 1956-57. A degradation product in cooked chlortetracycline-treated poultry. Antibiotics Annual. Pp. 843-48.

Stob, M., Andrews, F. N., Zarrow, M. X., and Beeson, W. M. 1954. Estrogenic activity of the meat of cattle, sheep and poultry following treatment with synthetic estrogens and progesterone. J. Anim. Sci. 13:138-51.

Taylor, B., Hale, W. H., and Burroughs, W. 1957. Growth and fattening stimulation in ewe lambs by certain androgenic and estrogenic compounds. J. Anim. Sci. 16:294-306.

Turner, C. W. 1956. Biological assay of beef steer carcasses for estrogenic activity following the feeding of diethylstilbestrol at a level of 10 mg. per day in the ration. J. Anim. Sci. 15:13-24.

Twombly, G. H. and Schoenewaldt, E. F. 1951. Tissue localization and excretion routes of radioactive diethylstilbestrol. Cancer 4:296-302.

Umberger, E. J., Gass, G. H., and Curtis, J. M. 1958. Design of a biological assay method for the detection and estimation of estrogenic residues in the edible tissues of domestic animals treated with estrogens. Endocrinol. 63:806-15.

Walker, S. M. and Stanley, A. J. 1941. Effect of diethylstilbestrol dipropionate on mammary development and lactation. Proc. Soc. Exper. Biol. Med. 48:50-53.

10

Incidental Additives:
Commentary and Discussion

WAYLAND J. HAYES, JR.

WAYLAND J. HAYES, JR.
COMMUNICABLE DISEASE CENTER
Atlanta, Georgia

COMMENTARY

ALL OF THE UNINTENTIONAL FOOD ADDITIVES are
poisonous in some degree. This is emphasized by the fact
that cases of human poisoning are known in connection with
most of them that have been used in significant tonnage (Hayes,
1960). The list of such compounds applied to crops includes inorganic insecticides such as cryolite,[1] lead arsenate, and Paris
green, botanical insecticides such as nicotine, pyrethrum, and
rotenone; chlorinated hydrocarbon insecticides such as aldrin,
BHC, chlordane, DDT, dieldrin, endrin, TDE, and toxaphene; organic phosphorus insecticides such as DDVP, demeton, Diazinon,[2]
Dipterex, malathion, methyl parathion, mipafox, parathion, Phos-
drin, schradan, TEPP, phorate, and Trithion; dinitro compounds
used as either insecticides or plant growth regulators; hormone-
type growth regulators; fungicides and seed dressings such as
hexachlorobenzene and ziram; several organic mercury compounds; and growth promoters such as antibiotics. Cases of poisoning are also known in connection with pesticides that may con-
taminate food although they are not intentionally used on food
crops. Such cases have involved the rodenticides such as ANTU,
arsenic trioxide, barium carbonate, thallium, sodium monofluoro-
acetate, and strychnine sulfate; fumigants such as hydrogen

---

[1] Terminology for insecticides is in accordance with the Committee on Insecticide
Terminology of the Entomological Society of America as indicated by Smith (1959a,
1959b, and 1961).

[2] Use of trade names is for identification purposes only and does not constitute
endorsement by the U. S. Public Health Service.

cyanide, dibromoethane, and methyl bromide; household insecticides such as Lethane, naphthalene, and sodium fluoride; and defoliating compounds such as sodium chlorate.

Contamination of Foods

Food may become contaminated in many ways. Poisoning has occurred in this country when leafy vegetables were sprayed with toxaphene entirely contrary to directions and then harvested within a day or so. The residue was 3315 ppm, and 3126 ppm still remained after washing (McGee et al., 1952). If toxaphene is used at all on such crops, it is supposed to be used on seedlings only so that no residue remains at harvest. The same sort of thing occurred with nicotine. Even two weeks after spraying the residue ranged from 69 to 123 ppm as compared with a permitted level of 2 ppm (Lemmon, 1956). The intentional addition of lindane to sugar by a child resulted in severe illness of four people. Poisoning by pesticides has been a much more serious problem in some other countries than in the United States. Thousands of people were made chronically sick and some were killed in Turkey when they ate grain intentionally treated with hexachlorobenzene (Schmid, 1960). In a somewhat similar way, prepared bait containing the rodenticide warfarin was used as food by a family in Korea. Each of the 14 persons in the family became sick, and two died (Lange and Terveer, 1954).

The most common and serious cause of outbreaks of nonoccupational poisoning has been the accidental contamination of grain or other staple food during shipment. Of the new compounds, parathion has been involved most frequently. An episode in the Kerala and Madras States in India followed the contamination of grain during ocean shipment by leakage of parathion from polyethylene carboys. The grain was distributed widely before the danger was detected, so that deaths continued to occur for about a month (Anonymous, 1958). At least 102 persons were killed (Karunakaran, 1958). There have been some similar occurrences in other countries. Although no such tragedy has occurred in this country, an outbreak of endrin poisoning in Wales reminds us that it can happen here. Fifty-nine persons were poisoned and 30 of them had fits. The trouble was traced to bread made from sacked flour that had been shipped in a freight car previously used to transport endrin. Much of the contamination probably came from the flour sacks, which were used to cover the loaves as they rose (Davies and Lewis, 1956).

Storage of DDT

At least one pesticide, DDT, is stored in essentially everybody in this country. This phenomenon, because it involves the whole population, probably deserves as much attention as the relatively rare instances in which pesticides produce recognized illness.

A trace of DDT has been found in every whole meal analyzed, although some individual foods contained no detectable insecticide. The average daily intake of DDT is about 0.2 mg/man/day (Walker et al., 1954; Hayes et al., 1956). The storage of DDT in the fat of people in the general population is shown in Table 10.1. It is almost certainly dietary in origin. Most of the DDT in our food is in meat and related products (Walker et al., 1954). However, Eskimos from isolated Alaskan villages where the native food contained no detectable DDT stored even less DDT than meat abstainers in the 48 contiguous states (Durham et al., 1961). Thus, it is the DDT content of the food rather than the amount of animal fat and protein that is important in determining the degree of storage.

People whose work brings them in contact with DDT store more of the compound than people in the general population (Hayes et al., 1958). The amount depends on the kind of work (see Table 10.1). The fat of one healthy formulator contained a concentration of 648 ppm (Hayes et al., 1956).

Table 10.1. Concentration (ppm) of DDT
and DDE in Fat of People*

Exposure Group	DDT	DDE
1. Died before DDT	0	0
2. Abstain from meat	2.3	3.2
3. General population	4.9	6.1
4. Environmental exposure	6.0	8.6
5. Agricultural applicators	17.1	22.3

*Adapted from Hayes, et al., 1958.

The relation between oral dosage of DDT and its storage in fat has been studied in volunteers (Hayes et al., 1956). An extension of this work permitted Ortelee (1958) to relate oral intake of DDT and the urinary excretion of DDA. He was able to show (Figure 10.1) that more than half of the workers in certain formulating plants excreted DDA at a rate equal to or greater than that in men taking a repeated oral dosage of 35 mg/man/day. The men had been employed at this kind of work for as long as 6.5 years.

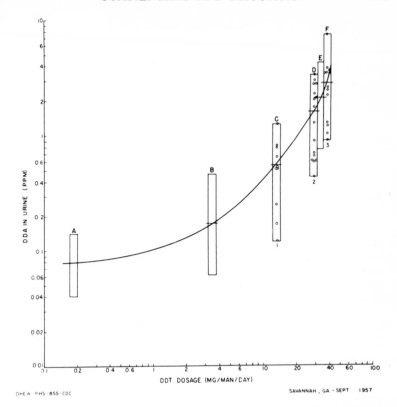

DHEW PHS BSS-CDC   SAVANNAH , GA - SEPT   1957

Fig. 10.1. Concentration of DDA in the urine of men with known daily oral
            intake of DDT (Bars A, B, and E) and of men with different de-
            grees of industrial exposure to the compound (Bars C, D, and F).
            Each bar indicates the range and mean for the group. In Bars
            C, D, and F, corresponding to Grades 1, 2, and 3 respectively of
            industrial exposure, the individual values are shown. (From
            Ortelee, 1958, reprinted with permission from the A.M.A. Ar-
            chives of Industrial Health.)

The fact that men can tolerate for years a dosage of DDT approximately 200 times that of the general population is reassuring.
However, it seems best to repeat the observations periodically
for at least one generation in order to detect any long-term effects
that may occur.

Mortality Caused by Pesticides

The annual mortality in the United States from all causes is
approximately 935 per 100,000, varying only slightly from year to

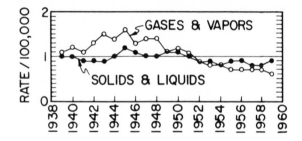

Fig. 10.2. Accidental deaths in the United States caused
by chemicals expressed as rate per 100,000
population. (Based on U. S. Public Health Serv.
Vital Statistics. U. S., 1936-1959 [1940-1960].

year. This rate may be compared to the rates of about 1.0 for
accidental acute poisoning by gases and vapors and by solids and
liquids, respectively. The exact rates over a period of several
years are shown in Figure 10.2. It may be seen that there has been
no dramatic change in the last two decades. The current over-
all rate for acute poisoning is only about one-fourth the rate at
the turn of the century. People have learned to use poisons, and
chemicals generally, with greater care. However, there is much
room for improvement.

Each year pesticides contribute approximately 10 percent of
the deaths caused by solid and liquid materials (Conley, 1958;
Hayes, 1960). The deaths caused by fumigants are very few.
Thus, pesticides are associated with an annual mortality rate of
about 0.1 per 100,000 population in this country.

More than half of the deaths caused by pesticides are in children. For example in 1956, 62 percent of these deaths were in
children less than 10 years old and 51 percent were in children
less than 4 years old (Hayes, 1960). It may be more unexpected
that, during 1956, the last year for which complete figures are
available, 68 percent of the deaths caused by pesticides were
caused by compounds introduced before DDT.

It seems that the greatest opportunity for improvement involves the older poisons and the unnecessary exposure of children. There can be no doubt that the careful labeling that has
been required for all pesticides since 1947 (U.S. Laws, 1947) has
been a powerful factor in educating the public in the use of these
compounds. The newer pesticides have been manufactured in
greater tonnage than the older ones for several years, and the
newer compounds are just as toxic on the average as the older
ones. It follows that the good record of the newer compounds

results from their better handling and not from any lack of hazard. In this connection, it is interesting that sodium arsenite was first declared on January 1, 1962, to be an "injurious material" under the California law that directly regulates the use of pesticides.

To be sure, as the relative importance of the newer compounds increases, there will come a time when the mortality caused by these newer materials will exceed that caused by the older poisons. This has already happened in California so far as the morbidity of occupationally exposed persons is concerned. Seventy-five percent of the systemic cases resulting from occupational exposure were associated with organic phosphorus insecticides (Kleinman et al., 1960). For several years, use of the more toxic organic phosphorus compounds in California has required a special permit (California Law, 1953). Now their use will not be permitted without medical supervision (California Law, 1961).

How Others See Us

Since we not only have sufficient food but even have surpluses, we can afford to be conservative about any contamination of food by chemical residues. The same is not true in all parts of the world. Another less obvious difference is that we can afford to carry out a great deal of systematic toxicological study. But the necessity to focus technically trained talent on real problems is not confined to the developing countries; it extends to some of the most technically advanced nations (Barnes, 1961).

Problems

Real problems with unintentional food additives include our lack of adequate data on storage of pesticides in the body and our incomplete understanding of the dynamics of storage in general. A problem that potentially involves pesticides, although it affects drugs more, is the lack of an adequate test for proving or disproving a relationship between a specific chemical and blood dyscrasia in individual cases (Hayes, 1961). Finally, the possibility must be kept in mind that the most important hazards of foods may concern what is conventionally called nutrition. If sodium chloride contributes to hypertension as it does in rats (Meneely et al., 1957) and if saturated fats contribute to arteriosclerosis (Bronte-Stewart, 1961), are these not toxic effects?

*   *   *

DISCUSSION

Question: What are the relative merits of electron capture and microcoulometric detection systems in gas chromatography?

Gunther: This is not an easy question to answer. The sensitivity of microcoulometric cells on the market now is ten times greater for sulfur and chlorine than it was a month ago. The new cells are thus in the range of sensitivity of electron capture techniques. The electron capture technique, which I prefer to call electron transfer depicting the actual process involved, can be incredibly sensitive under ideal conditions. But people who are applying these techniques to gas chromatography are finding there are hosts of naturally occurring plant compounds, presumably also animal compounds, that will cause interference.

Often as one goes down the decade scale of amounts detectable, interferences become more troublesome than if one is working in a less sensitive range. It may seem anomalous, but there is no piece of equipment, no matter what its cost, available to the analytical chemist which is free from background interference. In very low regions of detectability one may have an uncertainty factor increasing at twice the rate of detectability.

These factors are coming into prevalence in electron capture. It's a marvelous technique and has marvelous potentialities, but at the present time it should be regarded as are all other new techniques. That is, one should anticipate trouble in all applications and be prepared to compensate for it, realizing that background from the instrument itself may be seriously misleading if the technique is stretched to the utmost.

Question: Is chloramphenicol being used as a growth promoter for animals? If so, are there any reports of blood disorders resulting from its use?

Kastelic: This compound has not been used commercially as an animal growth factor to the best of my knowledge nor am I certain it has been demonstrated to cause blood disorders under certain circumstances.

Question: Would Dr. Kastelic elaborate on the possible sensitization of humans as a result of eating foods containing antibiotics?

Kastelic: To start with I would refer you to a well-written review on the subject of penicillin in milk (Albright, Tuckey and Woods, 1961). The possibility of sensitization seems to be present, but the difficulty is to determine how a particular individual became sensitized. I think most dairy farmers follow the recommended practice of avoiding the collection of milk from cattle that have been treated with massive doses of penicillin introduced in the teat to control mastitis. But possibly such milk, although it doesn't get into pooled supplies going to dairies, may be consumed at home and cause difficulty.

Hayes: I might mention that sensitivity to penicillin is not new. It was observed just as soon as there was enough penicillin to treat diseases. My impression is that there hasn't really been any change in sensitivity levels in recent years, but that there has been a great deal more attention given to the significance of sensitivity to the patient and to methods for detecting sensitivity prior to treatment (Brown, Simpson and Price, 1961).

Question: What is the current thinking on biological control of insects?

Vlitos: I am not an entomologist, but I have had experience the last few years with control of the frog hopper, a sugar cane pest. This insect annually costs sugar cane growers on a small island like Trinidad over a million dollars a year just for insecticide, so quite naturally we've been interested in biological control. It could well be that many of the natural predators have been destroyed by the organo-phosphates we've been using.

We have evaluated bacterial toxins applied to soil to control nymphs of the frog hopper — unfortunately these had very low residual activity in soil and disappear after the first heavy rain. As far as the tropics are concerned, there doesn't seem to be much use for natural toxins against this particular insect.

Question: What is the half life of DDT stored in man?

Hayes: The form of the question presumes that the rate of loss is directly proportional to the amount present, or if loss is graphed — log of the concentration vs. time — there will be a linear relationship. This is the conventional kind of reporting for loss of drugs from tissues and most drugs do obey such a rule, but DDT doesn't. There are two possible explanations. It could be that excretion or total loss becomes less efficient with time or the excretion may be more efficient at higher concentrations. One can distinguish between these two possibilities and the fact is that the excretion of DDT is more efficient at higher concentrations.

Question: Suppose the average intake of DDT is 0.2 mg/day. How much of this will be stored and how much excreted?

Hayes: Some people believe that of the compounds absorbed some are stored and some are excreted. Actually all compounds which are absorbed are stored, and the amount stored varies with the excretion rate. It's a quantitative difference not qualitative.

Of every dose that's absorbed, some fraction will be excreted completely that day and some fraction stored. One can feed an animal DDT for months or years, and then after a single feeding of radioactive DDT, detect radioactive material in the urine and feces the same day and decreasing amounts for as long as the sensitivity of the method will follow it. In the meantime the total quantity stored remains constant.

For man, the time required to reach a constant storage level of DDT is approximately one year (Hayes, Durham and Cueto, 1956). There's every evidence that the general population had reached equilibrium when storage of DDT was first measured and reported by scientists in the U.S. Food and Drug Administration in 1951 (Laug, Kunze and Prickett). We had studies at the same time which were never published but which were in complete agreement. The average storage of DDT in humans has not increased since 1951 (Hayes et al., 1958).

Question: Is there any evidence that individuals with high storage levels of DDT showed liver damage, and what is the relation to starvation?

Hayes: DDT given in massive doses will produce liver injury in any species. In some species it's difficult to keep the animal alive long enough to show liver damage since the major action of this compound is on the nervous system, and death may result before liver injury becomes apparent.

Rodents have a peculiar reaction in which relatively low levels of DDT will produce minor changes in the liver. This has been the subject of rather extensive discussion. It seems that the effect cannot be produced at will. Scientists from the U.S. Food and Drug Administration (Laug et al., 1950) and our own group (Ortega et al., 1956) have reported positive results whereas others (Cameron and Cheng, 1951) could not find the effect. Our group has now been able to get both positive and negative results, but we don't know what the factor is that makes the difference. When we used English food for the rats we couldn't produce liver damage, but we could with food available in this country about 1954. The effect of American food now is less than it was then. We have tried several hypotheses and are now working on another, but as yet we don't have the answer.

I don't think that there is any connection whatever between the liver pathology, which is minimal, and the phenomenon of producing symptoms in animals on starvation diets. As fat is utilized in a starving animal to maintain metabolism, the fat is used up faster than the animal can excrete stored DDT (Dale, Gaines and Hayes, 1962). The actual excretion of DDT increases when the dosage is cut off. The DDT is mobilized and excreted but not at a fast enough rate to prevent levels from increasing in the tissues. In order to produce the effect in rats, they have to be fed to the point at which they have a DDT concentration in their fat of about 2,000 ppm. Another factor which is important is that the rat has a metabolic rate approximately four times that of man, and consequently they starve faster than a man. Poisoning in starving rats due to DDT is a real phenomenon, but I don't think it's of any significance because even workers in formulating plants do not store the necessary high quantity of DDT. Furthermore, because of metabolic differences, man can't starve as fast as a rat.

Question: Since efficacy of herbicides seems to depend upon rainfall and solubility, with the efficacy decreasing after a rainfall, is there any evidence of a build-up of herbicides in subsurface waters?

Vlitos: Offhand I know of no published information on residual herbicides in the water table, but I would answer the question negatively. With an herbicide such as diuron, one of the substituted ureas which is fairly insoluble, even with 60 inches of rainfall a year, the most weed control we can hope for is not in excess of 6 to 8 weeks. This is a pretty good biological indicator of whether or not the compound is still present. It is possible with a more soluble herbicide that there would be traces in the water.

Question: What is known about the resistance of insects to various insecticides?

Vlitos: We are starting to use a rotational system for insecticides. Resistance to DDT and BHC was built up about six years ago. We then switched to Trithion. At this time we were treating fairly large acreages, about 20,000 acres, with one insecticide. I'm not sure we have resistance to Trithion but it was less effective after use for two years. We next introduced a carbamate which was used for two years. It was very successful the first year and not so good the second year.

Now we are breaking up areas into sections and numbering them. Section 1 will get a chlorinated hydrocarbon in 1961, a carbamate in 1962 and an organo-phosphate in 1963. I don't know whether this will confuse the insect or not. We're hoping it will.

Hayes: There is a tremendous amount of literature on this subject, and each of us, although not specifically in the field, could supply a number of references on it.

Question: With regard to detection of compounds using electron capture, it has been suggested that electron capture may give an evaluation of the biological activity of the compound. Would Dr. Gunther care to comment on this?

Gunther: We are sufficiently intrigued by this possibility to investigate it in connection with some of the carcinogens to confirm the original observations. It will be interesting to see how useful this concept might be. One must remember that many of the compounds involved are polynuclear with a high pi electron density, and these are ideally suited to electron capture evaluation.

LITERATURE CITED

Albright, J. L., Tuckey, S. L., and Woods, G. T. 1961. Antibiotics in milk — a review. J. Dairy Sci. 44:779-807.

Anonymous. 1958. Food poisoning blame fixed on Folidol Firm. False description of consignment. The Indian Express. Bombay, India, 26(241):1, July 10.

Barnes, J. M. 1961. Mode of action of some toxic substances with special reference to the effects of prolonged exposure. Brit. Med. J. II:1097.

Bronte-Stewart, B. 1961. Lipids and atherosclerosis. Federation Proc. 20(1)-III:127.

Brown, W. J., Simpson, W. G., and Price, E. V. 1961. Reevaluation of reactions to penicillin in venereal disease clinic patients. Pub. Hlth. Rep. 76:189-98.

California Agricultural Code Relating to Use and Application of Injurious Materials. 1953. Agricultural Code, Ch. 7, Art. 4, Par. 1080, Sacramento, California, As Amended.

California Industrial Safety Board Agricultural Safety Orders. 1961. Section 3298.14, Formulation and application of injurious materials. Department of Industrial Relations, Division of Industrial Safety.

Cameron, G. R. and Cheng, K. K. 1951. Failure of oral DDT to induce toxic changes in rats. Brit. Med. J. 2:819-21.

Conley, B. E. 1958. Morbidity and mortality from economic poisons in the United States. A.M.A. Arch. Ind. Hlth. 18:126-33.

Dale, W. E., Gaines, T. B., and Hayes, W. J., Jr. 1962. Storage and excretion of DDT in starved rats. Toxicol. & Appl. Pharmacol. 4:89-106.

Davies, G. M. and Lewis, Ieuan. 1956. Outbreak of food-poisoning from bread made of chemically contaminated flour. Brit. Med. J. II:393-98.

Durham, W. F., Armstrong, J. F., Upholt, W. M., and Heller, C.
1961. Insecticide content of diet and body fat of Alaskan na-
tives. Science 134(3493):1880.

Hayes, W. J., Jr. 1960. Pesticides in relation to public health.
Ann. Rev. Entomol. 5:379.

Hayes, W. J., Jr. 1961. Diagnostic problems in toxicology (agriculture). Arch. Environ. Hlth. 3:49.

_____, Durham, W. F., and Cueto, C., Jr. 1956. The
effect of known repeated oral doses of Chlorophenothane
(DDT) in man. J. Am. Med. Assoc. 162:890-97.

_____, Quinby, G. E., Walker, K. C., Elliott, J. W., and Up-
holt, W. M. 1958. Storage of DDT and DDE in people with
different degrees of exposure to DDT. A.M.A. Arch. Ind.
Hlth. 18:398-406.

Karunakaran, C. O. 1958. The Kerala food poisoning. J. Indian
Med. Assoc. 31:204.

Kleinman, G. D., West, I., and Augustine, M. S. 1960. Occupational disease in California attributed to pesticides and agricultural chemicals. Arch. Environ. Hlth. 1:118-24.

Lange, P. E. and Terveer, J. 1954. Warfarin poisoning. Report
of 14 cases. U. S. Armed Forces Med. J. 5(6):872.

Laug, E. P., Kunze, F. M., and Prickett, C. S. 1951. Occurrence
of DDT in human fat and milk. Am. Med. Assoc. Arch.
Indust. Hyg. 3:245-46.

_____, Nelson, A. A., Fitzhugh, O. G., and Kunze, F. M.
1950. Liver cell alteration and DDT storage in the fat of
the rat induced by dietary levels of 1 to 50 p.p.m. DDT. J.
Pharmacol. & Exper. Therap. 98:268-73.

Lemmon, A. B. 1956. Bureau of chemistry. Annual report for
the calendar year 1955. Thirty-sixth Annual Report, California Department of Agriculture XLV(2):128.

McGee, L. C., Reed, H. L., and Fleming, J. P. 1952. Accidental
poisoning by Toxaphene. J. Am. Med. Assoc. 149:1124-26.

Meneely, G. R., Ball, C. O. T., and Youmans, J. B. 1957.
Chronic sodium chloride toxicity: the protective effect of
added potassium chloride. Ann. Internal Med. 47:263-73.

Ortega, P., Hayes, W. J., Jr., Durham, W. F., and Mattson, A.
1956. DDT in the diet of the rat. Pub. Hlth. Mon. No. 43,
Pub. Hlth. Serv. Pub. No. 484, 27 pp.

Ortelee, M. F. 1958. Study of men with prolonged exposure to
DDT. A.M.A. Arch. Ind. Hlth. 18:433-40.

Schmid, R. 1960. Cutaneous porphyria in Turkey. New England
J. Med. 263:397-8.

Smith, C. N. 1959a. Common names of insecticides. J. Econ.
Entomol. 52:361-62.

_____. 1959b. Changes in common names of insecticides. J. Econ. Entomol. 52:1032.

_____. 1961. Designation of insecticides by trade marked names on code numbers in society publications. J. Econ. Entomol. 54:1066-67.

U. S. Laws, Statutes, etc. 1947. Federal Insecticide, Fungicide, and Rodenticide Act (Appr. June 25, 1947, 61 Stat. 163-73 Chap. 125; Publ. No. 104, 80th Congr.).

Walker, K. C., Goette, M. B., and Batchelor, G. S. 1954. Pesticide residues in foods. Dichlorodiphenyltrichloroethane and Dichlorodiphenyldichloroethylene content of prepared meals. J. Agr. Food Chem. 2:1034-37.

PART FOUR

Harmful and/or Pathogenic Organisms

11

Significance of Microorganisms in Foods

D. A. A. MOSSEL

CENTRAL INSTITUTE FOR NUTRITION AND FOOD RESEARCH T.N.O.
Utrecht, The Netherlands

A T FIRST SIGHT it may seem that sterile foods represent the ideal to be attained in Food Science and Technology, because no food-borne disease can be caused by such foods and neither will these foods be subject to microbial spoilage. For various reasons, however, this is far from true. First of all, such foods, when contaminated during handling may give rise to uninhibited growth of food poisoning organisms. In regular foods these bacteria have to compete with acid-forming and other antagonistic organisms (Winter, Weiser and Lewis, 1953; Brown, Vinton and Gross, 1960; Post, Bliss and O'Keefe, 1961; Oberhofer and Frazier, 1961; Johanssen, 1961; Peterson, Black and Gunderson, 1962) and are often prevented from reaching minimal infectious or toxic levels. Next, it is not yet possible to process all, or even the most important part of, our staple foods in such a way that they can be marketed in a sterile state without having thereby lost an essential part of their organoleptic and/or nutritive value. And, finally, if sterile foods were to become a reality, it would often be necessary to protect them thoroughly against various types of oxidative deterioration, because it is a regularly observed phenomenon that foods wherein all microbial activity is precluded are no longer protected from oxidation by the relatively low redox potential characteristic of metabolizing microbial cells. We have, therefore, to accept and to learn to live with quite a few infected foods (Hobbs, 1960).

This being the starting point it must of course be one of the most important tasks of the food scientist to keep our food supply safe by such measures as: limitation of the contamination of foods by microorganisms, elimination of certain organisms by a corrective heat treatment and especially inhibition of their subsequent growth in foods. High numbers of microorganisms in foods — and this may be considered to be above the order 10^4 per gram — are clearly undesirable for three reasons: 1) the risk of the food being, or rapidly becoming, unwholesome; 2) the actual or potential loss of organoleptic and nutritive quality; 3) esthetics. The first two points need hardly any further comments. However, we may devote one thought to microbiological esthetics of foods — so much overrated by the layman on emotional grounds, but certainly too much neglected by professional food bacteriologists.

It is generally accepted that the consumer has the right to be protected against unwarranted contamination of his foods with sand, grass, chalk and other, quite innocuous adulterations. However, it has only more recently become accepted — at least in Western Europe (Cockburn, 1960) — that foods have also to be kept free from microorganisms which, though in themselves not impairing the food's wholesomeness or quality, are technically to be considered unnecessary contaminants (Shelton, 1961). In this connection, one thing is obviously quite unjustified, i.e. to call a food unwholesome or spoiled, when it contains no pathogenic or toxinogenic bacteria and is also free from obvious signs of spoilage, but contains too high numbers of viable organisms or debris to be acceptable in the esthetic sense (Bond and Stauffer, 1955).

In the early era of bacteriology much attention was paid to detecting enteric pathogens in foods and in rejecting commodities containing them — whether they were encountered in 100 mg or in 1 kg (Levine, 1961). In the thirties, when the classical types of food-borne disease became more or less under control, a kind of "worship of numbers" (Tanner, 1949) became the behavioral pattern of food bacteriologists. Typing of the bacteria, harvested by the very popular "standard plate count," seemed to be an outdated or unknown procedure. Even "coliform" counts of foods often tended to be reported without any, or even tentative, identification of the organisms isolated and this soon led to disastrous consequences (Martinez and Appleman, 1949; Mundt, Shuly and McGarty, 1954; Wolford, 1955; Barber, 1955; van Uden and Sousa, 1957).

Already the indiscriminate use of the total aerobic plate count has given rise to much criticism by pathologists (Wilson, 1959), who very rightly pointed to the lack of specificity of such procedures, especially the hazard of overlooking dangerous anaerobes (Hobbs, 1955) and to the obvious risk of unjustified rejection of foods which were obtained by microbial processes. Included among such foods were cheese, fermented sausages and raw ham, or mixed foods containing these commodities as ingredients. It was then often forgotten, however, that a total viable count also

has the intrinsic merit of indicating the extent of microbial proliferation which occurred in a food sample due to faulty storage
— the most undesirable situation encountered in food hygiene.

All this may be summarized by stressing the necessity that
those who practice food microbiology should 1) give the results
of their examinations in both taxonomically and quantitatively
justified form; and 2) interpret their results in an ecologically
justified manner, i.e., by paying proper attention to the properties and mode of manufacture of the food wherein the organisms
have been detected (Buttiaux and Mossel, 1957). This chapter
will attempt to deal with the sweeping subject, "Significance of
Microorganisms in Foods," by clinging to these principles —
which are as old as the science of bacteriology itself.

A TAXONO-ECOLOGICAL REVIEW OF THE DOMINANT GROUPS OF MICROORGANISMS FOUND IN FOODS

Modern microbial taxonomy must seriously acknowledge as
well as respect the achievements of that flourishing branch of
microbiology which is called microbial genetics. It is, therefore,
quite acceptable to bring into one taxonomic unit, organisms of
which most, i.e., 85-95 percent, though not necessarily all, fit
exactly the type species, genus, etc. (Borman, Stuart and
Wheeler, 1944; Smith, Gordon and Clark, 1952; Shattock, 1955;
Ewing and Edwards, 1960; Shewan, Hobbs and Hodgkiss, 1960a
and b; Papavassiliou, 1962; Mossel, 1962a; Sharpe, 1962). Especially in food microbiology where taxonomy is not a purpose in
itself but only a requirement for a reliable system of determination, this is a quite realistic and, in fact, the only possible approach. With these limitations in mind, we shall now briefly review recent developments in the taxonomy of the most important
groups of food microorganisms.

Bacteria

Gram-Negative Rods. The Gram-negative rod-shaped bacteria where are of frequent occurrence in foods can be arranged in
five groups.

Generally peritrichous, nitrate-reducing and fermentative
bacteria: the Enterobacteriaceae. A very rational subdivision
of this most important group of bacteria has been suggested recently by Ewing and Edwards (1960). A determinative key for
Enterobacteriaceae, based on this scheme, is presented in Table
11.1.

Bacteria   E. Coli — Gram −ve rod shaped bacteria.

D. A. A. MOSSEL

Table 11.1. The classification of <u>Enterobacteriaceae</u> according to Ewing and Edwards (1960) and the biochemical pattern of the groups as shown in the series of tests known mnemonically as GNiLUMoCISKA (Mossel, 1961)

Division	Group	G	Ni	L	U	Mo	C	I	S	K	A
1	Shigella	A	+	-	-	-	-	d	-	-	-
	Alcalescens-Dispar	A	+	−or+	-	-	-	+	-	-	-
	Escherichia	AG or A	+	+or−	-	+or−	-	+or−	-	-	-
2	Salmonella	AG or A	+	-	-	+or−	+or−	-	+or−	-	-
	Arizona	AG	+	+or−	-	+	+	-	+	-	-
	Citrobacter	AG	+	+or−	-	+	+	-	+	+	-
	Bethesda-Ballerup (Citr. ballerupensis)	AG	+	-	−or+	+	+	-	+	+	-
3	Klebsiella	AG	+	+or−	+	-	+	d	-	+	-
	Enterobacter (Aerobacter)	AG	+	+	−or+	+	+	-	-	+	-
	Hafnia	AG	+	-	−or+	+	+	-	−or+	+	-
	Serratia	AG or A	+	−or+	-	+	+	-	-	+	-
4	Proteus vulgaris	AG	+	-	+	+	+or−	+	+	+	+
	mirabilis	AG	+	-	+	+	+or−	-	+	+	+
	morganii	AG	+	-	+or−	+	-	+	-	+	+
	rettgeri	AG or A	+	-	+	+	+	+	-	+	+
	Providencia	AG or A	+	-	-	+	+	+	-	+	+

Legend: $\overset{+}{\underset{-}{o}}$r = mostly positive; $\overset{-}{\underset{+}{o}}$r = mostly negative; d = both + and − reactions frequently occurring

A = acid   AG = acid and gas

The genera <u>Shigella</u>, <u>Salmonella</u> and <u>Arizona</u> of this group are the causal agents of classical enteric disease, while certain serotypes of <u>Escherichia coli</u> can also cause diarrhea. The minimal oral infectious dose, (MID) of these organisms varies widely with the organisms themselves as well as with the age and health condition of the consumer. However, because it is the commission of the food hygienist to cater to the weakest <u>tractus digestivus</u> likely to be exposed to foods, it is a wise policy to adopt the same attitude to all these organisms as to their normally considered harmless cohabitants of the digestive tract of man and animals: regular types of <u>Escherichia coli</u>, <u>Klebsiella</u> and <u>Proteus</u>, — they simply have no place in most foods.

<u>Generally polar and fermentative bacteria: Aeromonas.</u> The bacteria belonging to this genus can easily be mistaken for <u>Enterobacteriaceae</u> unless either the localization of their flagella

or their oxidase reaction is taken into account (Ewing and John-son, 1960; Eddy, 1960b; Steel, 1961). There is a definite ecologi-cal reason to distinguish Aeromonas from Enterobacteriaceae, because unlike the latter, aeromonads are very seldom found in human stools (Lautrop, 1961) or excreta of warm-blooded ani-mals; their habitat is usually the aqueous one, cold-blooded ani-mals included.

Polar and generally oxidative bacteria: Pseudomonas. Or-ganisms which, if attacking glucose, do this by oxidation only, were generally not recognized in routine bacteriology before Hugh and Leifson's classical paper of 1953. At present, Hugh and Leifson's test and its simplified modification (Mossel and Martin, 1961) are regularly used and, in combination with an oxi-dase test, (Steel, 1961) permit a very rapid identification of Pseudomonadaceae.

Although some Pseudomonas strains may occasionally give rise to rather serious human infections, the habitat of most pseudomonads is again soil, water and, in addition, proteinaceous foods stored at refrigeration temperature, where these bacteria, due to their psychrotrophic behavior soon dominate the microflora and eventually produce spoilage.

Morphologically related to the Pseudomonadaceae, but bio-chemically rather different (Unger, Rahman and DeMoss, 1961) are the bacteria of the Vibrio type. This group harbors both very dangerous enteric pathogens such as Vibrio comma and useful denitrifying organisms including Vibrio costicolus.

Nonmotile and sometimes oxidative bacteria: Achromobac-ter. It has been well established recently (Buttiaux and Gagnon, 1959; Thornley, 1960; Shewan, Hobbs and Hodgkiss, 1960a and b; Buttiaux, 1961) that the nonmotile, Neisseria-like, psychrotrophic bacteria which have for some time puzzled taxonomists are in fact genuine Achromobacter strains, in spite of the circumstance that the type species of this genus as described in Bergey's Man-ual allegedly possessed peritrichous flagella. Besides being nonmotile, these bacteria which have the same habitat as the pseudomonads, can be readily distinguished from the latter by their relative biochemical inertia: they are oxidase negative, do not decompose arginine and often do not oxidize glucose.

Nonmotile organisms of the biochemical types Alcaligenes and Flavobacterium as well as the rods formerly called Bacte-rium anitratum or group B5W organisms (Ferguson and Roberts, 1950) can all be considered Achromobacteriaceae now; the latter organisms are then of course to be indicated as Achromobacter anitratus (Mossel, 1962b).

Other Gram-negative rod-shaped organisms of interest in food

D. A. A. MOSSEL

microbiology. The genus Brucella need hardly be mentioned here, because these organisms belong to the classical pathogens spread by water, milk and food. The obligately anaerobic rods of the type Ristella (Bacteroides) occur in very high numbers in feces so that, in spite of their very rapid destruction by exposure to extraenteral conditions (Mossel, 1959) they may still occasionally be encountered in foods.

Finally, some attention must be given to the acetic acid bacteria, because they play a rather important role in the spoilage of fruit, fruit products and some fermented vegetable produce. Their taxonomy remained in a confused state until proper attention was paid to the morphological characters of these bacteria by Leifson in 1954. Later, Carr and Shimwell (1961) contributed greatly to the biochemical aspects of the classification of these organisms; Table 11.2, taken from their review (1961) and Leifson's work (1954), gives a clear picture of the taxonomic position of the various acetic acid bacteria to date.

Table 11.2. The classification of the acetic acid bacteria according to Leifson (1954) and Carr and Shimwell (1961)

| Characters | Genus | |
	Acetobacter	Acetomonas
Flagella, if any	Peritrichous	Polar
Type of nutrition		
Lactate preferring	+	
Glucose preferring	.	+
Type of oxidation		
Ethanol → CO_2	+	-
Lactate → CO_3^{--}	+	-
Amino acids attacked	+	-
Metabolic type		
Operation of citric acid cycle	+	-
Important species	aceti, rancens, oxidans, xylinum and mesoxydans	suboxydans, viscosum and melanogenum

Catalase Positive Cocci. The taxonomy of the catalase positive cocci, the "micrococci" of older editions of Bergey's Manual, and the tetrads has also long presented a puzzle. However, some newer evidence facilitates their arrangement in broad groups for determinative purposes.

First of all, it seems well established that many organisms grouped in the genus Pediococcus are catalase positive, as indicated in the recent study of Günther and White (1961), in which

26 out of 85 gave positive reactions. These organisms can there-fore be brought together with the other catalase positive cocci, although preferably with micrococci (Felton, Evans and Niven, 1953)rather than with the sarcinae, because pediococci do not often show the sarcinoform grouping of cells.

Further, our study of some 600 "micrococci" has shown that a distinction between the genera <u>Micrococcus</u> and <u>Staphylococcus</u> is indeed warranted. The former genus is obligate aerobic and, if attacking mannitol, does this by oxidation in the manner de-scribed by Hugh and Leifson (1953) and probably by a pathway similar to some pseudomonads (Mossel, 1962a).

The genus <u>Staphylococcus</u> is facultatively anaerobic and at-tacks mannitol either by fermentation (<u>i.e.</u>, following the glyco-lytic pathway) or not at all. Work in this laboratory also fully confirmed the correlation which was first noticed by Evans (1948) between anaerobic decomposition of mannitol and coagulase pro-duction as well as other signs of pathogenicity. This further clearly distinguishes <u>Staphylococcus aureus</u> from similar bac-teria.

With regard to ecology, the original habitat of <u>Staphylococcus aureus</u> is definitely limited to cutaneous lesions and stools of man and to certain animal lesions. <u>Staphylococcus saprophyticus</u> (Fairbrother, 1940) and micrococci dominate on healthy skin and by this route may gain access to various meat products. In some of these they are considered as essential to good quality (Niini-vaara and Pohja, 1957), although they frequently also cause spoil-age of other meat products. Sarcinae and pediococci occur natu-rally on many vegetable materials and may hence also be found in dairy products, horticultural produce, beer, etc.

<u>Catalase Negative Cocci</u>. The streptococci no longer present much of a taxonomic problem, since Lancefield's serotyping has become daily routine.

Group A streptococci are invariably associated with diseases such as sore throat, rheumatic fever and nephritis. Foods have fairly often been incriminated in the spread of these organisms (Moore, 1955; Boissard and Fry, 1955; Farber and Korff, 1958; Badiryan, Flexner and Voliman, 1959; Taylor and McDonald, 1959; Otte and Ritzerfeld, 1960). Various types of group D ("fe-cal") streptococci are found in the stools of man and many warm-blooded animal species (Winslow and Palmer, 1910; Ostrolenk and Hunter, 1946; Buttiaux, 1958; Bartley and Slanetz, 1960; Kenner, Clark and Kabler, 1960), and this seems to be their only primary habitat (Mundt, 1961).

Among the other catalase negative cocci, <u>Leuconostoc</u> species are much dreaded as slime-formers in sugar factories and in the manufacture of some vegetable products. Strains of <u>Aerococcus</u> (Williams, Hirch and Cowan, 1953; Deibel and Niven, 1960) often cause false positives in testing foods for Lancefield group D streptococci because they are quite resistant to azide, the inhibitor which is currently used in media for the detection and enumeration of fecal streptococci (Guthof and Daman, 1958; Mossel and Krugers Dagneaux, 1959). The genus <u>Aerococcus</u> is, as the name rightly indicates, mostly airborne; it can be readily distinguished from Lancefield's group D streptococci by its failure to grow at 45^0 C. The catalase negative strains of the <u>Pediococcus</u> group closely resemble aerococci.

<u>Catalase Negative Gram-Positive Rods</u>. The most important groups of these organisms are the lactic acid bacteria: the homofermentative genus <u>Lactobacillus</u> and the heterofermentative group <u>Betabacterium</u>. The former are generally more resistant to unfavorable external conditions than the latter; both types are rather frequently encountered in vegetable products, and <u>Lactobacillus</u> species also occur in cheese and fermented sausages.

Other Gram-positive, catalase negative rods of importance in food microbiology are <u>Brevibacterium</u> species, which dominate the flora of surface ripened cheeses. The obligately anaerobic rods of the catalase negative group are now indicated as <u>Bifidibacterium</u> (Lactobacillus bifidus) (Prévot, 1938). Although as numerous in feces as <u>Ristellae</u> they are also not normally found in fecally contaminated foods because of their low resistance to extraenteral conditions.

<u>The Sporeforming Bacteria: Bacillaceae</u>. This class of organisms is of paramount importance in food bacteriology because the spores which they form are unique in structure and composition and, therefore, in heat resistance. Although some spores are rather thermolabile, others resist heating in a medium of pH 6.5 for 15 min. at 120^0 C. Because spores of <u>Bacillaceae</u> occur widely in soil, on vegetable material, in feces, dust, etc., the manufacturer of sterile goods often has to process his low acid products on the assumption that the raw materials may have contained such highly heat resistant spores. The processing hence required generally impairs the nutritive and organoleptic quality of the commodities greatly and is often even prohibitive, <u>e.g</u>., in the case of large hams.

The aerobic, catalase positive genus of this group (<u>Bacillus</u>) has been very well studied, described and arranged in a determinative key (Smith, Gordon and Clark, 1952). The anaerobic, catalase negative genus <u>Clostridium</u> is not so well defined. Table 11.3 may therefore be helpful in identifying clostridia isolated from foods.

Table 11.3. The main characteristics of the most important clostridia occurring in foods
(Mossel, Bechet and Lambion, 1962)

Character	Cl. butyricum	Cl. pasteurianum	Cl. multifermentans	Cl. acetobutylicum	Cl. chauvoei	Cl. perfringens	Cl. botulinum	Cl. sporogenes	Cl. bifermentans	Cl. histolyticum	Cl. tertium	Cl. thermosaccharolyticum	Cl. pectinovorum	Cl. lentoputrescens	Cl. tetani
Spores and sporangia	s:d	s:d	c:d	s:d	s:d	c	s:d	s:d	c	s:d	t:d	t:d	t:d	t:d	t:d
Motility	+	+	+	+	+	+	+	+	+	+	+	+	+	+	+
Fermentation of															
glucose	+	+	+	+	+	+	+	+	+	-	+	+	+	-	-
sucrose	+	+	+	+	+	+	v	-	-	-	+	+	v	-	-
lactose	+	-	+	+	+	+	v	-	-	-	+	+	v	-	-
mannitol	+	+	-	+	-	-	-	-	-	-	+	-	+	-	-
starch	+	+	+	+	-	v	v	-	-	-	+	+	+	-	-
Production of AMC	-	-	-	+	-	+		-	-	-	+	-	-	-	-
Nitrate reduction	-	-	+ or -	-	+ or -	+	-	+	+ or -	-	+	-	-	-	+
Behavior in milk	sf	-	c	sf	c	sf	v	c,p (slow)	c,p	c,p	c	c,r	sf	c,p (slow)	c (slow)
Gelatin liquefaction	-	-	-	+	+ (slow)	+	+	+	+	+	-	-	+	+	+
H₂S-production	-	-	-	+	-	+	+	+	+	+	-	-	+	+	+
Indole-formation	-	-	-	-	-	-	-	-	+	-	-	-	+ (slow)	v	+ (slow)

Legend: s = subterminal   d = distended   c = central   t = terminal
v = various reactions possible   p = peptonization   sf = stormy fermentation   r = reduction
c = coagulation

The <u>Bacillaceae</u> group harbors, apart from active spoilage agents, a number of organisms which may cause food poisoning. In order of increasing minimal toxic or infectious dose these are certain types of <u>Clostridium botulinum</u> (Meyer, 1956; Dolman, 1961), <u>Clostridium perfringens</u> (Hobbs <u>et al.</u>, 1953; Dische and Elek, 1957) and <u>Bacillus cereus</u> (Hauge, 1955; Nygren, 1961).

<u>Miscellaneous Bacteria</u>. Among the <u>Corynebacteriaceae</u> the pathogenic species <u>Cor. diphtheriae</u>, <u>Listeria monocytogenes</u> and <u>Erysipelothrix insidiosa</u> are only exceptionally found in foods when contaminated by diseased persons or animals. The saprophytic genus <u>Microbacterium</u>, on the contrary, is frequently encountered especially in dairy and meat products (McLean and Sulzbacher, 1953; Drake, Evans and Niven, 1958; Miller, 1960); its metabolism is similar to that of <u>Lactobacilli</u>, but its heat resistance may be very much higher.

The rather isolated, rod-shaped Gram-positive, anaerobic genus <u>Propionibacterium</u> is only of limited significance in food microbiology; nevertheless, its constant occurrence in certain cheeses makes it of interest.

The acid fast, Gram-positive rods of the <u>Mycobacterium</u> group were once the most dreaded of the pathogens found in foods, especially dairy products. The progress made in the eradication of bovine tuberculosis, but especially the increasing acceptance of pasteurization as an obligatory step in dairy processing, has entirely changed this picture. In the few instances where pasteurization is not generally carried out, <u>i.e.</u>, in cheese manufacture, other raw milk pathogens, such as <u>Staphylococcus aureus</u>, <u>Salmonellae</u>, <u>Shigellae</u>, <u>Brucellae</u> and <u>Rickettsiae</u>, have received so much attention that it is likely that all dairy products will soon be subjected to a heat treatment in most countries (Thatcher, Simon and Walters, 1956; Enright, Sadler and Thomas, 1957; Gargani and Guerra, 1957; Carrère <u>et al.</u>, 1960; Del Vecchio, d'Arca Simonetti and d'Arca, 1960; Walker, Harmon and Stine, 1961; Mickelson <u>et al.</u>, 1961; Parry, 1962).

Molds and Yeasts

Yeasts are virtually never pathogenic when absorbed by the oral route. However, some molds may form compounds from certain foods and feeds, which are toxic when eaten (Mayer, 1953; Lancaster <u>et al.</u>, 1961; Schumaier <u>et al.</u>, 1961; Wilson and Wilson, 1962a and b). Molds and yeasts are yet most important as spoilage agents, especially in those foods which are either naturally less vulnerable to attack by bacteria or are stored under conditions more inhibitory to bacteria than to fungi.

Table 11.4. Equilibrium humidity (e.r.h.) as a decisive parameter
of microbial spoilage of foods (adapted from Mossel
and van Kuijk, 1955 and Money and Born, 1951)

E.r.h. range	Organisms inhibited by lower limit of range	Examples of foods showing the lower e.r.h. value of this range
1.00-c.0.95	Most Gram-negative rod-shaped bacteria	Foods containing up to 40 wt % of added sugar (sucrose) or 7% of salt
0.95-0.91	Most cocci, lactobacilli and bacilli	Foods with 55% of sugar of 12% of salt
0.91-0.88	Most yeasts	Foods with 65% of sugar or 15% of salt
0.88-0.80	Most molds	Flour, rice, peas, etc. with 17% moisture
0.80-0.75	Halophilic bacteria	Foods salted to saturation (26 wt%) of NaCl. Most jams and marmalades are in this range
0.75-0.65	Xerophilic molds	Dehydrated foods with 8-25 % of moisture, dependent on water binding capacity of the dry substance Marzipan generally e.r.h. = 0.70
0.65-0.60	Osmophilic yeasts	Saturated (82 wt%) fructose solution

In the first category are foods having a reduced equilibrium humidity (cf. Table 11.4), or a pH well below 5, as well as those which are preserved by the addition of antibacterial antibiotics, especially the broad spectrum types like the tetracyclines. Also commodities normally stored in the temperature range between -5 and +5°C often become moldy (Brown, 1922; Brooks and Hansford, 1923; Schwartz and Kaess, 1934) or are slowly attacked by yeasts (Drake, Evans and Niven, 1959; Lawrence, Wilson and Pederson, 1959). In fact, yeasts and molds together with certain Gram-negative rods, micrococci and Lactobacteriaceae are the most important groups of psychrotrophic microorganisms.

In considering the significance of yeasts in foods, attention should also be given to the importance of these organisms in the manufacture of bread and fermented beverages.

Viruses

In spite of the increasing amount of rather convincing epidemiological data which has become available during the last two decades (Read et al., 1946; Mathews, 1949; Kaufmann, Sborov and Havens, 1952; Roos, 1956; McCollum, 1961; Nauman, Bagley and Schlang, 1962; Mason and McLean, 1962) it is still not clear whether foods play an important role in the spread of viral diseases as infectious hepatitis, viral diarrheas and poliomyelitis. Recently improved methods of isolation of most of the viruses concerned (Status report, 1961) may soon provide some much needed information.

Pending the accumulation of these data, the policy of the public health bacteriologist can only be the rejection of all foods which show the classical signs of having been subjected to contamination by material of fecal origin, because it has been well established that the viruses concerned are secreted with stools. Because these viruses often show a much higher resistance to unfavorable extra-enteral conditions than most of the Enterobacteriaceae occurring in feces (Gilcreas and Kelly, 1955; Kabler et al., 1961) it is imperative not to rely on a negative "coliform" test as a criterion for freedom from enteric viruses but rather to attribute much significance to the absence or presence of the more persistent Lancefield group D streptococci (Buttiaux and Mossel, 1961).

THE QUANTITATIVE ASPECTS OF FOOD MICROBIOLOGY

The Necessity of Quantification

It is generally recognized (Cheftel, 1955; Wilson, 1955; Robertson, 1960; Fay, 1960; Buchbinder, 1961; Duffy, 1961; Hands, Symons and Vineall, 1961) that a control system based on the examination of samples can never be effective in safeguarding the supply of perishable foods. Preventive measures are required for this purpose, that is, formulating codes for the industrial preparation of foods as well as the cooking, storing and serving of meals in restaurants and carrying out sufficient inspections to be sure that these codes are followed. Part of such a system is the periodic sampling of foods for microbiological examination, as a check on the reliability of the modes of processing and storage (Wilson, 1955; Thatcher, 1958; Goresline, 1962).

Even if laboratory testing of samples is carried out only in this connection, the problem of sampling is a very difficult one, because taking a truly representative sample would require 1) a large number of packs or servings per given item; 2) taking this large number of samples from all the items produced or prepared in the period to which the sampling pertains (Mossel and Drion, 1958; Bischoff, 1961). If this were to be done, even the largest laboratories would be able to deal only with the samples taken on one or a few premises per day and this, in turn, would make reasonable frequencies of sampling for each company impossible. The situation is even more difficult in the sampling of goods to be imported and therefore prepared under conditions which are only vaguely known — or not at all.

It is mainly for this reason that there is a general tendency to design tests for the microbiological examination of foods in such a way, that a general impression of the way of preparation, handling and storage of the material under investigation is obtained. In other words, it is believed advisable to base the microbiological evaluation of foods on methods of examination giving figures indicative of the absence of potentially dangerous handling rather than exclusively on proofs of the absence of specific pathogenic or toxinogenic organisms (Heller, 1952). Testing of foods according to this principle is not only born from necessity, but also has an intrinsic merit. The repeated failure to find numbers and types of organisms indicative of dangerous handling renders it very unlikely that foods or meals prepared in the same way will ever reach the consumer in a dangerously contaminated state. This system of microbiological examination of foods makes strict adherence to two principles necessary, viz. 1) taxonomically, ecologically and physiologically warranted choice of indicator organisms; 2) a strict quantitative approach.

As to the first requirement, where, classically, certain Enterobacteriaceae are chosen as one of a group of indicators of the bacteriological condition of foods (Schardinger, 1892; Smith, 1895; Swenarton, 1927; McCrady and Langevin, 1932; Seeliger, 1952; Henriksen, 1955; Mossel, 1957; Buttiaux and Mossel, 1961) it is necessary to consider whether all Enterobacteriaceae are of interest or fecal types only. If a food has been subjected to heat treatment in the course of its preparation the detection of virtually any organisms belonging to the Enterobacteriaceae will indicate potentially dangerous recontamination. In raw foods, or heated foods to which raw components (egg, cheese, sugar, spices, etc.) are added, nonfecal organisms may regularly occur without giving any reason for alarm (Mossel, 1956). With reference to staphylococci, the situation is easier: because Staphylococcus aureus, Staphylococcus saprophyticus and mesophilic micrococci have similar habitats and tolerances, large numbers of any of these organisms, though they may be harmless in themselves, must be regarded as indicating the occurrence of gross mishandling of the items under examination (Evans, Buettner and Niven, 1950; Ingram, 1960). The same consideration is valid for all Lancefield group D streptococci (Mossel, Bechet and Lambion, 1962) and for the whole group of sulfite-reducing clostridia (Ghysen, 1962).

The need of justified quantification in food microbiology has been stressed for more than 75 years (Malapert-Neufville, 1886; Smith, 1895; von Freudenreich, 1895). But, although counts of viable as well as of dead organisms are regularly carried out, the limitations of the procedures applied have alternately been underrated as well as exaggerated. Because the significance of finding given organisms in food is, as we have already seen, entirely dependent on the order of magnitude of the numbers found, attention must be given to the accuracy of quantitative microbiology in general.

Accuracy of Current Procedures

Direct Microscopic Counts. The enumeration of dead plus viable microorganisms in foods can be performed only by microscopic methods, following Breed's classical procedure (1911). The methods of staining which have been recommended for this purpose in the course of years are virtually as numerous as there are laboratories where this procedure is routinely applied. Adequate staining of cells whose cyto-receptors have been damaged by heat treatment of the food in which they occur presents an additional problem, which is not yet entirely solved (Moats, 1961).

Nevertheless, if the methodology is duly standardized and if allowance is made for the high coefficient of variation of this technique (Ziegler and Halvorson, 1935) the microscopic count may be a useful tool in food microbiology (Mossel and Zwart, 1959; Mossel and Visser, 1960).

Enumeration by Dilution Methods. Enrichment of series of decimal dilutions of a food in suitable liquid media in order to determine bacterial load is as old as bacteriology itself, because it was the procedure originally applied by Pasteur as well as by Beijerinck. Soon after Robert Koch had introduced cultivation on and in solid media, colony counts using such media were also employed for the microbiological examination of water and foods. It has since been demonstrated repeatedly that the latter procedure is generally to be preferred, because its intrinsic coefficient of variation is considerably lower (Ziegler and Halvorson, 1935; Clegg and Sherwood, 1947; McCarthy, Thomas and Delaney, 1958; Pretorius, 1961; Lear, 1962). Also, in solid media every viable bacterial cell has a chance to develop into a colony without being inhibited by antagonistic influences exerted by bacteria growing in the same medium (Prescott and Baker, 1904; Etinger-Tulczynska, 1958; Habs and Langeloh, 1960).

There is one instance wherein dilution counts are required, because plate counts are not sufficiently sensitive for practical purposes, i.e., in the detection of very low levels ($\leq 10^{-1}$ per 1 gram) of Salmonellae and Arizonae in foods. The alternative of filtration through membranes and cultivating the latter on appropriate media — although it is the method of choice for the examination of beverages (Lindberg, 1960) — presents difficulties when it comes to the examination of large amounts of solid foods (Anderson and Woodruff, 1961).

Counts on Plates or in Tubes of Solid Media. At present, these are the methods of choice for the reasons given in the preceding paragraph.

Nevertheless it should be stated that in properly carried out counts the variation coefficient may still be of the order of 10 percent (Mossel, 1956) because of the errors involved in sampling, dilution, breaking up of clumps, etc. Higher coefficients of variation up to 30 percent can be expected when selective media are used (Mossel, Bechet and Lambion, 1962) because here often a compromise must be accepted between selectivity and productivity. When badly prepared or insufficiently pretested media are used, results may become entirely unreliable.

Accuracy Reducing Factors not Inherent to Media. In many, if not most foods, the distribution of microorganisms over the commodity or within a given consignment is heterogeneous (Rahn and Boysen, 1929; Rishbeth, 1947; Johns, 1951; von Gavel, 1957; Michener, Thompson and Dietrich, 1960; Hartman and Huntsberger, 1960; Bradshaw, Dyett and Herschdoerfer, 1961; Hoeke, 1962). This makes it imperative that a sufficiently large sample be taken to be representative; this may be as much as a few hundred grams. Impairment of the accuracy of bacterial counts because of this heterogeneous distribution is not always controlled in current practice.

Next, after the sample is taken it may have to be stored and shipped. If this is done at too high a temperature, microbial growth will occur and an entirely false and unfavorable impression of the food's microbiological condition will be obtained. But if during storage or shipping the samples are accidentally frozen, deceivingly favorable counts are obtained, because most asporogenous bacteria die off slowly when held at a temperature a few degrees below 0°C (Haines, 1938; Michener, Thompson and Dietrich, 1960; Woodburn and Strong, 1960; Raj and Liston, 1961).

Because of the heterogeneous distribution of microorganisms in foods and meals, it is necessary to homogenize every sample carefully before subjecting it to microbiological examination.

Finally, foods usually have to be diluted before being examined bacteriologically. The phenomenon of spontaneous death of bacteria in some current diluents has been known for many years (Butterfield, 1932) — yet only relatively few investigators seem to take account of this by using a suitable protective diluent (Straka and Stokes, 1957; Neilson, MacQuillan and Campbell, 1957;

Table 11.5. Review of quantitative methods for the microbiological examination of foods
(Mossel, Bechet and Lambion, 1962)

Group of organisms sought	Presumptive counting procedure	Confirmation	Completion or identification
Salmonella and Arizona	MPN-enrichment procedure with at least 10 grams using simultaneously: (1) Muller or Muller-Kauffmann's tetrathionate broth; (2) selenite cystine broth of North and Bartram; (3) mannitol pre-enrichment broth of Taylor.	Plating simultaneously on (1) brilliant green phenol red sulfapyridine agar;[1] (2) desoxycholate citrate agar.[2]	Subsequently: (1) LUMoSKAD; (2) agglutinations; (3) phage test.
Shigella	Plating of at least 100 mg on SS-agar.	cf. completion tests for Salmonella	
Enterobacteriaceae	For perishable foods: plating in violet red bile glucose agar, incubated 20 h at 37°C. For low count foods: MPN-enrichment procedure with at least 1 gram using buffered brilliant green bile glucose broth without paying attention to gas formation.	GNiOx

Plating on violet red bile glucose agar. | LUMoCISKAD-reaction with the necessary supplements and if required agglutinations. GNiOx followed by LUMoCISKAD and, if necessary, Eijkman's test or agglutinations. |
Gram-negative rods	Plating on Olson's 1 p.p.m. crystal violet plate count agar.	GNiLMoOx	The necessary biochemical tests, incl. test for pigment formation on the agar of King et al.
S. aureus	Plating on Chapman's mannitol salt phenol red agar.	CoMaAnGel	Phage typing.
Lancefield group D streptococci	Plating in Packer's crystal violet sodium azide blood agar, incubated at 37°C.	TAzECa	Sherman characters plus So.
Clostridium	Counts with and without prior pasteurization for one min. at 80°C in glucose-free sulfite-iron agar contained in Miller-Prickett-tubes.	AnCaSu	The necessary morphological and biochemical tests; cf. Table 11.3.
B. cereus	Counts with and without prior pasteurization in infusion agar.	GAnNiLec	-
Molds and yeasts	Counts in yeast extract glucose oxytetracycline agar, incubated at 22°C.	Microscopic - but hardly necessary	-

Legend:
A = formation of phenylpyruvic acid from phenylalanin;
An = growth under anaerobic conditions;
Az = azide tolerance;
C = assimilation of citrate;
Ca = catalase activity;
Co = coagulase activity;
D = lysin decarboxylase activity;
E = tolerance of azide + ethyl violet;
G = glucose metabolism;
Gel. = gelatin liquefaction;
I = formation of indole from tryptophan;
K = KCN-tolerance;

L = lactose metabolism;
Lec. = lecithinase activity;
Ma = mannitol metabolism;
Mo = motility;
Ni = nitrate reduction;
Ox = oxidase activity;
S = cysteine desulfhydrase and thiosulphate reductase activity;
So = sorbitol metabolism;
Su = sulphite reduction;
T = growth at 45°C;
U = hydrolysis of urea.

[1] Osborne and Stokes.
[2] Leifson.

Heller, 1957; Heather and Vanderzant, 1958; Bretz and Hartsell, 1959; Schmidt-Lorenz, 1960).

Conclusion. Proper sampling and treatment of the sample followed by expert choice of media and application of reliable and taxonomically justified methods can reduce the variation coefficients of microbial counts of foods to an acceptable minimum of the order of 10 percent and confer a high degree of significance to such counts.

A review of the procedures found to be reliable during about 10 years of experience in the author's laboratory is given in Table 11.5.

THE SIGNIFICANCE OF THE RESULTS OF MICROBIOLOGICAL EXAMINATION OF FOODS

General Principles

The significance of a given number of given organisms in a given food depends on various factors characteristic for the food in question, such as: its composition, the treatments to which it has been subjected in the course of its manufacture, the way in which it is stored before being consumed and whether this is done before or after a culinary heat treatment. The wholesomeness and quality of foods is generally endangered only by high numbers of organisms and these can stem from any of the following causes, or a combination of these:

1) a high initial infection with dangerous organisms or spoilage agents;
2) an inadequate heat treatment;
3) a low initial infection followed by microbial proliferation during subsequent storage.

As a rule, only viable organisms are considered undesirable, but there are at least two exceptions. When high levels (i.e. 10^6 or more per gram) of enterotoxic strains of Staphylococcus aureus have occurred in a food, the product will remain toxic, even if the viable bacteria have been destroyed by autolysis, pasteurization, or cooking, because staphylococcal enterotoxin has a high thermostability. When very high total counts, i.e. $> 10^7$ per gram, have once existed in a food not prepared by fermentation, even if these organisms have been destroyed by heating, the food is undesirable in the esthetic sense. Such counts imply that approximately 0.01 percent of the food consists of bacterial debris — a quite considerable level, e.g., the same as the vitamin C content of potatoes.

Factors Influencing Significance of Numbers

Logistics. The first principle in interpreting bacterial
counts is that the numbers which are decisive are those likely to
be present at the moment of consumption (Dack, 1956). Therefore, it is important to consider the "future" of a given food
sample: the way in which it will be stored, the period of time
that storage will last, and whether any heating of the food is anticipated.

Specificity as to Types of Organisms. The tolerances defined
are primarily determined by the minimal infectious or toxic dose
(MID or MTD) or the level at which spoilage becomes apparent.
These, as has been said before, vary widely for the different
types of organisms so that a high degree of specificity is required in interpreting microbial counts.

Classical pathogens like Streptococcus pyogenes, Mycobac-
terium tuberculosis and the enterics Salmonella typhi, Salmonella
paratyphi, Shigella flexneri and Vibrio comma have very low
MID's. For other Salmonella serotypes it seems more cautious,
in the light of the recent experience of Angelotti et al. (1961), to
fix the MID not above 10^6 organisms. For Staphylococcus aureus
viable counts above 10^5/g are suspect (Dolman, 1943; Googins et
al., 1961) which shows that the MTD corresponds to the equivalent
of the order of 10^7 organisms. It is interesting to note that
Hobbs' (1953 and 1955) classical analysis of the status of numerous samples of foods incriminated in outbreaks of "unspecific
food poisoning," i.e., cases in which none of the classical pathogens could be isolated, has revealed that such foods always had
total counts of 10^5/g or more. Figures recently published by
Nikodemusz et al. (1962) pertaining to approximately 50 outbreaks
of food poisoning attributed to Bacillus species in Hungary, fully
confirm the older observations of Hobbs. Hence, either the MTD
of these "unspecific agents" is not very much higher than those of
many of the classical causes of food-borne disease, or counts
above 10^5/g may be indicative of a previous dangerous development of enterotoxinogenic Staphylococcus aureus strains, later
outgrown by a saprophytic flora (Dolman, 1943). However this
may be, foods not prepared by microbial fermentation and showing a total count of mesophiles well over 10^5/g at the moment of
consumption must generally be considered as potential health
risks. It is of interest to note that this level is well below the
one at which spoilage of foods occurs in several cases (Mossel
and Ingram, 1955) which accounts for the fact that these toxic
foods have been eaten at all.

The Limited Significance of Low Counts. The third principle in dealing with numerical tolerances is that the inverse of the second is not valid, i.e., foods showing low counts are not ipso facto safe or acceptable.

It has been shown repeatedly that pathogenic bacteria may also be isolated from foods having relatively low counts (Sutton and McFarlane, 1947; Hobbs, 1955; Hobbs and Wilson, 1959; Silverstolpe et al., 1961; Thatcher and Montford, 1962). Therefore, it is necessary under all circumstances to examine foods for the classical pathogens, as shown in Table 11.5, even though our scientific credo contains the indicator organism concept.

There is a second important reason why foods showing a count below Hobbs' limit of $10^5/g$ are not always acceptable. It is fairly certain that commodities containing about $10^4/g$ only of viable Escherichia freundii or Staphylococcus saprophyticus are in themselves harmless. Yet, such levels of these organisms have, as Hunter stated in 1939, no proper place in most foods under the environmental conditions existing in the area. With a few exceptions, viz., foods obtained by microbial fermentation, viable counts of these organisms well over $10/g$ result either from heavy contamination due to unsanitary preparation (Thatcher, Coutu and Stevens, 1953; Thatcher, Simon and Walters, 1956; Abrahamson and Clinton, 1960) or, much more frequently, from bacterial proliferation during faulty storage of foods which initially had a satisfactory bacterial status (Thatcher and Montford, 1962). Both conditions involve obvious reasons for refusal of a commodity.

Psychrotrophic Behavior of Organisms. The fourth principle is a refinement of the conception that viable counts of well over $10/g$ of organisms with a human or animal habitat often indicate inadequate storage of the commodity from which such organisms have been isolated.

If most of the organisms isolated from a food sample are psychrotrophic, i.e., capable of relatively rapid growth at about $5^\circ C$ (Mossel and Zwart, 1960; Eddy, 1960a; Kereluk, Peterson and Gunderson, 1961), their occurrence in relatively large numbers does not necessarily prove that the food was stored at too high a temperature. It will therefore sometimes be necessary to test a representative number of isolated bacteria for their growth rate at 5°. If isolates do show visible growth of a considerable part of the seeded population within 5 days at that temperature, their occurrence in considerable numbers in the sample under investigation may have been due to development at refrigeration temperature.

Such psychrotrophes, which include various Gram-negative rods, micrococci, lactobacilli, molds and yeasts, may eventually spoil the food. Although undesirable, this is a less serious defect than finding numbers of mesophilic organisms which are clearly indicative of the fact that the temperature during storage of the food has not been sufficiently low to prevent growth of organisms causing food-borne disease — or may permit this in the future.

Specificity as to Foods.   After all that has been said previously on this point, this important aspect may be simply summarized by stressing that the food bacteriologist should never fail to consider whether a given number of given organisms is encountered in fresh versus fermented, low versus high acid, or raw versus heated foods.

Application to Daily Routine

To illustrate the principles given in the previous section, a few practical cases taken from the author's experience will be summarized in historical order. They may serve as a guide to the interpretation of the results of bacteriological examinations of foods.

Significance of Bacteria in Various Types of Sausage.   Early in the past decade, a small, but excellent, meat products manufacturer in our country received a cable saying that a rather important consignment of sausage shipped by him had been rejected in a foreign country, because a few samples taken at random from this consignment had proved to contain considerable numbers of mannitol-fermenting staphylococci. Because the greater part of the consignment consisted of cooked sausage of the bologna type, this had the appearance of being a rather serious affair and we were rushed to the foreign country concerned.

The first thing which soon became evident was that the county health veterinarian who had carried out the examination had sampled only a few of the fifty or so salami type sausages which happened to be located in the rear of the van and, at that stage had not checked on the cooked bologna sausage at all. Yet, even in fermented sausages, we thought at that time staphylococci had no proper place, because lactobacilli were believed to predominate.

We succeeded in isolating significant numbers of cocci from the salami and these bacteria produced acid from mannitol under aerobic conditions; but this is a normal phenomenon also for saprophytic micrococci (Dickscheit, 1961). However, they did not do this under anaerobiosis, the key criterion for S. aureus indicated by Evans (1948) and since fully confirmed by Mossel (1962a). Also we found that quite a few types of freshly prepared and sound fermented sausages contained large numbers

Table 11.6. Review of the bacteriological condition of 60 samples
of cooked sausage of standard Dutch quality

Type of bacteria counted; cf. Table 11.5	Percent of samples in range					
	< 10	$10 - 10^2$	$10^2 - 10^3$	$10^3 - 10^4$	$10^4 - 10^5$	$> 10^5/g$
Enterobacteriaceae	100	-	-	-	-	-
Lancefield group D streptococci		81 $(< 10^2)$	7	2	7	3[*]
Presumptive Staph. aureus		97 $(< 10^2)$	2	1	-	-
Sulfite-reducing clostridia		87 $(< 10^2)$	8	3	2	-
Aerobic plate count		36 $(< 10^3)$		28	25	11
Anaerobic total count		70 $(< 10^3)$		13	8	9

[*] Sausage prepared from cooked ham, which is mostly very rich in Lancefield group D streptococci.

of such micrococci (Mossel, 1952). The classical paper by Niini-vaara and Pohja which came out a few years later (1957) was in perfect agreement with this chance observation.

The cooked sausages which were examined subsequently showed what Madelung (1953) rightly defined as an excellent bacteriological condition: a total count of most samples below $10^5/g$ and absence of significant numbers of all of the usual dangerous species and indicator organisms (Table 11.6; Mossel and Eijgelaar, 1956).

Enterobacteriaceae in Fruit Juices and on Vegetables. Like others (Hahn and Appleman, 1952; Dack, 1955; Wolford, 1956) we have occasionally found confirmed counts (cf. Table 11.5) up to about 10^2 per gram of viable Enterobacteriaceae in fruit juices marketed or served in our country. This indicates that infected products of this type were stored whereby the rather heavy initial contamination was reduced by the acids present in the juice (Mossel and de Bruin, 1960). If these juices are stored for a longer time above refrigeration temperature they will soon be almost free of Enterobacteriaceae.

Obviously, a high initial infection with organisms other than Erwinia species points to improper manufacturing practices in need of correction. There is, however, no health risk involved in the consumption of such fruit juices because the enteric bacteria will definitely never increase to MID-levels, which may be expected to occur in vegetable juice having pH values well over 4.5 (Thomann, 1953).

As a contrary example, to find about 10^3/g viable <u>Escherichia coli</u> per gram of a vegetable like spinach (Slocum and Boyles, 1941; Steiniger, 1961) is again not very serious, because this vegetable is invariably cooked before being consumed and this treatment will destroy virtually all viable Gram-negative rods (Angelotti, Foter and Lewis, 1961). The only exception known (<u>Alcaligenes tolerans</u>) presents no risk for the consumer because it is nonpathogenic (Abd-El-Malek and Gibson, 1952).

The Problem of Potentially Infected Raw Proteinaceous Staple Foods. The rather large loads of microorganisms on fresh meats (Lundbeck, Plazikowski and Silverstolpe, 1955; Hobbs and Wilson, 1959; Kampelmacher et al., 1961; Galbraith, Archer and Tee, 196 Granville, 1961) and egg products (Thatcher and Montford, 1962) in general makes it difficult to decide what destination to recommend for such materials, though they may appear to be free from Salmonellae.

If the examination of a sample of meat, fish or whole egg shows the presence of, e.g., 10^4/g of viable <u>Pseudomonas</u>, <u>Achromobacter</u> and psychrotrophic species of <u>Enterobacteriaceae</u> (Eddy and Kitchell, 1959; Mossel and Zwart, 1960; Schultze and Olson, 1960) spoilage of the product will be inevitable when stored for more than about 3 additional days at 5 to 10^0C. Even if the food is to be eaten without a storage period, cooking before use might be strongly recommended, which, fortunately, is the rule rather than the exception with such commodities.

However, this same bacterial count does not mean very much for a product which is to be frozen, dehydrated, or incorporated into a suitably prepared meat, fish or egg salad, because both freezing, dehydration and pickling (Mossel and van der Meulen, 1960) tend to limit and even reduce the numbers of Gram-negativ rod-shaped bacteria to harmless levels.

The Significance of Lancefield Group D Streptococci in Dehydrated Milk. The occurrence of staphylococcal enterotoxin in dehydrated milk products (Anderson and Stone, 1955; Armijo et al., 1957) once more emphasized that the rather labile coliform bacteria are not under all circumstances the only reliable group of indicator organisms. Obviously, the remedy against toxin food poisoning caused by dehydrated milk products is prevention by starting with low count milk followed by proper quality control in the dry milk plant (Heinemann, 1958). Once in a while the bacteriologist may still be faced with the necessity of examining food samples and interpreting the data obtained. In spite of all progress made (Heinemann, 1960; Moats, 1961) the Breed count on heat-treated milk products still has its limitations and it was therefore thought worthwhile to attempt an estimation of the

Table 11.7. Frequency distribution of Lancefield group D ("fecal")
streptococci in 200 samples of Dutch pasteurized milk

Viable counts	Percent of samples in range			
	< 10	$10\text{-}10^2$	$10^2\text{-}10^3$	$< 3 \times 10^4/\text{ml}$
Total plate count	-	-	-	100
Enterobacteriaceae	98	2	-	-
Lancefield group D streptococci	91	5	4	-

value of Lancefield group D streptococci as indicator organisms
in such products.

The underlying idea was, that, while these streptococci doubtless "occur" in most pasteurized milks due to their relatively
high heat resistance (White and Sherman, 1944; Mieth, 1961), they
may normally do so in relatively low numbers. These organisms
may increase greatly when milk or milk products are mishandled
in some way or another, involving potential development of S. aureus. Also, Lancefield group D streptococci will certainly survive
subsequent heat treatment (Angelotti, Foter and Lewis, 1961; Silliker
et al., 1962) as well as storage of the dehydrated final product
(Brown and Gibbons, 1950; Higginbottom, 1953) much better than
staphylococci and therefore be useful potential indicator organisms.

Because proper evaluation of the product depends on the numbers of such streptococci occurring in pasteurized milk of acceptable quality, approximately 200 Dutch samples taken strictly
at random during 6 months were examined. Table 11.7 gives the
results of this investigation. Over 90 percent of the samples
showed a viable count of Lancefield group D streptococci of less
than $10/\text{ml}$. Hence proper handling of freshly pasteurized milk
may guarantee a liquid product with less than 10 of these organisms per ml, corresponding to less than $10^2/\text{g}$ of dehydrated
product. These results are in agreement with the conclusion of
Mattick, Hiscox and Crossley (1945) that with improvement in
cleanliness and efficiency of a milk plant the incidence of fecal
streptococci in its products declines.

The Significance of the Microflora of Minor Components of
Canned Foods. It has long been recognized that spores of Bacill-
aceae and other relatively heat resistant organisms, such as
Lancefield group D streptococci and Microbacterium species, occurring in ingredients such as cereals, spices, sugar, curing
salts, gelatin, etc., may impair the keeping quality of heat-
preserved foods in which they are incorporated, because they
may exert the decisive influence on the initial microbial load of

such commodities. A justified principle of examining these ingredients will have to take account both of the qualitative aspect,
i.e., the heat treatment to which an appropriate dilution of the ingredient is to be exposed prior to plating in order to eliminate
insignificant contaminants, and of quantitation: the acceptable
infection rate.

A suitable ecological approach to the former aspect has been
suggested by Knock and Baumgartner (1947). It introduces the
concept of "potentially process-resistant organisms," defined as
the types of organisms resisting the heat treatment applied to the
commodity to which the ingredient is to be added. This process,
for example, will be of the order of 5 to 30 min. at 63°C for ingredients used in various types of canned large size hams (Clarenburg, 1955; Coretti, 1957) and 30 min. at 110°C for "commercially sterile" other low acid packs.

As to the significance of numbers, we have suggested (Mossel
1955) that canning ingredients should not essentially influence the
initial load of potentially process-resistant organisms generally
encountered in the major component of the commodity. Thus it
makes a great difference whether meat or vegetable packs are
considered and whether we have to judge spices, used at a level
of about 0.1 percent or sugars often used in concentrations of a
few percent.

Requirements for Dehydrated, Instant Baby Formulae and
Cereals.   The three salmonellosis outbreaks involving dried food
products in the United States (Slocum, 1959) and especially the
Swedish epidemic of infantile salmonellosis due to infection of a
dehydrated baby food with Salmonella muenchen (Silverstolpe et
al., 1961) have caused much alarm and raised many questions as
to suggested standards for this type of product because it is intended for a class of consumers with exceptional susceptibility to
enteric disease (Dack, 1961).

Clearly the approach must primarily again be a preventive
one, in which the manufacturing industries check carefully every
consignment of raw materials, maintain a very high standard of
plant hygiene and adhere strictly to the requisite terminal heating
procedure. Further, these products must be free from enteric
pathogens when reasonably large samples (ca. 50 g) are examined
properly.

As to other justified requirements for viable organisms, an
ecological approach has led to the following suggested standard.
A representative sample is reconstituted using tap water and
thereupon stored for 6 hours at 25-30°C. The beverage or porridge so obtained has to meet the specifications for freshly pasteurized milk, which, in almost every country (Kästli, 1957), are:

total count of 10^4/ml as a maximum and absence of "coliform bacteria" — which we prefer to substitute by <u>Enterobacteriaceae</u> — in numbers over 10/ml.

This requirement is very reasonable because it is well established that the greater part of the infection existing in the freshly prepared product will die off rather quickly during storage of such a dehydrated commodity (Crossley and Johnson, 1942; Haines and Elliot, 1944; Rishbeth, 1947; Brown and Gibbons, 1950; van der Schaaf, van Zijl and Hagens, 1962), whereas the residual flora needs a considerable lag-time before vigorous growth will start (Silverstolpe <u>et al.</u>, 1961). The consumer is fully protected by this approach involving a storage period of 6 hours as a safety margin, because keeping perishable foods, especially those intended for babies, for more than a few hours at ambient temperature is widely discouraged.

<u>Enterobacteriaceae as Indicator Organisms for General Use.</u> In spite of the favorable experience, apart from an occasional exception, obtained by the use of "coliform bacteria" as indicator organisms for improper manufacture or storage of foods, the wisdom of maintaining this group of bacteria as indicators of sanitation or contamination has recently been questioned again (Machala, 1961). This question is not unreasonable because <u>indiscriminate</u> use of these organisms is certainly to be discouraged. However, once more the ecologically justified use of well-chosen and also properly identified organisms for this purpose is advocated.

Limiting this choice to the lactose positive organisms is not warranted, because most of the lactose-negative <u>Enterobacteriaceae</u> with the exception of <u>Erwiniae</u> are at least of the same health or ecological significance as the coliform group (Seeliger, 1952; Mossel, 1956; Mossel, 1957; Kretzchmar, 1959; Schönherr, 1961; Mossel, Mengerink and Scholts, 1962).

Next, the approach must again be strictly quantitative; to reject foods indiscriminately just because they contain e.g., one <u>Aerobacter aerogenes</u> per 10 grams is certainly not justified from the point of view of consumer protection nor is such a standard attainable even in the best sanitized food plant (Dack, 1955).

Finally, in judging the significance of encountering a certain number of <u>Enterobacteriaceae</u> in a particular low acid food, one must consider the question whether or not the product under review has been heat-treated in the course of its industrial preparation.

In the case of heat treated foods, the presence of numbers of <u>Enterobacteriaceae</u> of <u>any</u> kind of the order 10/g indicates either insufficient heat processing of the product (because, as shown before, virtually all such rods are very heat-labile), or adequate treatment followed by post-pasteurization recontamination. As either of these conditions is highly undesirable, such a food must be rejected. If counts of Enterobacteriaceae in heated foods are even well above $10/g$, this points to either 1) a gross under-pasteurization; or 2) contamination from a severely infected focus (Abrahamson and Clinton, 1960; Fee, 1961); or 3) slight under-processing or slight pollution, followed by subsequent growth of the residual or initial contaminating flora due to stor-age of the product at elevated temperature. These, clearly, are further reasons for refusal of the commodity.

For foods not subjected to any heat treatment in the course of their manufacture, the situation is entirely different. The natural association of many of these foods may well comprise some types of fermentative Gram-negative rods. Therefore, only if the num-bers of these organisms exceed the normal level (indicating growth in the food) and/or if the isolated bacteria belong to patho-genic groups, or are of unquestionable fecal origin (Escherichia coli, urea positive Klebsiellae and Proteus; Buttiaux and Mossel, 1961) is there a valid reason to reject a sample. This illustrates anew the absolute necessity of identifying bacteria isolated from foods and of not worshipping numbers only (Tanner, 1949; Wilson, 1959).

This plea to retain Enterobacteriaceae, chosen after proper ecological study, as indicator organisms does not attempt to say that the use of such organisms leaves nothing to be desired. On the contrary, it has been shown repeatedly in this paper, that a negative outcome of this test may not be of too much significance. One of the most valuable indicator organisms of this group, Escherichia coli, has relatively little resistance to various extra-enteral conditions and therefore is easily outgrown by other less specific Enterobacteriaceae and even quite different bacteria (Mossel, 1962b). Hence, more resistant and therefore more val-uable enteric indicator organisms, especially the Lancefield group D streptococci must certainly be used simultaneously (Penna, 1959; Buttiaux and Mossel, 1961; Wilkerson, Ayres and Kraft, 1961).

The Significance of Microbiological Findings in Margarine. The significance of the results of the microbiological examina-tion of a sample of margarine may be difficult to interpret unless the composition of the product is known in detail.

If a margarine contains milk solids, has a pH well above 4.5 in its aqueous phase and contains no preservatives, e.g., as is the case in France, practically all microorganisms may grow.

Such a product must therefore be virtually sterile, and should definitely be free from significant numbers of pathogenic and lipolytic organisms, unless its structure is such that autosterilization of its very small water droplets will occur during storage (Winkle and Adam, 1959).

On the contrary, where milk solids are not used and the pH of the aqueous phase is generally around 4.5 (e.g., in The Netherlands) the occurrence of most organisms, with the exception of molds, yeasts and a few acid-tolerant bacteria, is of no significance. Where such products also contain benzoic acid, even the highest microbial count is no indication of an impaired keeping quality — yet, it reveals bad hygiene during preparation and is therefore a reason for refusal.

NEEDS FOR FURTHER STUDY

It would be quite unwise to pretend that the food bacteriologist of 1962 has so much published experimental data at his disposition that he may in every case be able to attribute proper significance to finding certain numbers of taxonomically well-identified organisms in a food of well-known ecological history. It appears more prudent and wiser to point out a few essential deficiencies in our knowledge — thereby deliberately risking future criticism of our decisions.

Pathogenesis of Food-Borne Disease Outbreaks

It has been said repeatedly, although most explicitly by Dauer (1961), that there is need for really representative information of the frequency and etiology of food-borne disease outbreaks.

This involves: 1) a much more adequate and immediate reporting of such disease outbreaks; 2) a much swifter and more complete bacteriological investigation of virtually all, and not merely of an arbitrary selection of, outbreaks of food-borne diseases; 3) reporting all details — especially the results of counts of various significant organisms, which may lead to better quantification of MID's — in a meticulous way. The British excel in this in their reports on "Food poisoning in England and Wales," published annually in the Monthly Bulletin of the Ministry of Health, London.

Food Stability Studies

Next, it may not be forgotten that food spoilage is an important ailment of our society too, albeit perhaps less speculative than food poisoning.

Therefore, previous investigations on the stability pattern of foods (Mossel and Ingram, 1955) should be extended, with particular reference to the influence exerted on such patterns by freezing, dehydration, acid preservation and also radiation. In this connection a search for inhibitors of microbial growth which are not expected to exhibit any chronic toxicity for the consumer at the level necessary for food preservation, and whose antimicrobial activity is not influenced by pH, may be very useful too, especially for those countries where cold chains are not yet very frequent. That such a line of research may be promising is demonstrated by the development of the macrolide antibiotic Tylosin, which seems effective (Greenberg and Silliker, 1962a and 1962b) as well as reasonably safe from the toxicological point of view (Berkman et al., 1960).

Referee Panel Procedure Studies

Panel studies carried out under the auspices of the Association of Official Agricultural Chemists of the United States in the field of chemical examination of foods as laid down in the successive editions of the "Official Methods" are gratefully used by food chemists all over the world.

In the opinion of many microbiologists, the A.O.A.C. should definitely undertake a similar endeavor in the field of microbiological examination of foods, starting from the material already collected by its sister organization, The American Public Health Association, in its "Recommended Methods" (1958).

Post Doctoral and Refresher Courses

What Rutstein recently (1961) stated so admirably about the difficulty of maintaining contact with progress in medical knowledge, also fully pertains to food microbiology.

About twenty years of experience has taught the author two things:

1) It is practically impossible for an average graduate, Ph.D., D.V.M. or M.D., to start practicing food

microbiology without a proper postdoctoral training. There is not at all a need for a long apprenticeship, but it should be spent in a competent institute aware of the current status of this special field.

2) The period of education, so defined, must extend from completing such a postdoctoral course until retirement. The necessary refresher courses need not be long or very frequent. The experience obtained at the International Center for Food Bacteriology at the Pasteur Institute at Lille (Buttiaux and Mossel, 1957) suggests that an annual or biannual seminar of about two weeks fully assures that contact with food microbiology will be maintained — provided that the teaching content is selected and presented by a competent board of professors.

It might be worthwhile to consider whether, with our well-known "shrinking distances," such refresher courses might be alternately held in the United States and in European centers such as the Pasteur Institute at Lille or the Low Temperature Research Station at Cambridge.

ACKNOWLEDGEMENTS

The author is greatly indebted to his friends of long standing Dr. M. Ingram (Cambridge, England) and Dr. W. L. Sulzbacher (Beltsville, Md.), for numerous valuable discussions pertaining to specific aspects of this paper.

LITERATURE CITED

Abd-El-Malek, Y. and Gibson, T. 1952. Studies in the bacteriology of milk. IV. The Gram-negative rods of milk. J. Dairy Research 19:294-301.

Abrahamson, A. E. and Clinton, A. F. 1960. The control of bacterial populations in foods. Quart. Bul. Assoc. Food and Drug Offic. U. S. 24:31-38.

_____, Caputo, G., Kellerman, W. B., and Weiner, L. 1961. The use of automatic machine methods to control and schedule sanitary inspections and process related data. Quart. Bul. Assoc. Food and Drug Offic. U. S. 25:69-77.

Allen, J. R. and Foster, E. M. 1960. Spoilage of vacuum-packed sliced processed meats during refrigerated storage. Food Research 25:19-25.

American Public Health Association, Inc. 1958. Recommended methods for the microbiological examination of foods. New York.

Anderson, K. and Woodruff, P. 1961. The bacteriological screening of desiccated coconut. Med. J. Australia I:856-58.

Anderson, P. H. R. and Stone, D. M. 1955. Staphylococcal food poisoning associated with spray-dried milk. J. Hygiene 53: 387-97.

Angelotti, R., Foter, M. J., and Lewis, K. H. 1961. Time-temperature effects on Salmonellae and staphylococci in foods. III. Thermal death time studies. Applied Microbiol. 9:308-15.

_____, Bailey, G. C., Foter, M. J., and Lewis, K. H. 1961. Salmonella infantis isolated from ham in food poisoning incident. Public Health Repts. Wash. 76:771-76.

Armijo, R., Henderson, D. A., Timothée, R., and Robinson, H. B. 1957. Food poisoning outbreaks associated with spray-dried milk — An epidemiologic study. Am. J. Public Health 47: 1093-1100.

Badiryan, L. G., Flexner, S. Y., and Voliman, I. B. 1959. An outbreak of a streptococcal alimentary infection. Gigiena i. Sanitariya 24, nr. 5:58-59.

Barber, F. W. 1955. The value of the coliform test applied to fruit flavored ice cream. J. Dairy Sci. 38:233-35.

Bartley, C. H. and Slanetz, L. W. 1960. Types and sanitary significance of fecal streptococci isolated from feces, sewage and water. Am. J. Public Health 50:1545-52.

Berkman, R. N., Richards, E. A., van Duyn, R. L., and Kline, R. M. 1960. The pharmacology of tylosin, a new antibiotic, in the chicken. Antimicrobial Agents Annual:595-604.

Bischoff, J. 1961. Vorschläge zur Änderung der Verordnung zum Schutze gegen Infektion durch Erreger der Salmonellagruppe in Eiprodukten vom 17. Dezember 1956. Berl. münch. tierärztl. Wschr. 74:70-71.

Boissard, J. M. and Fry, R. M. 1955. A food-borne outbreak of infection due to Streptococcus pyogenes. J. Applied Bact. 18: 478-83.

Bond, R. G. and Stauffer, L. D. 1955. Food sanitation and/or the infectious process. J. Am. Diet. Assoc. 31:993-96.

Borman, E. K., Stuart, C. A., and Wheeler, K. M. 1944. Taxonomy of the family Enterobacteriaceae. J. Bact. 48:351-67.

Bradshaw, N. J., Dyett, E. J., and Herschdoerfer, S. M. 1961. Rapid bacteriological testing of cooked or cured meats, using a tetrazolium compound. J. Sci. Food Agr. 12:341-44.

Breed, R. S. 1911. The determination of the number of bacteria in milk by direct microscopic examination. Zentralbl. Bakteriol. Parasitenk. Abt. II, 30:337-40.

Bretz, H. W. and Hartsell, S. E. 1959. Quantitative evaluation of defrosted Escherichia coli. Food Research 24:369-375.

Brooks, F. T. and Hansford, C. G. 1923. Mould growths upon cold-store meat. Trans. Brit. Mycol. Soc. 8:113-42.

Brown, H. J. and Gibbons, N. E. 1950. Enterococci as an index of fecal contamination in egg products. Canad. J. Research 28F:107-17.

Brown, W. 1922. On the germination and growth of fungi at various temperatures and in various concentrations of oxygen and carbon dioxide. Ann. Botany 36:257-83.

Brown, W. L., Vinton, C. A., and Gross, C. E. 1960. Heat resistance and growth characteristics of microorganisms isolated from semi-perishable canned hams. Food Research 25:345-50.

Buchbinder, L. 1961. Current status of food poisoning control. Public Health Repts. Wash. 76:515-20.

Butterfield, C. T. 1932. The selection of a dilution water for bacteriological examinations. J. Bact. 23:355-68.

Buttiaux, R. 1958. Les streptocoques fecaux des intestins humains et animaux. Ann. Inst. Pasteur 94:778-82.

_____. 1961. Pseudomonas non pigmentés et Achromobacter. Ann. Inst. Pasteur 100, Suppl. au no. 6:43-58.

_____, and Gagnon, P. 1959. Au sujet de la classification des Pseudomonas et des Achromobacter. Ann. Inst. Pasteur Lille 10:121-49.

_____, and Mossel, D. A. A. 1957. L'analyse bactériologique des produits alimentaires perissables et conserves. Ann. Inst. Pasteur Lille 9:138-75.

_____. 1961. The significance of various organisms of faecal origin in foods and drinking water. J. Applied Bact. 24:353-64.

Carr, J. G. and Shimwell, J. L. 1961. The acetic acid bacteria, 1941-61. A critical review. Antonie van Leeuwenhoek 27:386-400.

Carrère, L., Lafenètre, H., Quatrefages, H., and Noronha, F. de. 1960. Durée de la survie des Brucella dans le fromage de Roquefort. Bul. Acad. Vét. France 33:469-73.

Cheftel, H. 1955. Remarques à propos du contrôle bactériologique des jambons conservés en boîtes. Ann. Inst. Pasteur Lille 7:256-62.

Clarenburg, A. 1955. Factors determining survival and development of Salmonella and other Enterobacteriaceae in semi-preserved meats. Ann. Inst. Pasteur Lille 7:124-32.

188   D. A. A. MOSSEL

Clegg, L. F. L. and Sherwood, H. P. 1947. The bacteriological examination of molluscan shellfish. J. Hygiene 45:504-21.

Cockburn, W. C. 1960. Food poisoning. (a) Reporting and incidence of food poisoning. Roy. Soc. Health J. 80:249-53.

Coretti, K. 1957. Eine Schnellmethode zum Nachweis der ausreichenden Erhitzung in Dosenschinken. Fleischwirtschaft 9:113-15.

Crossley, E. L. and Johnson, W. A. 1942. Bacteriological aspects of the manufacture of spray-dried milk and whey powders, including some observations concerning moisture content and solubility. J. Dairy Research 13:5-44.

Dack, G. M. 1955. Significance of enteric bacilli in foods. Am. J. Public Health 45:1151-56.

_____. 1956. Evaluation of microbiological standards for foods. Food Technol. 10:507-9.

_____. 1961. Why microbiological standards for foods? Proc. 13th Research Conf. Am. Meat Inst. Foundation:29-33.

_____, Wheaton, E., Mickelson, M. N., and Schuler, M. N. 1960. Public health significance of microorganisms in frozen pot pies. Quick Frozen Foods 22, nr. 11:44-45; 160; 162.

Dauer, C. C. 1961. 1960 summary of disease outbreaks and a 10-year résumé. Public Health Repts. Wash. 76:915-22.

Davis, G. H. G. and Park, R. W. A. 1962. A taxonomic study of certain bacteria currently classified as Vibrio species. J. Gen. Microbiol. 27:101-19.

Deibel, R. H. and Niven, C. F. 1960. Comparative study of Gaffkya homari, Aerococcus viridans, tetrad-forming cocci from meat curing brines and the genus Pediococcus. J. Bact. 79:175-80.

Dickscheit, R. 1961. Beiträge zur Physiologie und Systematik der Pediokokken des Bieres. I. Mitteilung. Zentralbl. Bakteriol. Parasitenk. Abt. II, 114:270-84.

_____. 1961. Beiträge zur Physiologie und Systematik der Pediokokken des Bieres. II. Mitteilung. Zentralbl. Bakteriol. Parasitenk. Abt. II, 114:459-74.

Dische, F. E. and Elek, S. D. 1957. Experimental food-poisoning by Clostridium welchii. Lancet 273:71-74.

Dolman, C. E. 1943. Bacterial food poisoning. Canad. J. Public Health 34:97-111; 205-35.

_____. 1961. Further outbreaks of botulism in Canada. Canad. Med. Assoc. J. 84:191-200.

Drake, S. D., Evans, J. B., and Niven, C. F. 1958. Microbial flora of packaged frankfurters and their radiation resistance. Food Research 23:291-96.

_____. 1959. The identity of yeasts in the surface flora of packaged frankfurters. Food Research 24:243-46.

Duffy, M. P. 1961. California pure Foods and Drugs Acts and related laws. Food Drug Cosmetic Law J. 16:443-65.

Eddy, B. P. 1960a. The use and meaning of the term "psychrophilic." J. Applied Bact. 23:189-90.

_____. 1960b. Cephalotrichous, fermentative Gram-negative bacteria: the genus Aeromonas. J. Applied Bact. 23: 216-49.

_____ and Kitchell, A. G. 1959. Cold-tolerant fermentative Gram-negative organisms from meat and other sources. J. Applied Bact. 22:57-63.

Enright, J. B., Sadler, W. W., and Thomas, R. C. 1957. Pasteurization of milk containing the organism of Q fever. Am. J. Public Health 47:695-700.

Etinger-Tulczynska, R. 1958. The effect of various water bacteria on the growth of Escherichia coli I in MacConkey's broth at 37° and 42°. J. Applied Bact. 21:174-79.

Evans, J. B. 1948. Studies of staphylococci with special reference to the coagulase-positive types. J. Bact. 55:793-800.

_____, Buettner, L. G., and Niven, C. F. 1950. Evaluation of the coagulase test in the study of Staphylococci associated with food poisoning. J. Bact. 60:481-84.

Ewing, W. H. and Edwards, P. R. 1960. The principal divisions and groups of Enterobacteriaceae and their differentiation. Internat. Bul. Bact. Nomenclat. and Taxon. 10:1-12.

_____ and Johnson, J. G. 1960. The differentiation of Aeromonas and C27 cultures from Enterobacteriaceae. Internat. Bul. Bact. Nomenclat. and Taxon. 10:223-30.

Fairbrother, R. W. 1940. Coagulase production as a criterion for the classification of staphylococci. J. Pathol. Bact. 50: 83-88.

Farber, R. E. and Korff, F. A. 1958. Foodborne epidemic of group A beta hemolytic streptococcus. Public Health Repts. Wash. 73:203-9.

Fay, A. C. 1960. Reappraisal of the quality control of milk supplies from farm bulk tanks. J. Dairy Sci. 43:116-19.

Fee, W. M. 1961. An unusual vehicle of infection. Health Bul. Scotland 19:10-12.

Felton, E. A., Evans, J. B., and Niven, C. F. 1953. Production of catalase by the Pediococci. J. Bact. 65:481-82.

Ferguson, W. W. and Roberts, L. F. 1950. A bacteriological and serological study of organism B5W (Bacterium anitratum). J. Bact. 59:171-83.

Freudenreich, Ed. von. 1895. Ueber den Nachweis des Bacillus coli communis im Wasser und dessen Bedeutung. Zentralbl. Bakteriol. Parasitenk. Abt. I, 18:102-5.

Galbraith, N. S., Archer, J. F., and Tee, G. H. 1961. Salmonella saint-paul infection in England and Wales in 1959. J. Hygiene 59:133-41.

Gargani, G. and Guerra, M. 1957. Sulla sopravvivenza della Br. abortus e della Br. melitensis nel burro. Igiene mod. 50: 528-38.

Gavel, L. von. 1957. Die räumliche Verteilung des Mikroorganismenwachstums in der Butter. Milchwissenschaft 12:276-80.

Ghysen, J. 1962. De microbiologische gesteldheid van hulpstoffen voor vleesprodukten in verband met de volksgezondheid. Conserva 10:138-42.

Gilcreas, F. W. and Kelly, S. M. 1955. Relation of coliform-organism test to enteric-virus pollution. J. Am. Water Works Assoc. 47:683-94.

Googins, J. A., Collins, J. R., Marshall, A. L., and Offutt, A. C. 1961. Two gastroenteritis outbreaks from ham in picnic fare. Public Health Repts. Wash. 76:945-54.

Goresline, H. E. 1962. Role of sanitation in dehydration of foods. J. Milk Food Technol. 25:11-13.

Granville, A. 1961. Le problème de la dissémination des Salmonella par les viandes de boucherie et les produits de charcuterie. Ann. Med. Vet. 105:82-105.

Greenberg, R. A. and Silliker, J. H. 1962a. The effect of tylosin on coagulase-positive staphylococci in food products. J. Food Sci. 27:60-63.

_____. 1962b. The action of tylosin on spore-forming bacteria. J. Food Sci. 27:64-68.

Günther, H. L. and White, H. R. 1961. The cultural and physiological characters of the Pediococci. J. Gen. Microbiol. 26: 185-97.

Guthof, O. and Dammann, G. 1958. Ueber die Brauchbarkeit von Enterokokken-Testen zur Beurteilung von Trinkwasser und Oberflächenwasser. Arch. Hygiene Bakteriol. 142:559-68.

Habs, H. and Langeloh, U. 1960. Versuche über den Antagonismus zwischen E. coli und E. freundii bei der bakteriologischen Trinkwasseruntersuchung. Arch. Hygiene Bakteriol. 144:277-86.

Hahn, S. S. and Appleman, M. D. 1952. Microbiology of frozen orange concentrate. I. Survival of enteric organisms in frozen orange concentrate. Food Technol. 6:156-58.

Haines, R. B. 1938. The effect of freezing on bacteria. Proc. Royal Soc. London 124B:451-63.

_____ and Elliot, E. M. L. 1944. Some bacteriological aspects of dehydrated foods. J. Hygiene 43:370-81.

Hands, A. H., Symons, H. W., and Vineall, A. J. P. 1961. Hygiene in frozen food factories. J. Roy. Inst. Public Health and Hygiene 24:217-25.

Hartman, P. A. and Huntsberger, D. V. 1960. Sampling procedures for bacterial analysis of prepared frozen foods. Applied Microbiol. 8:382-86.

Hauge, S. 1955. Food poisoning caused by aerobic sporeforming Bacilli. J. Applied Bact. 18:591-95.

Heather, C. D. and Vanderzant, C. 1958. Effect of metabolites on the viability of heat-treated Pseudomonas fluorescens. Food Research 23:126-29.

Heinemann, B. 1958. Quality control in the dry milk plant. J. Dairy Sci. 41:1114-17.

_____. 1960. Factors affecting the direct microscopic clump count of nonfat dry milk. J. Dairy Sci. 43:317-28.

Heller, C. L. 1952. Bacteriological standards for perishable foods. (b) Manufactured meats. J. Royal Sanit. Inst. 72: 396-404.

_____. 1957. The pH values of food samples diluted with various diluents. Lab. Practice 6:388-89.

Henriksen, S. D. 1955. A study of the causes of discordant results of the presumptive and completed coliform tests on Norwegian waters. Acta Pathol. Microbiol. Scand. 36:87-95.

Higginbottom, C. 1953. The effect of storage at different relative humidities on the survival of microorganisms in milk powder and in pure cultures dried in milk. J. Dairy Research 20:65-75.

Hobbs, B. C. 1953. The intensity of bacterial contamination in relation to food poisoning with special reference to Clostridium welchii. Att VI Congr. Internaz. Microbiol. 7:280-82.

_____. 1955. The laboratory investigation of non-sterile canned hams. Ann. Inst. Pasteur Lille 7:190-200.

_____. 1960. Food poisoning. (d) Staphylococcal and Clostridium welchii food poisoning. Roy. Soc. Health J. 80: 267-71.

_____, Smith, M. E., Oakley, C. L., Warrack, G. H., and Cruickshank, J. C. 1953. Clostridium welchii food poisoning. J. Hygiene 51:75-101.

_____, and Wilson, J. G. 1959. Contamination of wholesale meat supplies with Salmonellae and heat-resistant Clostridium welchii. Monthly Bul. Min. Health London 18:198-206.

Hoeke, F. 1962. Microbiologische gesteldheid van gedroogd kippeeiwit en gemalen kokos als bakkerijgrondstof. Conserva 10:186-94.

Hugh, R. and Leifson, E. 1953. The taxonomic significance of fermentative versus oxidative metabolism of carbohydrates by various Gram negative bacteria. J. Bact. 66:24-26.

Hunter, A. C. 1939. Uses and limitations of the coliform group in sanitary control of food production. Food Research 4: 531-38.

Ingram, M. 1960. Bacterial multiplication in packed Wiltshire bacon. J. Applied Bact. 23:206-15.

_____, and Shewan, J. M. 1960. Introductory reflections on the Pseudomonas-Achromobacter group. J. Applied Bact. 23:373-78.

Johanssen, A. 1961. SIK Rapport nr. 100. Svenska Institutet for Konserveringsforskning, Götebarg, Sweden.

Johns, C. K. 1951. Bacteriological analysis of edible gelatin. Food Research 16:281-87.

Kabler, P. W. and Clark, H. F. 1960. Coliform group and fecal coliform organisms as indicators of pollution in drinking water. J. Am. Water Works Assoc. 52:1577-79.

_____, Clarke, N. A., Berg, G., and Chang, S. L. 1961. Viricidal efficiency of disinfectants in water. Public Health Repts. Wash. 76:565-70.

Kästli, P. 1957. L'influence du traitement thermique du lait sur la viabilité des germes pathogènes et sur l'activité de leurs toxines. Le Lait 37:241-53; 404-17.

Kampelmacher, E. H., Guinée, P. A. M., Hofstra, K., and Keulen, A. van. 1961. Studies on Salmonella in slaughter-houses. Zentrabl. Vet. Med. 8:1025-42.

Kaufmann, G. G., Sborov, V. M., and Havens, W. P. 1952. Outbreak of infectious hepatitis — presumably food-borne. J. Am. Med. Assoc. 149:993-95.

Kenner, B. A., Clark, H. F., and Kabler, P. W. 1960. Fecal streptococci. II. Quantification of streptococci in feces. Am. J. Public Health 50:1553-59.

Kereluk, K., Peterson, A. C., and Gunderson, M. F. 1961. Effect of different temperatures on various bacteria isolated from frozen meat pies. J. Food Sci. 26:21-25.

Knock, G. G. and Baumgartner, J. G. 1947. The estimation of thermophile spores in canning ingredients. Food Manufact. 22:11-15.

Kretzchmar, W. 1959. Die Bedeutung der Differenzierung von Coli und coliformen Keimen für die bakteriologische Wasseruntersuchung. Z. ges. Hygiene Grenzgeb. 5:73-91.

Lancaster, M. C., Jenkins, F. P., Philip, J. M., Sargeant, K., Sheridan, A., O'Kelly, J., and Carnaghan, R. B. A. 1961. Toxicity associated with certain samples of groundnuts. Nature 192:1095-97.

Lautrop, H. 1961. Aeromonas hydrophila isolated from human faeces and its possible pathological significance. Acta Pathol. Microbiol. Scand. 51, Suppl. 144:299-301.

Lawrence, N. L., Wilson, D. C., and Pederson, C. S. 1959. The growth of yeasts in grape juice stored at low temperatures. II. The types of yeast and their growth in pure culture. Applied Microbiol. 7:7-11.

Lear, D. W. 1962. Reproducibility of the most probable numbers technique for determining the sanitary quality of clams. Applied Microbiol. 10:60-64.

Leifson, E. 1954. The flagellation and taxonomy of species of Acetobacter. Antonie van Leeuwenhoek 20:102-10.

_____. 1958. Identification of Pseudomonas, Alcaligenes and related bacteria. Zentralbl. Bakteriol. Parasitenk. Abt. I, Orig., 173:487-88.

Levine, M. 1961. Facts and fancies of bacterial indices in standards for water and foods. Food Technol. 15, nr. 11:29-38.

Lindberg, R. B. 1960. Observations on the relative sensitivity of enrichment broth and membrane filter techniques in the detection of coliform organisms in bottled carbonated beverages. Ann. Inst. Pasteur Lille 11:77-82.

Lundbeck, H., Plazikowski, U., and Silverstolpe, L. 1955. The Swedish Salmonella outbreak of 1953. J. Applied Bact. 18:535-48.

McCarthy, J. A., Thomas, H. A., and Delaney, J. E. 1958. Evaluation of the reliability of coliform density tests. Am. J. Public Health 48:1628-35.

McCollum, R. W. 1961. An outbreak of viral hepatitis in the Mediterranean fleet. Military Med. 126:902-10.

McCrady, M. H. and Langevin, E. 1932. The coli-aerogenes determination in pasteurization control. J. Dairy Sci. 15:321-29.

McLean, R. A. and Sulzbacher, W. L. 1953. Microbacterium thermosphactum, spec. nov.; a non heat resistant bacterium from fresh pork sausage. J. Bact. 65:428-33.

Machala, W. E. 1961. A bacteriological investigation of frozen foods in the Oklahoma City area. J. Milk Food Technol. 24:323-27.

Madelung, P. 1953. Efter hvilke principer bør man anlaegge den bakteriologiske undersøgelse af kogepølser og andre varmebehandlede tilberedte kødvarer? Nord. Vet. Med. 5, Suppl. 2-16.

Malapert-Neufville, R. von. 1886. Bakteriologische Untersuchung der wichtigsten Quellen der städtischen Wasserleitung Wiesbadens sowie einer Anzahl Mineralquellen zu

Schlangenbad, Schwalbach, Soden i.T. und Bad Weilbach. Z. anal. Chem. 25:39-88.

Martinez, J. and Appleman, M. D. 1949. Certain inaccuracies in the determination of coliforms in frozen orange juice. Food Technol. 3:392-94.

Mason, J. O. and McLean, W. R. 1962. Infectious hepatitis traced to the consumption of raw oysters. An epidemiologic study. Am. J. Hygiene 75:90-111.

Mathews, F. P. 1949. Poliomyelitis epidemic, possibly milk-borne, in a naval station, Portland, Oregon. Am. J. Hygiene 49:1-7.

Mattick, A. T. R., Hiscox, E. R., and Crossley, E. L. 1945. The effect of temperature of pre-heating, of clarification and of bacteriological quality of the raw milk on the keeping properties of whole-milk powder dried by the Kestner spray-process. Part II. The effect of the various factors upon the bacterial (plate) count of the intermediate products and of the final powder. J. Dairy Sci. 14:135-44.

Mayer, C. F. 1953. Endemic panmyelotoxicosis in the Russian grainbelt. Mil. Surgeon 113:173-89; 295-315.

Meyer, K. F. 1956. The status of botulism as a world health problem. Bul. World Health Org. 15:281-98.

Michener, H. D., Thompson, P. A., and Dietrich, W. C. 1960. Time-temperature tolerance of frozen foods. XXII. Relationship of bacterial population to temperature. Food Technol. 14:290-94.

Mickelsen, R., Foltz, V. D., Martin, W. H., and Hunter, C. A. 1961. The incidence of potentially pathogenic staphylococci in dairy products at the consumer level. II. Cheese. J. Milk Food Technol. 24:342-45.

Mieth, H. 1961. Untersuchungen über das Vorkommen von Enterokokken bei Tieren und Menschen. II. Mitteilung: Ihr Vorkommen in Stuhlproben von gesunden Menschen. Zentralbl. Bakteriol. Parasitenk. Abt. I, Orig., 183:68-89.

Miller, W. A. 1960. The microbiology of self-service, packaged square slices of cooked ham. J. Milk Food Technol. 23:311-14.

Moats, W. A. 1961. Chemical changes in bacteria heated in milk as related to loss of stainability. J. Dairy Sci. 44:1431-39.

Money, R. W. and Born, R. 1951. Equilibrium humidity of sugar solutions. J. Sci. Food Agr. 2:180-85.

Moore, B. 1955. Streptococci and food poisoning. J. Applied Bact. 18:606-18.

Mossel, D. A. A. 1952. De betekenis van de zgn. bederf-associatie voor de levensmiddelenmicrobiologie. Tijdschr. over Plantenziekten 58:267-68.

_____. 1955. The importance of the bacteriological condition of ingredients used as minor components in some canned meat products. Ann. Inst. Pasteur Lille 7:171-85.

_____. 1956. Aufgaben und Durchführung der modernen hygienischbakteriologischen Lebensmittelüberwachung. Wien. tierärztl. Monatsschr. 43:321-40; 596-610.

_____. 1957. The presumptive enumeration of lactose negative as well as lactose positive Enterobacteriaceae in foods. Applied Microbiol. 5, 379-81.

_____. 1958. Die Verhütung der Verbreitung von Salmonellosen und sonstigen Enterobacteriosen durch Lebensmittel. Zentralbl. Bakteriol. Parasitenk. Abt. I, Ref., 166, 421-32.

_____. 1959. The suitability of certain obligately anaerobic nonsporeforming enteric bacteria, as part of a more extended bacterial association, as indicators of 'faecal contamination' of foods. J. Applied Bact. 22:184-92.

_____. 1961. Eine Schnellidentifizierung von Gramnegativen stäbchenförmigen Bakterien isoliert bei der hygienischbakteriologischen Lebensmitteluntersuchung. Arch. Lebensm. Hyg. 12:180-82.

_____. 1962a. An attempt of classification of the catalase positive cocci: Staphylococci and Micrococci. Bact. Proc. 39.

_____. 1962b. An ecological investigation on the usefulness of two specific modifications of Eijkman's test as one of the methods used for the detection of faecal contamination of foods. J. Applied Bact. 25:20-29.

_____, Bechet, J., and Lambion, R. 1962. La prévention des infections et des toxi-infections alimentaires. Le contrôle hygiénique des industries alimentaires et l'analyse bactériologique de leurs produits. Cooperative d'Editions pour les Industries Alimentaires, Bruxelles, Belgium. 279 p.

_____ and Bruin, A. S., de. 1960. The survival of Enterobacteriaceae in acid liquid foods stored at different temperatures. Ann. Inst. Pasteur Lille 11:65-72.

_____ and Drion, E. F. 1958. Sampling of canned foods for bacteriological analysis. Food 27:333-37.

_____ and Eijgelaar, G. 1956. Noodzakelijkheid en vervangbaarheid van boorzuur als conserveermiddel in eiwitrijke voedingsmiddelen. Conserva 5:7-12.

_____ and Ingram, M. 1955. The physiology of the microbial spoilage of foods. J. Applied Bact. 18:232-68.

_____ and van Kuijk, H. J. L. 1955. A new and simple technique for the direct determination of the equilibrium relative humidity of foods. Food Research 20:415-23.

_____ and Krugers Dagneaux, E. L. 1959. Bacteriological requirements for and bacteriological analysis of precooked ('instant') cereals and similar foods. Antonie van Leeuwenhoek 25:230-36.

_____ and Martin, G. 1961. Milieu simplifié permettant l'étude des divers modes d'action des bactéries sur les hydrates de carbone. Ann. Inst. Pasteur Lille 12:225-26.

_____, Mengerink, W. H. J., and Scholts, H. H. 1962. A McConkey type agar medium for the selective growth and enumeration of all Enterobacteriaceae. J. Bact. 84:381.

_____ and Meulen, H. S., van der. 1960. Untersuchung des Absterbens von Salmonella-Arten in feinster Mayonnaise und in Salat-Mayonnaise bei einer Aufbewahrungstemperatur von ungefähr 15°C. Arch. Lebensm. Hyg. 11:245-48.

_____ and Visser, M. 1960. The estimation of small numbers of micro-organisms in opalescent non-alcoholic drinks by applying centrifugation. Ann. Inst. Pasteur Lille 11:193-202.

_____ and Zwart, H. 1959. Die quantitative bakterioskopische Bewertung von Gemüse/Fleischkonserven. Arch. Lebensm. Hyg. 10:229-32.

_____. 1960. The rapid tentative recognition of psychrotrophic types among Enterobacteriaceae isolated from foods. J. Applied Bact. 23:185-88.

Mundt, J. O. 1961. Occurrence of enterococci: bud, blossom and soil studies. Applied Microbiol. 9:541-44.

_____, Shuly, G. A., and McGarty, I. C. 1954. The coliform bacteria of strawberries. J. Milk Food Technol. 17:362-65.

Nauman, R., Bagley, C., and Schlang, H. 1962. Food-borne hepatitis — Florida. Morbidity and Mortality Weekly Rept. 11, nr. 8:58-59.

Neilson, N. E., MacQuillan, M. F., and Campbell, J. J. R. 1957. The enumeration of thermophilic bacteria by the plate count method. Canad. J. Microbiol. 3:939-43.

Niinivaara, F. P. and Pohja, M. S. 1957. Ueber die Reifung der Rohwurst. II. Mitteilung. Aus Rohwurst isolierte Bakterienstämme und ihre Bedeutung beim Reifungsprozess. Z. Lebensm. Unters. Forsch. 106:187-96.

Nikodémusz, I., Bodnár, S., Boján, M., Kiss, M., Kiss, P., Laczkó, M., Molnár, E., and Pápay, D. 1962. Aerobe Sporenbildner als Lebensmittelvergifter. Zentralbl. Bakteriol. Parasitenk. Abt. I, Orig., 184:462-70.

Nygren, B. 1961. Lecithinase-producing bacteria and food poisoning. Acta Pathol. Microbiol. Scand. 51, Suppl. 144:297-98.

Oberhofer, T. R. and Frazier, W. C. 1961. Competition of Staphylococcus aureus with other organisms. J. Milk Food Technol. 24:172-75.

Official methods of analysis of the Association of Official Agricultural Chemists. 1960. Ninth Ed., Washington.

Ostrolenk, M. and Hunter, A. C. 1946. The distribution of enteric Streptococci. J. Bact. 51:735-41.

_____, Kramer, N., and Cleverdon, R. C. 1947. Comparative studies of Enterococci and Escherichia coli as indices of pollution. J. Bact. 53:197-203.

Otte, H. J. and Ritzerfeld, W. 1960. Massenerkrankung an Angina durch Streptokokken in Lebensmitteln. Deut. medizin. Wschr. 85:1625-28.

Papavassiliou, J. 1962. Species differentiation of group D streptococci. Applied Microbiol. 10:65-69.

Parry, W. H. 1962. A milk-borne outbreak due to Salmonella typhimurium. Lancet I:475-77.

Penna, R. 1959. Importanza dei fagi nel giudizio di potabilita di un'acqua in relazione agli aetri indici microbiologici di inquinabilita. Riv. Ital. Igiene 19:51-86.

Peterson, A. C., Black, J. J., and Gunderson, M. F. 1962. Staphylococci in competition. I. Growth of naturally occurring mixed populations in precooked frozen foods during defrost. II. Effect of total numbers and proportion of staphylococci in mixed cultures on growth in artificial culture medium. Applied Microbiol. 10:16-22; 23-30.

Post, F. J., Bliss, A. H., and O'Keefe, W. B. 1961. Studies on the ecology of selected food poisoning organisms in foods. I. Growth of Staphylococcus aureus in cream and a cream product. J. Food Sci. 26:436-41.

Prescott, S. C. and Baker, S. K. 1904. On some cultural relations and antagonisms of Bacillus coli and Houston's sewage streptococci; with a method for the detection and separation of these micro-organisms in polluted waters. J. Infect. Diseases 1:193-210.

Pretorius, W. A. 1961. Investigations on the use of the roll tube method for counting Escherichia coli I in water. J. Applied Bact. 24:212-17.

Prévot, A. R. 1938. Études de systématique bactérienne. III. Invalidité du genre Bacteroides Castellani et Chalmers. Démembrement et reclassification. Ann. Inst. Pasteur 60:285-307.

Rahn, O. and Boysen, H. H. 1929. Die Verteilung der Bakterien in der Butter. Milchwirtsch. Forsch. 7:214-32.

Raj, H. and Liston, J. 1961. Survival of bacteria of public

health significance in frozen sea foods. Food Technol. 15: 429-34.

Read, M. R., Bancroft, H., Doull, J. A., and Parker, R. F. 1946. Infectious hepatitis — presumedly food-borne outbreak. Am. J. Public Health 36:367-70.

Rishbeth, J. 1947. The bacteriology of dehydrated vegetables. J. Hygiene 45:33-45.

Robertson, A. H. 1960. Frozen foods. J. Assoc. Offic. Agricult. Chemists 43:176-81.

Roos, B. 1956. Hepatitepidemi, spridd genom ostron. Svenska Läkartidn. 53:989-1003.

Rutstein, D. D. 1961. Maintaining contact with medical knowledge. New England J. Med. 265:321-24.

Schaaf, A., van der, Zijl, H. J. M., van, and Hagens, F. M. 1962. Diermeel en salmonellosis. Tijdschr. Diergeneesk. 87:211-21.

Schardinger, F. 1892. Ueber das Vorkommen Gährung erregender Spaltpilze im Trinkwasser und ihre Bedeutung für die hygienische Beurtheilung desselben. Wien. klin. Wschr. 5: 403-5; 421-23.

Schmidt-Lorenz, W. 1960. Ueber den Einfluss der Verdünnungslösungen auf das Ergebnis von Bakterienzählungen bei bestrahltem und gefrorenem Fisch. Arch. Lebensm. Hyg. 11: 60-63.

Schönherr, W. 1961. Zur Abhängigkeit des Nachweisses coliformer Keime in der Milch von Methodik, Keimart und Keimmenge. Zentralbl. Vet. Med. 8:473-82.

Schultze, W. D. and Olson, J. C. 1960. Studies on psychrophilic bacteria. II. Psychrophilic coliform bacteria in stored commercial dairy products. J. Dairy Sci. 43:351-57.

Schumaier, G., Panda, B., de Volt, H. M., Laffer, N. C., and Creek, R. D. 1961. Hemorrhagic lesions in chickens resembling naturally occurring "hemorrhagic syndrome" produced experimentally by mycotoxins. Poultry Sci. 40:1132-34.

Schwartz, W. and Kaess, G. 1934. Das Wachstum von Schimmelpilzen auf gekühltem Fleisch bei verschiedenen Luftzuständen. Arch. Mikrobiol. 5:157-84.

Seeliger, H. 1952. Die Keimzahlbestimmung und Differenzierung coliformer Bakterien in Milch und Speiseeis. Milchwissenschaft 7:389-94.

Sharpe, M. E. 1962. Taxonomy of the Lactobacilli. Dairy Sci. Abstr. 24:109-18.

Shattock, P. M. F. 1955. The identification and classification of Streptococcus faecalis and some associated streptococci. Ann. Inst. Pasteur Lille 7:95-100.

Shelton, L. R. 1961. Frozen precooked foods – plant sanitation and microbiology. J. Am. Diet. Assoc. 38:132-34.

Shewan, J. M., Hobbs, G., and Hodgkiss, W. 1960a. A determinative scheme for the identification of certain genera of Gram-negative bacteria, with special reference to the Pseudomonadaceae. J. Applied Bact. 23:379-90.

_____. 1960b. The Pseudomonas and Achromobacter groups of bacteria in the spoilage of marine white fish. J. Applied Bact. 23:463-68.

Silliker, J. H., Jansen, C. E., Voegeli, M. M., and Chmura, N. W. 1962. Studies on the fate of staphylococci during the processing of hams. J. Food Sci. 27:50-56.

Silverstolpe, L., Plazikowski, U., Kjellander, J., and Vahlne, G. 1961. An epidemic among infants caused by Salmonella muenchen. J. Applied Bact. 24:134-42.

Slocum, G. G. 1959. Food sanitation and food poisoning. Quart. Bul. Assoc. Food and Drug Off. U. S. 23:3-10.

_____ and Boyles, W. A. 1941. Incidence of coliform bacteria on fresh vegetables and efficiency of lactose broth, brilliant green bile two percent and formate ricinoleate broth as presumptive media for the coliform group. Food Research 6:377-85.

Smith, N. R., Gordon, R. E., and Clark, F. E. 1952. Aerobic sporeforming bacteria. USDA, Agr. Mono. No. 16, Washington, 148 p.

Smith, T. 1895. Notes on Bacillus coli communis and related forms, together with some suggestions concerning the bacteriological examination of drinking water. Am. J. Med. Sci., new series, 110:283-302.

Status report on tissue-culture cultivated hepatitis virus. 1961. J. Am. Med. Assoc. 177:671-82.

Steel, K. J. 1961. The oxidase reaction as a taxonomic tool. J. Gen. Microbiol. 25:297-306.

Steiniger, F. 1961. Wie lange halten sich Salmonellen aus verregnetem Abwasser auf Pflanzen? Berl. müch. tierärztl. Wschr. 74:389-92.

Straka, R. P. and Stokes, J. L. 1957. Rapid destruction of bacteria in commonly used diluents and its elimination. Applied Microbiol. 5:21-25.

Sutton, R. R. and McFarlane, V. H. 1947. Microbiology of spray-dried whole egg. III. Escherichia coli. Food Research 12:474-83.

Swenarton, J. C. 1927. Can B. coli be used as an index of the proper pasteurization of milk? J. Bact. 13:419-29.

Tanner, F. W. 1949. Quality in foods. Food Technol. 3:179.

Taylor, P. J. and McDonald, M. A. 1959. Milkborne streptococcal sore throat. Lancet I:330-33.

Thatcher, F. S. 1958. Microbiological standards for foods. II. What may microbiological standards mean? Food Technol. 12:117-22.

_____, Coutu, C., and Stevens, F. 1953. The sanitation of Canadian flour mills and its relationship to the microbial content of flour. Cereal Chem. 30:71-102.

_____ and Montford, J. 1962. Egg-products as a source of Salmonellae in processed foods. Canad. J. Public Health 53: 61-69.

_____, Simon, W., and Walters, C. 1956. Extraneous matter and bacteria of public health significance in cheese. Canad. J. Public Health 47:234-43.

Thomann, O. 1953. Prinzipien bei der hygienischen Beurteilung von Lebensmitteln, insbesondere roher Frucht- und Gemüsesäfte. Mitt. Geb. Lebensm. Unters. Hygiene 44:308-32.

Thornley, M. J. 1960. The differentiation of Pseudomonas from other Gram-negative bacteria on the basis of arginine metabolism. J. Applied Bact. 23:37-52.

Uden, N., van and Sousa, L. do C. 1957. Presumptive tests with liquid media for coliform organisms in yoghurt in the presence of lactose fermenting yeasts. Dairy Ind. 22:1028-29.

Unger, L., Rahman, A. K. M. M., and DeMoss, R. D. 1961. Anaerobic dissimilation of glucose by vibrio comma. Canad. J. Microbiol. 7:844-47.

Vecchio, V., Del, Arca Simonetti, A., d', and Arca, S., d'. 1960. Indagine sullo stato igienico della "crema di latte" e delle "panna montata" in vendita nella città di Roma. Nuov. Ann. Igiene Microbiol. 11:465-82.

Walker, G. C., Harmon, L. G., and Stine, C. M. 1961. Staphylococci in Colby cheese. J. Dairy Sci. 44:1272-82.

White, J. C. and Sherman, J. M. 1944. Occurrence of enterococci in milk. J. Bact. 48:262.

Wilkerson, W. B., Ayres, J. C., and Kraft, A. A. 1961. Occurrence of enterococci and coliform organisms on fresh and stored poultry. Food Technol. 15:286-92.

Williams, R. E. O., Hirch, A., and Cowan, S. T. 1953. Aerococcus, a new bacterial genus. J. Gen. Microbiol. 8:475-80.

Wilson, B. J. and Wilson, C. H. 1962a. Hepatotoxic substance from Penicillium rubrum. J. Bact. 83:693.

_____. 1962b. Studies on toxic substances produced on feedstuffs by fungi including isolates from moldy feed. Bact. Proc. 28.

Wilson, G. S. 1955. Symposium on food microbiology and public health: general conclusions. J. Applied Bact. 18:629-30.
_____. 1959. Faults and fallacies in microbiology. J. Gen. Microbiol. 21:1-15.
Winkle, S. and Adam, W. 1959. Ueber die Lebensdauer von Enterobacteriaceen, insbesondere Salmonellen, in künstlich infizierter Margarine. Zentralbl. Bakteriol. Parasitenk., Abt. I, Orig., 174:364-83.
Winslow, C. E. A. and Palmer, G. T. 1910. A comparative study of the intestinal streptococci from the horse, cow and man. J. Infect. Diseases 7:1-16.
Winter, A. R., Weiser, H. H., and Lewis, M. 1953. The control of bacteria in chicken salad. II. Salmonella. Applied Microbiol. 1:278-81.
Wolford, E. R. 1955. Significance of the presumptive coliform test as applied to orange juice. Applied Microbiol. 3:353-54.
_____. 1956. A source of coliforms in frozen concentrated orange juice. Fruit surface contamination. Applied Microbiol. 4:250-53.
Woodburn, M. J. and Strong, D. H. 1960. Survival of Salmonella typhimurium, Staphylococcus aureus and Streptococcus faecalis frozen in simplified food substrates. Applied Microbiol. 8:109-13.
Ziegler, N. R. and Halvorson, H. O. 1935. Application of statistics to problems in bacteriology. IV. Experimental comparison of the dilution method, the plate count and the direct count for the determination of bacterial populations. J. Bact. 29:609-34.

12

Microbial Inhibitors

H. L. A. TARR

FISHERIES RESEARCH BOARD OF CANADA

T HE FOOD INDUSTRY is the largest business on this continent, and in the United States alone the retail value of
food has been estimated as at least $70 billion per year. It
is impossible to state how much of this food is rendered unfit for
human consumption through microbiological attack; but even if it
were only 0.1 per cent, a very significant sum of money would be
involved. When considered on an international basis, food losses
caused by microorganisms must reach staggering proportions.
Successful preservation of foods is, therefore, an all-important
problem, especially in a world which is largely peopled by individuals who, if they do not go to bed hungry, are at least inade-
quately fed.

Two of the important problems of the food microbiologist are
to preserve food so that it is safe and edible, and to use his
knowledge to improve the quality of foods for the consumer. His
success will depend on his ability to apply known or new techniques of inhibiting microbial development in foods, or, in some
instances, to make sure that a desirable microbial flora develops
to the exclusion of undesirable types. Several common methods
of inhibiting growth of microorganisms in foods were employed
long before man knew the causes of food spoilage; most of these
are still in use today. However, in recent years there have been
new developments, and there is little doubt that in the years to
come even better methods of preserving foods will be found.

Food preservation techniques may be roughly divided into two
broad categories, namely 1) physical methods and 2) chemical
methods. Physical methods include the removal of water, application of heat (as in pasteurization and canning); storage under
cold or frozen conditions; application of radiations (ultra-violet
light, penetrating radiations); and adjustment of the foods to acid

or alkaline pH values which are incompatible with bacterial development. In practice it is not easy to separate physical and chemical techniques of preservation, for the successful application of many chemical methods often depends upon maintenance of adequate physical conditions. The purpose of this chapter is to discuss some recognized chemical methods of preserving foods. For convenience the application of chemical substances will be considered individually as applied to each of a number of different foods, rather than discussing each food separately. Fellers reviewed this subject briefly in 1955.

To the layman, and more especially to the neurotic food faddist, the term "chemical" as applied to food all too frequently implies or suggests that some harmful substance has been added. It is difficult, and all too often impossible, to overcome this unfortunate attitude. In general, there is some agreement among civilized peoples regarding chemical additives which should be permitted in foods intended for human consumption. However, some appear to be absurdly conservative and others absurdly liberal in their attitudes towards food additives. British regulations tend to be rather strict regarding the use of chemicals in foods (Banfield, 1952; Ingram, 1959). In general there are comparatively few substances which are in reasonably constant use as food additives intended to retard or prevent microbiological deterioration. Some are employed as bacteriostats, others as fungistats, and some because they possess both these properties. Though salt, sucrose, glucose, vinegar, wood smoke, spices, etc. are considered additives under certain food and drug regulations, their use has been so general that it hardly warrants discussion at the present time. For the present discussion a chemical preservative will be regarded as, "any substance which is added to a food in (usually) less than 0.1 per cent concentration with the intention of inhibiting or preventing microbiological spoilage of that food." Comparatively few chemical substances are permitted for inclusion in foods within the scope of this definition. The use of these substances, and of a few others which show considerable promise, will be discussed.

PRESERVATIVES

Benzoic Acid

The position regarding benzoic acid was discussed in 1944 by Tanner and has not altered appreciably. It occurs widely in nature and is present in many fruits and vegetables. It is readily excreted in conjugation with glycine in the form of hippuric acid and as benzoylglucuronide. According to early workers cumulative poisoning is of no significance as far as this compound is concerned. It has been known for at least 50 years that it can be ingested by man over extended periods without harmful effects. Since the early work of Cruess and his collaborators (see Tanner, 1944), it has been recognized that the preservative action of benzoic acid is a function of the undissociated molecules rather than of the ions, and that high acidity in the food and resulting suppression of dissociation enhances its activity. Benzoic acid, together with certain of its salts and simple derivatives, has largely or entirely taken the place of boric acid and salicylic acid. These latter compounds are considered toxic since there is some danger of accumulation in the body, and salicylates at least have marked physiological activity. On this continent, food and drug regulations usually permit 0.1 per cent benzoic acid in certain foods. In Great Britain, levels of from 0.012 to 0.06 per cent are permitted in different instances, and Ingram (1959) has pointed out that this creates difficulties with certain foods.

The methyl, ethyl, and propyl esters of parahydroxybenzoic acid were first introduced as preservatives in Germany by Sabalitschka about 1926 according to von Schelhorn (1951). Since this time considerable evidence has accumulated showing that these esters are, for fruit products at least, somewhat more effective than benzoic acid itself. The molecules remain more undissociated at higher pH values. Parachlorbenzoic acid has been used for fruit preservation in Germany. Both the methyl and propyl esters may be used instead of benzoic acid according to the Canadian Food and Drug regulations.

Benzoic acid and its various derivatives are widely used in certain fruit products such as candied fruits intended for baking purposes, certain syrups, acidic sweet beverages, in jams, and in margarine.

From time to time attempts have been made to use benzoic acid or its derivatives in preservation of flesh foods and more particularly for fish. The reaction of flesh foods is normally not sufficiently acidic for benzoic acid to exert any significant antimicrobiological activity. As far as fresh fish are concerned, Tarr and Bailey (1939) pointed out that the only useful action of benzoic acid in preservation of fresh fish is its ability to inhibit trimethylamine formation.

In conclusion it appears reasonable to assume that benzoic acid and its derivatives will continue to be employed as harmless and fairly effective preservatives for certain classes of foodstuffs, particularly acid fruits and vegetable products, and possibly for acid-cured marinated fish products in certain countries. Sorbic acid (see p. 206) may take its place in certain operations.

Sulfur Dioxide

There is little doubt that sulfur dioxide is singly the most versatile food preservative presently in use. It is employed either as a gas, as sulfurous acid, or in the form of its sulfites, bisulfites or metabisulfites. Several excellent reviews are available on the use and properties of sulfurous acid in foods (Joslyn and Braverman, 1954; von Schelhorn, 1951; Vaughn, 1955). When sulfur dioxide is consumed in small amounts, it has no appreciable toxicity for it is readily oxidized in the body to sulfates, which are excreted. Very high concentrations would undoubtedly be toxic but these would have such a deleterious effect upon color and flavor of foods that it is unlikely that such levels would be used. Another important factor in use of sulfur dioxide in foods is that cooking and other heating procedures normally result in extensive loss or removal of the compound. It also has valuable protective properties which improve retention of ascorbic acid, carotene and certain other oxidizable, biologically active compounds. Its use as an agent to prevent Maillard browning reactions is well known. As with benzoic acid, only the undissociated acid has significant preservative properties, and it becomes increasingly ineffective as the pH rises. Thus, two to four times as much sulfur dioxide are required for successful preservation effects at pH 3.5 as at pH 2.5. Sulfur dioxide is rapidly bound by sugars and aldehydes and in this form is only from 1/30 to 1/60 as active bacteriostatically as the free acid.

Fairly high concentrations of SO_2 have been used, and are still used routinely in many countries, for preserving fruit pulps for preparation of preserves and jams. Either raw or cooked fruit pulps can be so treated. However, the wine industry probably benefits most from the use of sulfur dioxide. Its principal function in this industry is to hinder and prevent development of acetic acid bacteria and various undesirable aerobic yeasts. It also has important selective action on many of the true wine yeasts which can often be trained to grow in fairly high concentrations of SO_2. Indeed, it is doubtful if the wine industry could survive if the use of SO_2 was forbidden.

Another industry in which the use of fairly high levels of SO_2 is almost mandatory if a satisfactory product is to be obtained is that of the preparation of dried or dehydrated fruits. Usually

from about 1,000 to 2,500 parts per million of SO_2 are necessary
if satisfactory dried bleached raisins, apples, peaches and apricots are to be obtained. Certain vegetables are also treated with
sulfites. The sulfiting of fruits and vegetables not only prevents
certain undesirable types of microbiological deterioration but is
all-important from the point of view of retention of color. It is
unlikely that an overall increase will occur in the types of acidic
foods which are treated with SO_2. As with benzoates, treatment
of flesh foods with sulfites is neither a satisfactory procedure
nor is it usually permitted by Food and Drug regulations. The
only exception appears to be the permitted use of bisulfite ice for
icing shrimp in Louisiana (Fieger and Novak, 1961).

Sorbic Acid, Propionic Acid and Sodium Diacetate

Sorbic acid, propionic acid and sodium diacetate have found
use in the food industry mainly as inhibitors of yeast and mold
growth. Propionic acid (usually 0.2 per cent) has been used for
many years, while the other substances have been introduced
somewhat more recently. Sodium diacetate (0.3 percent) is permitted in some foods such as bread and cheese. A survey of the
recent literature indicates that sorbic acid has been studied much
more extensively than the other compounds in recent years.
Though the excellent fungistatic action of this substance was recognized and patented 17 years ago (Gooding, 1945) several years
elapsed before thorough investigations of its properties were
undertaken and its potential in the food industry was recognized
(Phillip and Mundt, 1950; Deuel et al., 1954). In recent years, the
value of sorbic acid in preservation of a number of foods has
been studied, and some of the available information will be discussed.

Ferguson and Powrie (1957) found that the addition of 0.05 per
cent sorbic acid to unpasteurized apple juice inhibited yeast development, and that the simultaneous addition of ascorbic acid
(0.035 per cent) inhibited development of aerobic acetobacter
species. This combination enabled apple juice to be held two
weeks without refrigeration. Becker and Roeder (1957) found that
sorbic acid (0.05 to 0.1 per cent) was more satisfactory than benzoic acid in preventing spoilage of margarine. Pasteurized,
cooled, bulk grape juice could be stored for considerable periods
at 22 to 28^0C. without spoilage if 0.01 to 0.015 per cent sorbic
acid was added. Sorbic acid in concentrations of about 0.05 to 0.1
per cent markedly delayed development of yeasts and molds in
brine-smoked fish and of the organisms occasioning "dun" in

salted fish. Moreover, the acid was found to be comparatively stable in the stored fish (Boyd and Tarr, 1954, 1955). Saller and Kolewa (1957) claimed that the addition of sorbic acid (200 mg. per liter) was more useful than SO_2 to preserve wines of high sugar content and improved the organoleptic characteristics. Tanaka et al. (1956) used 0.1 to 0.2 per cent sorbic acid successfully to preserve sausages. Niven and Chesbro (1957) used sorbic acid successfully to control yeasts on microbiological spoilage of meats. The value of sorbic acid as a fungistatic agent for cheese was recognized early (Melnick, Luckmann and Goodman, 1954; Smith and Rollin, 1954). According to Deane (1961) potassium sorbate solutions are very satisfactory for preventing mold growth on rindless Swiss-type cheese. The cheese was painted with 15 per cent to 20 per cent solutions of potassium sorbate (1 oz. per 20-lb. block). These and many other papers would appear to confirm the fact that sorbic acid is a very useful and versatile food fungistat.

The nontoxicity and digestibility of sorbic acid are well recognized and permission has been received for its use in several different types of foods by pure food and drug divisions. Further studies with this substance from the point of view of applications to other foods and in combination with other preservatives might be rewarding. Studies by Hansen and Appleman (1955) suggest that sorbic acid does not favor the growth of strict anaerobes such as C. botulinum in foods.

Nitrates and Nitrites

Nitrates have long been used in meat curing and are considered practically indispensable. The whole position was admirably reviewed by Jensen about 20 years ago (Jensen, 1942). Evidently, sodium nitrate, especially in concentrations in excess of 0.2 or 0.5 per cent, retarded growth of anaerobes in canned ham and probably affected the thermal death time of anaerobic spores. The inhibitory action of nitrate in meat curing was believed to be due to inhibition of catalase by hydroxylamine formed during nitrate reduction, thereby permitting accumulation of hydrogen peroxide to which clostridia are very sensitive. At that time, the bacteriostatic action of nitrites in meats was not considered significant for various reasons.

The value of nitrites as a fresh fish preservative was first recognized in 1939 (Tarr and Sunderland, 1939, 1940). Nitrite was found to be more effective for some fish than for others. This was traced to the fact that nitrite in approximately 200 parts per million concentration was only effective at pH levels below 7 and preferably below pH 6.5. This finding precipitated the first thorough investigations of the effect of pH and other conditions on bacteriostatic action, of nitrites (Tarr, 1940; 1941; 1942; 1944a). This effect was later rediscovered by meat microbiologists (Castellani and Niven, 1955; Eddy, 1957; Ingram, 1959). Silliker and his collaborators have recently demonstrated that 78 ppm of nitrite inhibits growth of putrefactive anaerobes and thermoduric enterococci in canned meats (Silliker, Greenberg and Schack, 1958; Greenberg and Silliker, 1961). These studies show that nitrite possesses an inhibitory effect on bacterial growth, that it is usually more effective anaerobically than aerobically and that undissociated nitrous acid is necessary for optimum activity. In conclusion it is interesting to note that for over ten years nitrite was used successfully for fresh fillet preservation in Canada on a very large scale; permission for such use was only withdrawn when tetracycline antibiotics were permitted in 1956 (Tarr, 1961). The United States regulations were relaxed last year to permit use of nitrite in certain cured fish products and in certain canned fish pet foods. Extensive investigations concerning preservation of Baltic cod with ice containing sodium nitrite have been carried out in recent years (Borowik et al., 1957).

ANTIBIOTICS

Historical

No chemical method of food preservation has ever received as much publicity as has that involving antibiotics. This publicity has probably done more harm than good. The whole subject has been reviewed several times in recent years (Campbell and O'Brien, 1955; Deatherage, 1956; Hawley, 1957a and b; Farber, 1960; Wrenshall, 1959; Tarr, 1956, 1960a and b, 1961). Early attempts to preserve flesh foods (Tarr, 1944b, 1946, 1947; Tarr and Deas, 1948; Hounie, 1950), and milk (Curran and Evans, 1946; Foley and Byrne, 1950) with antibiotics were not very successful and it was not until 1950 that tests with Aureomycin (CTC) and Terramycin (OTC) indicated that these were remarkably effective in very low concentrations in preserving chilled fish (Tarr et al., 1950). The addition of rimocidin, an antifungal antibiotic, was found to inhibit yeast development which tended to occur on prolonged storage of CTC-treated fish and meat (Tarr et al., 1952). About this time, independent experiments by Goldberg et al. (1953) with red meats confirmed and extended these observations.

Also at this time, the first promising attempts to reduce the rather severe heating conditions normally required to sterilize canned vegetables were made by use of the antibiotic subtilin (Anderson and Michener, 1950).

Since these initial small-scale trials, research and development concerning use of antibiotics in a number of different foods has proceeded at a somewhat accelerated pace. Some of the significant findings will be reviewed. Most of the references will be found in the reviews mentioned above and only a few of the more recent ones will be discussed.

Preservation of Fish

Since fish is one of the most perishable of all foods, considerable effort has been expended in application of antibiotics for preservation of fish and seafoods in different countries. To obtain optimum effectiveness in retarding bacterial spoilage of fish with antibiotics, it was necessary to apply these as soon as possible after capture. Therefore some of the early experiments involved incorporation of antibiotics into ice used for icing fish, or their addition to chilled sea water in which the fish were to be transported. Chlortetracycline (CTC) has been used in most instances for preservation purposes, since, as will be pointed out later, it is normally found more effective than OTC and tetracycline.

One of the first problems encountered was that of obtaining uniform distribution of CTC in block ice. For various reasons, flake-type ice is little used in the fishing industry. The problem of distributing CTC in ice blocks was overcome fairly satisfactorily by addition of certain hydrocolloids to the water; the most useful substances were carrageenin, carboxymethylcellulose and certain alginates. In early tests in which CTC was used in manufacturing flake ice the antibiotic was extremely labile in hard water, but addition of citric acid sufficient to bring the pH to about 3.5 satisfactorily prevented destruction.

Hard waters present another problem in that the calcium, iron and magnesium salts present form relatively water-insoluble complexes between CTC and the hydrocolloids used in making block ices. When the block ice thaws, curdy precipitates result, which, though they contain antibiotic activity, remove much of the readily available soluble CTC from solution (Moyer et al., 1958). A serious problem in certain areas is that of destruction of CTC by free chlorine in water supplies. Demineralization of hard and heavily chlorinated waters by means of ion exchange resin treatments is undoubtedly the most logical means of rendering them satisfactory for use with CTC. Chlorine residues are, of course, readily removed by a brief immersion of fish in the water before adding the CTC.

During the past 8 years there have been extensive experimental and practical applications of antibiotics in the fishing industries in different countries. In British Columbia, where experimental work on preservation of fish with tetracycline antibiotics started as long ago as 1950, there have been no extensive industrial applications other than those which were requested on an experimental basis by the fishing industry. On Canada's Atlantic seaboard, during the past 3 or 4 years, application of CTC to fillets has become a major successful operation. In 1961 about 8 million pounds of fillets were treated (Anon., 1961). The Halifax Station of the Fisheries Research Board of Canada deserves considerable credit for the success of this operation; Castell and his collaborators (see Tarr, 1961) have carried out extensive experimental work, much of which has been designed to indicate to the fishing industry the conditions necessary for a satisfactory dipping operation. Experimental work has also shown that Canadian fresh-water fish can be preserved with CTC as successfully as sea water fish.

In Japan, Tomiyama and his collaborators have carried out extensive experiments concerning preservation with CTC of both round and eviscerated fish (see Tarr, 1961). They have published more than 20 papers on their large-scale experiments, most of which have been carried out at sea. These include studies of diverse species and application of the antibiotics incorporated in ice or in dipping solutions. Presently there is a large-scale industrial application of CTC in the Japanese fishing industry, and more especially, in connection with fish which are used in the preparation of kamoboko. Attempts to obtain information concerning the exact size of this antibiotic operation have not been successful, though indirect assessment based on the dollar value of the antibiotic sold for this purpose would indicate that it is very extensive. The Japanese consume a good deal of raw fish and more extensive applications in certain of their fisheries probably have been delayed until the CTC residue problem has been cleared with the appropriate authorities.

In the United States practically all the work on the use of antibiotics in fish preservation has been carried out by a few small groups of investigators in New York, California, Louisiana and on the Atlantic seaboard. In Great Britain, Shewan and his collaborators carried out, over a two- or three-year period, a very thorough investigation of the value of CTC in retarding bacterial spoilage of Atlantic ground fish. Presently CTC is not permitted in Great Britain. There have also been extensive studies of fish preservation by means of CTC in Russia (Dubrova, 1961). Successful investigations have also been made in India, South Africa, Germany and the French Cameroons. Considerable success has attended experimental work in connection with preservation of shellfish such as shrimp, crab and oysters with CTC. However, this work has been characterized by the fact that these shellfish have almost invariably required higher concentrations of CTC for successful preservation than do eviscerated fish or fish fillets. The reason remained obscure until recent work in our own laboratory suggested that divalent ions such as magnesium and calcium are probably responsible for this effect (Southcott and Tarr, 1961). Indeed, more recent results have shown that shrimp meat and shells strongly chelate CTC, rendering it unavailable for preservative purposes. Undoubtedly bacterial spoilage of shrimp and other shellfish is markedly retarded by CTC, but it is important to remember that these shellfish have characteristic delicate flavors which rapidly disappear during chilled storage. Therefore every effort should be made to overcome the somewhat undesirable tendency for holding them for many days at $32°$ F. before freezing or otherwise processing them.

About 6 years ago a marked improvement in keeping quality of whale carcasses post-mortem was first demonstrated when about 50 or 100 grams of either CTC or OTC was dissolved in water and injected into the visceral cavity of the animals. Though not all tissues were preserved to the same extent, research indicated a marked general improvement in bacteriological quality and in the free fatty acid values of the oil obtained from treated whales. These results have been recently verified in Russia.

Comprehensive investigations in a number of different laboratories in Canada, Great Britain, the United States, Japan, and Russia have dealt with the residues of antibiotics in raw fish and shellfish treated with CTC or OTC by a number of different methods, and of their destruction by heat as in cooking. CTC tends to concentrate in fish skins and to penetrate the flesh slowly. In general antibiotic residues in the flesh of fish which have been iced with ice containing 5 ppm of CTC rarely exceed 0.2 ppm. The residues found in fillets given a brief dip in a 10 ppm solution of CTC would be somewhat higher than this as a rule. The amount taken up varies with the size and condition of the fillets. Most cooking procedures destroy all, or all but traces, of active antibiotic so that after cooking there is usually no demonstrable CTC or less than 0.1 ppm of the antibiotic. CTC has had an enviable record as a preservative of fresh fish. When

failures have been reported, in nearly all instances they have bee
traced to inexcusable, unclean procedures. There has so far bee
no indication that CTC encourages development of pathogenic or
other undesirable food-poisoning organisms in fish.

Preservation of Poultry

Five years after the discovery that CTC and OTC were successful preservatives for fresh fish, reports appeared from two
different laboratories in the United States indicating that CTC
was an excellent preservative for eviscerated chilled poultry and
was by far the most effective of a number of antibiotics studied
(Ziegler and Stadelman, 1955; Kohler et al., 1955). During the
next 3 years about 35 papers dealing largely with the effect of
tetracycline antibiotics on the preservation of eviscerated chilled
poultry appeared in leading food or poultry journals. With few
exceptions there was a general agreement among the various authors that tetracycline antibiotic treatment extended shelf life of
chilled eviscerated poultry very markedly, especially if a storage
temperature of about 32° F. was employed.

Some confusion has existed regarding the comparative effectiveness of the various tetracycline antibiotics in preservation of
eviscerated poultry. Stadelman et al. (1957) stated there were no
statistically significant differences in shelf life of birds treated
with OTC or CTC when held at 41°F. Also, Vaughn et al. (1957)
found little or no difference in preservative power of the three
tetracyclines studied. However, Wells et al. (1957) stated that
under all conditions CTC was more effective in prolonging storage time than OTC or TC. These differences probably can be explained on the basis of an early observation by Ayres and his collaborators (Ayres et al., 1956) that, while CTC was more effectiv
than OTC when applied to poultry at a 3 ppm level, the two antibiotics were comparable at a 30 ppm level. Magnesium and certain other polyvalent cations chelate and bind OTC to a greater
extent than they do CTC (Southcott and Tarr, 1961). Thus, at low
levels OTC would be preferentially removed and there would be
insufficient amounts present to cause proper bacteriostatic actior
At higher levels, there would be an excess of both antibiotics. Mc
Vicker et al. (1958) also observed that CTC was more effective
than OTC. Several competent groups of investigators have shown
that antibiotic resistant organisms already present on eviscerate
poultry very rapidly gain ascendancy if unsanitary conditions prevail in processing plants. In fact, antibiotic treatment can becom
valueless after a few weeks under such conditions.

Failure to improve storage life of eviscerated poultry in packing plants by CTC treatment could possibly be used as an indicator of poor processing conditions in such plants. The danger that pathogenic fungi such as Candida parapsilosis might in time gain ascendancy in antibiotic-treated chilled poultry has been pointed out, but so far apparently such a condition has not arisen in practice (Njoku-Obi et al., 1957). Recent studies in Germany (Janke et al., 1958) have shown that eviscerated poultry treated with 10 ppm of CTC suffered no visible deterioration for 3 weeks at 3°C. Provided proper processing conditions are adhered to and strict sanitation is employed, CTC treatment of eviscerated poultry yields satisfactory results in plants even after many years use (Abbey et al., 1960).

Vaughn et al. (1960) found, as had been earlier reported by Tarr and Boyd (Tarr, 1956) that the level of resistance attained after a given number of transfers of a bacterium was dependent upon the initial resistance of the culture in CTC-treated poultry. It is generally agreed that resistant organisms on CTC-treated flesh foods usually arise as a result of preferential development of strains of organisms already resistant to CTC and that development of CTC-resistance in CTC-sensitive strains is a slow stepwise process. Nagel et al. (1960) reported that the location of processing facilities and use of antibiotics has no influence on the distribution of the various genera of spoilage organisms developing in chilled eviscerated poultry. However, Thatcher and Loit (1961) found a selective development of CTC-resistant microorganisms, and particularly of psychophilic bacteria and yeasts on eviscerated poultry commercially treated with CTC. Of particular significance was their observation that there was no evidence of an increased health hazard due to development of staphylococci, enterococci, coliforms or pathogenic yeasts; that recovery of Salmonella organisms was reduced markedly in presence of CTC; and that Salmonella resistant to concentrations of CTC in excess of 7 ppm were not found in treated poultry.

Preservation of Red Meats

Hounie (1950) was apparently the first to attempt to preserve red meat with antibiotics. He claimed that the most satifactory results were obtained with a combination of subtilin and streptomycin. Shortly afterwards, experiments carried out independently in two different laboratories (Tarr et al., 1952; Goldberg et al., 1953) showed for the first time the outstanding preservative power of CTC and OTC for preserving fresh beef at temperatures between 0 and 21°C. Subsequent experiments in the same and other laboratories verified these initial results. Since meats intended for human consumption are often hung at chill temperatures for considerable periods for tenderizing purposes, the desirability of introducing antibiotics by infusion techniques became apparent. Indeed, the practicability of infusing beef rounds and whole animals with CTC was soon demonstrated. CTC-treated beef not only kept much better than untreated beef, but such treated meat could be kept for 48 hours at room temperature thereby greatly accelerating ageing (Weiser et al., 1953). Other investigators obtained satisfactory preservation of sheep and hogs by intraperitoneal injection of OTC into the animals one or two hours before slaughter. In these experiments a quaternary complex of the antibiotic solubilized with tartaric acid was injected at a level of approximately 3 milligrams of OTC per lb. of body weight (Sacchi, et al., 1956; Downing, McMahan and Baker, 1956). These experiments were followed by similar work in South America, Russia and elsewhere. Jay et al. (1957) found that CTC-resistant strains of Proteus vulgaris required concentrations of CTC as high as 100 ppm for inhibition in broth, but were inhibited by only 3 ppm in beef. Since CTC inhibition in beef was reversed by divalent ions such as manganese, molybdate, tungstate, magnesium, calcium, etc., the suggestion was made that CTC competed with the bacterial cells for essential cations. More recent work in our laboratory has indicated that a more probable explanation is chelation of the CTC itself rendering it insoluble and unable to perform its normal inhibitory function. Other methods of applying tetracycline antibiotics in preservation of beef have included intraperitoneal or intramuscular injections two to three hours prior to slaughter (Wilson et al., 1960), ante-mortem aqueous injections of OTC in the tail followed by an external OTC spray (Sleeth and Naumann, 1960), intravenous injection of animals followed by an external CTC spray post-mortem (Firman et al., 1959), and finally the use of CTC coated films for packaging beef for storage in refrigerated cabinets (Ayres, 1959). These and many other investigations, have proved beyond doubt that tetracycline antibiotics when applied under proper conditions have an enormous potential for preservation of beef and other red meats where the need is proven and could prove very valuable in controlling surface and other microbiological spoilage during ageing.

Special Problems in Preservation of Flesh Foods

During early studies on preservation of fish and beef with tetracycline antibiotics it was observed that if storage at chill temperatures was prolonged beyond about 10 days, yeast-like organisms developed rapidly. These organisms could be checked to some extent, at least, by use of the antifungal antibiotic, rimocidin, in about 10 ppm concentration (Tarr et al., 1952). A few years later Niven and Chesboro (1957) showed that sorbic acid was quite effective in delaying yeast and mold growth in CTC-treated ground beef. They, and more recently Phillips et al. (1961), have shown that tetracycline antibiotics plus treatment with comparatively low concentrations of penetrating radiations offer promise in extending storage life of meats. Sleeth et al. (1960) found that a spray consisting of 20 ppm each of OTC and nystatin was more effective in controlling surface microflora of CTC-injected beef carcasses than a similar combination of OTC and sorbic acid. Possibly higher concentrations of sorbic acid with OTC could have caused more beneficial results. Yacowitz et al. (1957) observed that nystatin, at levels of 5 and 10 ppm in dip solutions containing 10 ppm of CTC, effectively prevented the growth of yeasts and molds and the development of fungal odors on stored chicken wings and drumsticks, and that nystatin did not penetrate the chicken skin or muscle and was rapidly inactivated by boiling. Herold et al. (1959) suggested, from results obtained on studies with isolated cultures, that combinations of CTC with benzalkonium chloride with and without fungicidin offered possibilities in controlling yeasts and molds on CTC-treated meats. These results indicate that fungistatic compounds can cause significant inhibition of yeast and mold growth on CTC-treated flesh foods.

Potential Use in Canning

Canning is singly the most widely used method of food preservation but in certain instances the products suffer from undesirable flavor defects or vitamin losses which could be overcome if less severe heating conditions could be employed. Since the original work of Anderson and Michener (1950), a number of competent research groups have investigated the possibility of reducing processing times and temperatures of certain canning procedures by use of antibiotics such as subtilin, nisin and tylosin. During the ensuing seven years at least 20 scientific papers appeared in which various aspects of the problem were outlined (Wrenshall,

1959). These antibiotics seem to be quite safe as far as ingestion is concerned, and in most cases they are effective in enabling the processing conditions to be reduced without spoilage. Therefore, the processing times and temperatures could be reduced for canned foods by use of the above antibiotics; this has been found to be true when foods have been inoculated with certain heat resistant spores of organisms such as <u>Bacillus coagulans,</u> <u>Clostridium botulinum</u> and certain putrefactive anaerobes. However, it is difficult to prove that this procedure would be reliable under all circumstances. This is undoubtedly the reason it has not as yet been adopted on this continent. In view of this it is interesting that a recent report (Anon., 1960) has stated that: "Nisin has been recommended for canned foods (in Great Britain) with no statutory limits." If this antibiotic finds successful application in enabling processing conditions to be reduced without development of microbiological spoilage and toxin formation in canned foods a useful application of antibiotics in canning technology may yet be developed. In Great Britain, nisin is apparently not regarded legally as an antibiotic or food preservative as it is produced by lactic acid bacteria which are normally found in dairy products (Hawley, 1958). Hawley (1957a and b) has exhaustively reviewed the many possible uses of nisin in food processing.

Potential Use of Antibiotics in Preservation of Other Foods

Milk was one of the first foods studied with respect to the possibility of employing antibiotics in its preservation. Extensive work has shown that tetracycline and other antibiotics preserve milk quite effectively. However, except in localities where pasteurizing facilities are not available, or preservation is required as a purely emergency measure, no good reason can be proposed for the use of antibiotics for this purpose. On the other hand, nisin, or a starter culture of streptococci which produces nisin, has been used for some years in the European cheese industry. Both original cheese and processed cheese are very susceptible to deterioration resulting from growth of various clostridia. Both nisin producing cultures of streptococci and the isolated crystalline antibiotic are used commercially with considerable success in controlling butyric acid and other undesirable fermentations in cheese (Hawley, 1957a and b, 1958). Relaxation of regulations on this continent may yet occur regarding use of nisin in cheese.

There have been a number of studies concerning prevention of microbiological deterioration of vegetables and fruits with antibiotics. Studies by Bonde (1953) indicated that streptomycin, tetracyclines and certain other antibiotics controlled bacterial decay of potatoes. Smith (1953) found that streptomycin and OTC were effective in controlling bacterial soft rot of packaged spinach. Several publications dealing with this subject have appeared since the above reports in which the results have been confirmed and extended to other vegetables including lettuce, broccoli, radishes, etc. (Becker et al., 1960; Shapiro and Holder, 1960). Nystatin delayed onset of brown rot of harvested peaches markedly (DiMarco and Davis, 1957). Ayres and Denisen (1958) found that spoilage of strawberries and raspberries was delayed by treatment with myprozine and rimocidin, but that nystatin, ascosin and candidin were unsatisfactory for this purpose.

In general, the concentrations of antibiotics used have been comparatively high and the residues in raw and in cooked products have been significant. This fact has undoubtedly delayed industrial application, especially since consumption of uncooked vegetables and fruits is not uncommon. As with antibiotic treatments of flesh foods, experts in this field have repeatedly stressed that success depends upon intelligent application of the antibiotic under sanitary conditions.

LITERATURE CITED

Abbey, A., Kohler, A. R., and Hopper, P. F. 1960. Effectiveness of chlortetracycline in poultry preservation after long-term commercial use. Food Technol. 14:609-12.

Anderson, A. A. and Michener, H. D. 1950. Preservation of foods with antibiotics. I. The complimentary action of subtilin and mild heat. Food Technol. 4:188-89.

Anonymous. 1960. Great Britain, Food Standards Committee Preservatives. Food 29:1.

_____. 1960-61. Annual report of the Fisheries Research Board of Canada, 11, 131.

Ayres, J. C. 1959. Use of coating materials or film impregnated with chlortetracycline to enhance color and storage life of fresh beef. Food Technol. 13:512-15.

_____, and Denisen, E. L. 1958. Maintaining freshness of berries using selected packaging materials and antifungal agents. Food Technol. 12:562-67.

_____, Walker, H. W., Fanelli, M. J., King, A. W., and Thomas, F. 1956. Use of antibiotics in prolonging storage life of dressed chicken. Food Technol. 10:563-68.

Banfield, F. H. 1952. Problems arising from the use of chemicals in food. Preservatives, including anti-mold agents. Chem. and Ind. 114-19.

Becker, E. and Roeder, I. 1957. Sorbic acid as a preservative for margarine. Fette, Seifen, Anstritchmittel 59:321-28.

Becker, R. F., Goodman, R. N., and Goldberg, H. S. 1960. Factors affecting decay of prepackaged spinach. Food Technol. 14:127-30.

Bonde, R. 1953. The control of bacterial decay of the potato with antibiotics. Am. Potato J. 30:143-47.

Borowik, J., Fischer, E., Ostrowski, St., and Trezesinski, P. 1957. Investigations on the influence of sodium nitrited ice on the keeping qualities of Baltic cod. Prace Morskiego Instyutut Ryrbackiego No. 9:633. (Translation available from the Office of Tech. Services, U. S. Dept. Comm., Wash. 25, D. C.)

Boyd, J. W. and Tarr, H. L. A. 1954. Inhibition of mold development in fish products. Fisheries Research Bd. Canada, Prog. Repts. Pacific Coast Stas. 99:22.

_____. 1955. Inhibition of mold and yeast development in fish products. Food Technol. 9:411-12.

Campbell, L. L., Jr. and O'Brien, R. T. 1955. Antibiotics in food preservation. Food Technol. 9:461-65.

Castellani, A. G. and Niven, C. F., Jr. 1955. Factors affecting bacteriostatic action of sodium nitrite. Appl. Microbiol. 3: 154-59.

Curran, H. R. and Evans, F. R. 1946. The activity of penicillin in relation to bacterial spores and the preservation of milk. J. Bact. 52:89-98.

Deane, D. D. 1961. Potassium sorbate solutions satisfactorily prevent mold growth on rindless-type Swiss cheese. J. Dairy Sci. 44:457-65.

Deatherage, F. E. 1956. The use of antibiotics in preservation of foods other than fish. Proc. 1st Internat. Conf. on Use of Antibiotics in Agriculture. Natl. Acad. Sci., Natl. Research Counc. Publ. 397:211.

Deuel, H. J., Jr., Alfin-Slater, R., Weil, C. S., and Smyth, H. F., Jr. 1954. Sorbic acid as a fungistatic agent. I. Harmlessness of sorbic acid as a dietary component. Food Research 19:1-12.

DiMarco, G. R. and Davis, B. H. 1957. Prevention of decay of peaches with post-harvest treatments. Plant Disease Reporter 41:284-88.

Downing, H. E., McMahan, J. R., and Baker, C. 1956. Antibiotic preservation of meats. IV. Intraperitoneal injection of oxytetracycline in hogs. Antibiotics Ann. 1955-56:737-38.

Dubrova, G. B. 1961. The use of antibiotics in the preservation of food products. Gostorgizdat, Moscow, 88 (in Russian).

Eddy, B. P. 1957. The inhibitory action of nitrite. In The microbiology of fish and meat curing brines. Dept. Sci. Ind. Research, Food Investigation. London, 79-86.

Farber, L. 1959. Antibiotics in food preservation. Annual Review Microbiol. 13:125.

Fellers, C. R. 1955. In Handbook of Food and Agriculture. Ed. F. C. Blanck. Reinhold Pub. Co., New York, 331.

Ferguson, W. E. and Powrie, W. D. 1957. Studies on the preservation of fresh apple juice with sorbic acid. Appl. Microbiol. 5:41-43.

Fieger, E. A. and Novak, A. F. 1961. Microbiology of shellfish deterioration, p. 561-611. In Fish as Food. Ed. G. Borgstrom, Academic Press Inc., New York.

Firman, M. C., Bachmann, H. J., Heyrich, F. J., and Hopper, P. F. 1959. A comparison of the methods of applying Acronize chlortetracycline to beef. Food Technol. 13:529-33.

Foley, E. J. and Byrne, J. V. 1950. Penicillin as an adjunct to the preservation of quality of raw and pasteurized milk. J. Milk and Food Technol. 13:170-74.

Goldberg, H. S., Weiser, H. H., and Deatherage, F. E. 1953. Studies on meat. IV. Use of antibiotics in preservation of fresh beef. Food Technol. 7:165-66.

Gooding, C. M. 1945. Process for inhibiting growth of molds. U. S. Patent, No. 2,379, 294 (assigned to Best Foods Inc.).

Greenberg, R. A. and Silliker, J. H. 1961. Evidence of heat injury in enterococci. J. Food Sci. 26:622-25.

Hansen, J. D. and Appleman, M. D. 1955. The effect of sorbic, propionic and caproic acids on the growth of certain clostridia. Food Research 20:92-96.

Hawley, H. B. 1957a. Nisin in food technology. Food Manufacture 32:370, 430.

_____. 1957b. Nisin. A survey of its development and applications in food technology. Mimeo. 34 pp.

_____. 1958. The permissibility and acceptability of nisin as a food additive. Milchwissenschaft 13:253-57.

Herold, M., Hoffman, J., Capkova, J., and Necasek, J. 1959. Suppression of growth of fungal microflora in the conservation of meat by chlortetracycline. Antibiotics Annual 1958-59:935.

Hounie, E. 1950. Meat preservation by antibiotics. Food Manufacture 25:508-9.

Ingram, M. 1959. Technical aspects of the commercial use of anti-microbial chemicals in food preservation. Chem. and Ind. 552-57.

Janke, A., Janke, R. G., Bauer, K., Heilman, H., Kirchmeyer, H., and Kostjak, W. W. 1958. Increased storage life of hen carcasses by treatment with chlortetracycline. Zbl. f. Bakt. 711 (Abt. II):291.

Jay, J. M., Weiser, H. H., and Deatherage, F. E. 1957. The effect of chlortetracycline on the microflora of beef and studies on the mode of action of this antibiotic in meat preservation. Antibiotics Annual 1955-57:954-65.

Jensen, L. B. 1942. Microbiology of Meats. The Garrard Press, Champaign, Ill., 242 pp.

Joslyn, M. A. and Bravermann, J. B. S. 1954. The chemistry and technology of the pre-treatment and preservation of fruit and vegetable products with sulfur dioxide and sulfites. Adv. in Food Research 5:97-160.

Kohler, A. R., Miller, W. H., and Broquist, H. P. 1955. Aureomycin-chlortetracycline and the control of poultry spoilage. Food Technol. 9:151-54.

McVicker, R. J., Dawson, L. E., Mallman, W. L., Walters, S., and Jones, E. 1958. Effect of certain bacterial inhibitors on shelf life of fresh fryers. Food Technol. 12:147-49.

Melnick, D., Luckmann, F. H., and Gooding, C. M. 1954. Sorbic acid as a fungistatic agent for foods. VI. Metabolic degradation of sorbic acid in cheese by molds and the mechanism of mold inhibition. Food Research 19:44-58.

Moyer, R. H., Southcott, B. A., and Tarr, H. L. A. 1958. Distribution of chlortetracycline (CTC) antibiotic in ice made from hard waters. Fisheries Research Bd. Canada, Prog. Repts. Pacific Coast Stas. 112:21-22.

Nagel, C. W., Simpson, K. L., Ng, H., Vaughn, R. H., and Stewart, G. F. 1960. Microorganisms associated with spoilage of refrigerated poultry. Food Technol. 14:21-23.

Niven, C. F., Jr. and Chesboro, W. R. 1957. Complementary action of antibiotics and irradiation in preservation of fresh meats. Antibiotics Annual 1956-57:855-59.

Njoku-Obi, A. N., Spencer, J. V., Sauter, E. A., and Eklund, M. W. 1957. A study of the fungal flora of spoiled chlortetracycline treated chicken meat. Applied Microbiol. 5: 319-21.

Phillips, A. W., Newcomb, H. R., Robinson, T., Back, F., Clark, W. L., and Whitehill, A. R. 1961. Preservation of fresh beef with antibiotics and radiation. Food Technol. 15:13-15.

Phillips, G. F. and Mundt, J. O. 1950. Sorbic acid as an inhibitor of scum yeast in cucumber fermentation. Food Technol. 4:291-93.

Sacchi, E. M., McMahan, J. R., Ottke, R. C., and Kersey, R. C. 1956. Antibiotic preservation of meats. II. Intraperitoneal injection of oxytetracycline in beef cattle. Antibiotics Ann. 1955-56:731-33.

Saller, W. and Kolewa, S. R. 1957. Sorbic acid in the preservation of wine. Mitt. Klosterneuberg Ser. A. 7:21.

Schelhorn, M., von. 1951. Control of microorganisms causing spoilage in fruit and vegetable products. Adv. in Food Research 3:429-82.

Shapiro, J. E. and Holder, I. A. 1960. Oxytetracycline and citric acid dips for packaged salad mix. Appl. Microbiol. 8:341-45.

Silliker, J. H., Greenberg, R. A., and Schack, W. R. 1958. Effect of individual curing ingredients on the shelf stability of canned comminuted meats. Food Technol. 12:551-54.

Sleeth, R. B., Armstrong, J. C., Goldberg, H. S., and Naumann, H. D. 1960. Antibiotic preservation of beef with subsequent feeding to experimental animals. Food Technol. 14:No. 6 (special abstracts section).

_____ and Naumann, H. D. 1960. Efficacy of oxytetracycline for ageing beef. Food Technol. 14:98.

Smith, D. P. and Rollins, N. J. 1954. Sorbic acid as a fungistatic agent for foods. VIII. Need and efficacy in protecting packaged cheese. Food Technol. 8:133-35.

Smith, W. L. 1952. Antibiotic treatments for the bacterial soft rot of spinach. Phytopathology 42:475.

Southcott, B. A. and Tarr, H. L. A. 1961. Magnesium as a factor in determining the comparative sensitivity of marine bacteria to four tetracycline antibiotics. Canadian J. Microbiol. 7: 284-86.

Stadelman, W. J., Marion, W. W., and Eller, M. L. 1957. Antibiotic preservation of fresh poultry meat. Antibiotics Annual 1956-57:839-42.

Tanaki, S., Kiyota, S., Takao, H., and Murakani, Y. 1956. Effect of antiseptics on the preservation of sausage. J. Fermentation Technol. (Japan) 34:485.

Tanner, F. W. 1944. The Microbiology of Foods. Garrard Press, Champaign, Ill., 1-58.

Tarr, H. L. A. 1940. Bacteriostatic action of nitrites. Nature 147:417.

_____. 1941. The action of nitrites on bacteria. J. Fisheries Research Bd. Canada 5;265-75.

_____. 1942. The action of nitrites on bacteria: further experiments. J. Fisheries Research Bd. Canada 6:74-89.

_____. 1944a. Action of nitrates and nitrites on bacteria. J. Fisheries Research Bd. Canada 6:233-49.

_____. 1944b. Chemical inhibition of growth of fish spoilage bacteria. J. Fisheries Research Bd. Canada 6:257-66.

Tarr, H. L. A. 1946. Germicidal ices. Fisheries Research Bd. Canada Prog. Repts. Pacific Coast Stas. 67:36-40.

_____. 1948. Comparative value of germicidal ices for fish preservation. J. Fisheries Research Bd. Canada 7:155-61.

_____. 1956. Control of bacterial spoilage of fish with anti-biotics. In Proc. 1st Internat. Conf. on Use of Antibiotics in Agri. Publ. No. 397, Natl. Acad. Sci., Natl. Research Counc. 119.

_____. 1960a. Antibiotics as a preservative measure. Ca-nadian Food Ind. 31:39.

_____. 1960b. Antibiotics in fish preservation. Fisheries Research Bd. Canada Bul. No. 124, 24.

_____. 1961. Chemical control of microbiological deterio-ration. In Fish as Food. Ed. G. Borgstrom, Academic Press Inc., New York, 639.

_____ and Bailey, B. E. 1939. The effectiveness of benzoic acid ice for fish preservation. J. Fisheries Research Bd. Canada 4:327-34.

_____ and Deas, C. P. 1948. Action of sulpha compounds, antibiotics and nitrites on growth of bacteria in fish flesh. J. Fisheries Research Bd. Canada 7:221.

_____, Southcott, B. A., and Bissett, H. M. 1950. Effect of several antibiotics and food preservatives in retarding bacte-rial spoilage of fish. Fisheries Research Bd. Canada, Prog. Repts. Pacific Coast Stas. 83:35-38.

_____. 1952. Experimental preservation of flesh foods with antibiotics. Food Technol. 6:363-66.

_____ and Sunderland, P. A. 1939. The role of preserva-tives in enhancing the keeping quality of fresh fillets. Fish-eries Research Bd. Canada, Prog. Repts. Pacific Coast Stas. 39:13.

_____. 1940. The comparative value of preservatives for fresh fillets. J. Fisheries Research Bd. Canada 5:148-63.

Thatcher, F. S. and Loit, A. 1961. Comparative flora of chlor-tetracycline-treated and untreated poultry. Appl. Microbiol. 9:39-45.

Vaughn, R. H. 1955. Bacterial spoilage of wines with special reference to California conditions. Adv. in Food Research 6:67-108.

_____, Nagel, C. W., Sawyer, F. M., and Stewart, G. F. 1957. Antibiotics in poultry meat preservation: a compari-son of the tetracyclines. Food Technol. 11:426-29.

_____, Ng, H., Stewart, G. F., Nagel, C. W., and Simpson, K. L. 1960. Antibiotics in poultry meat preservation,

development in vitro of bacterial resistance to chlortetracycline. Appl. Microbiol. 8:27-30.

Wells, F. E., Fry, F. L., Marion, W. W., and Stadelman, W. J.
1957. Relative efficacy of three tetracyclines with poultry
meat. Food Technol. 11:656-58.

Weiser, H. H., Goldberg, H. S., Cahill, V. R., Kunkle, L. E., and
Deatherage, F. E. 1953. Observations on fresh meat processed by the infusion of antibiotics. Food Technol. 7:495-99.

Wilson, G. D., Brown, P. D., Pohl, C., Weir, C. E., and Clesbro,
W. R. 1960. A method for the rapid tenderization of beef
carcasses. Food Technol. 14:186-89.

Wrenshall, C. L. 1959. Antibiotics in food preservation. In Antibiotics, Their Chemistry and Nonmedical Uses. Ed. H. S.
Goldberg. Van Nostrand Co., Princeton, N. J., pp. 449-527.

Yacowitz, H., Pansy, F., Wind, S., Stander, H., Sassaman, H. L.,
Pagano, J. F., and Trejo, W. H. 1957. Use of nystatin (My-
costatin) to retard yeast growth on chlortetracycline-treated
chicken meat. Poultry Sci. 36:843-49.

Ziegler, F. and Stadelman, W. J. 1955. The effect of Aureomycin
treatment on the shelf life of fresh poultry meat. Food Technol. 9:107-8.

13

Salmonellae

BETTY C. HOBBS
CENTRAL PUBLIC HEALTH LABORATORY
London, England

D ATA FOR FOOD POISONING in the United Kingdom are summarized year by year in reports published in the Monthly Bulletin of the Public Health Laboratory Service and the Ministry of Health. The Report for 1960 (Report 1961a) shows, as usual, that the predominant cause of gastroenteritis attributed to food is salmonellosis, of which a high proportion of notified incidents are sporadic cases. It shows also that the contaminated foods responsible for general and family incidents of salmonella food poisoning are found in a minority of instances only, whereas the foods responsible for incidents of Clostridium perfringens and staphylococci are more frequently traced. Thus, in 1960, of 165 outbreaks where the food vehicle was known, 31.5 per cent and 44.8 per cent were due to Cl. perfringens and staphylo cocci respectively and 7.5 per cent only were attributed to salmonellosis.

There are many reasons for lack of information on food sources of sporadic cases of salmonella food poisoning, such as late notifications of isolated cases, the failure to correlate cases occurring simultaneously in different areas or too long a delay in doing so. There is apathy with regard to the collection of fecal samples, and a delayed diagnosis of salmonellosis diminishes the chance of finding the relevant foodstuff; there is often too narrow a search for the food vehicle or source of infection. Some of these factors may be responsible also for the lack of information about family and general incidents.

Not many years ago the human carrier was considered to be the precursor of food contamination, although Savage (1932) had pointed out that animals were the main reservoirs of salmonellae, and in 1956 he said that more extensive surveys of animals as reservoirs of salmonella strains and their association with food poisoning was needed. In 1947 the data required to confirm this statement began to accumulate. A Report (1947) described the isolation of salmonellae from dried whole egg imported from the United States of America during the war years, and also the many outbreaks of salmonellosis due to serotypes isolated from the dried egg but hitherto unknown in the United Kingdom. Furthermore, pigs fed dried whole egg became symptomless excretors and, after slaughter, the offal and meat were found to be contaminated to a greater or lesser extent according to the hygiene of the abattoir. Ten more years elapsed before it was realized that frozen whole egg and dried egg albumen from China were bringing Salmonella schottmuelleri* (see p. 226) and many other salmonella serotypes into the country. In fact, all egg products, whether liquid, frozen or dried whole egg, frozen or dried albumen or dried yolk, and from whatever country they originated, were contaminated with salmonellae (Report, 1958).

Outbreaks of paratyphoid fever and salmonellosis following the consumption of cream cakes were suspected to be due to the cross-contamination of imitation cream from egg products in bakeries (Newell, Hobbs and Wallace, 1955); when symptomless excretors were found in the bakery they were considered to be the victims of the situation rather than the cause (Cameron, 1959).

Meanwhile the veterinary profession was proving poultry to be one of the most prolific sources of salmonellae (Buxton, 1957, 1958; Gordon, 1959; Garside, Gordon and Tucker, 1960), and later investigations at broiler stations showed that chicken and turkey carcasses were leaving stations contaminated with salmonellae (Galton et al., 1955; Sadler et al., 1961; Dixon and Pooley, 1961, 1962). The development of the phage typing of certain salmonella serotypes and its use in the study of salmonella infection created the possibility of discriminating organisms on a much finer level than has been possible hitherto.

Gradually, the evidence incriminating food from animal sources increased. Meat and meat products were investigated and the results showed that imported frozen boneless meats, such as beef, veal and mutton (Hobbs and Wilson, 1959) were frequently contaminated with salmonellae and that salmonellae could be found in 50 per cent or more of samples of imported frozen horsemeat (Hobbs, 1961). Home produced pork products from retail and manufacturing establishments were contaminated to the extent of approximately 2 per cent of samples; it was believed that most of the salmonellae came from the ingredients, either directly or by cross contamination from raw to cooked

*The designation S. schottmuelleri is used in this chapter to include both the classical S. paratyphi B and the java variety.

products. Earlier investigations showed that the same salmonella serotypes could be isolated from feed, pigs (rectal swabs), post-mortem cecal swabs and sausage meat (Newell et al., 1959). Nottingham and Urselmann (1961) found that 5 per cent of 7,497 samples from 2,755 animals slaughtered at meat works and from the pasture and water supply of 5 farms contained salmonellae. Fifteen per cent of cows and sheep, 13 per cent of calves and 4 per cent of beef cattle were found to have salmonellae in one or more samples examined. Leistner et al. (1961) found a relatively low incidence of salmonellae in animal feeds and pig intestines at the farm level. There was an increasing incidence in the pig intestine after the animals had arrived at the terminal market and after they had been held in pens; holding pen deposits were heavily infected. There was a direct relationship between the incidence of salmonellae in the colons and mesenteric glands of pigs and the length of holding time in pens prior to slaughter. This work agrees with that of Galton et al. (1954). Leistner et al. (1961) found a relatively high incidence but low numbers of salmonellae in rendered animal by-products used as feed constituents.

The Proceedings of the 65th Annual Meeting of the U.S. Livestock Sanitary Association 1961 give much useful information on the isolation of salmonellae from processing plants, swine and cattle and animal by-products, feeds and feed ingredients. The incidence of salmonellae and the significance of international spread are also considered.

Desiccated coconut, used uncooked on and within various popular cakes, sweets, cookies and sweetmeats prepared commercially and in the home, was found to be another source of salmonellae including S. schottmuelleri (S. paratyphi B) (Galbraith et al., 1960). By means of phage typing, Anderson (1960) showed that a type of the paratyphoid bacillus isolated from coconut was the same as that responsible for an incident of enteric fever. It was thought that the origin of the paratyphoid bacilli in desiccated coconut was likely to be human excretors. As far as other salmonella serotypes were concerned, pollution with infected animal or bird excreta could not be ruled out.

These investigations have raised problems with regard to the bulk sampling and laboratory examination of large numbers of samples of imported foods. The first consideration is to find the extent of the problem, i.e., the proportion of contaminated samples, and secondly, the distribution of contaminated batches should be prevented (Wilson, 1962). Furthermore, research has been initiated on methods to render the foods safe by eliminating the salmonellae either during processing or immediately before distribution for sale or use.

Meringue powder containing egg albumen, cake-mix (Thatcher and Montford, 1962) containing dried whole egg, albumen or coconut, raw and processed meats and desiccated coconut, are used daily in food handling establishments and homes, and contamination from these products to other foods may complicate the search for sources of salmonella infections. These substances may have been used before an outbreak occurred but not necessarily as an ingredient of the food which actually gave rise to the symptoms; intermediate agents such as hands, utensils, mixing bowls and cloths all play their part in the transmission of infection.

The increased use of phage typing particularly for S. typhimurium (Callow, 1959) has resulted in the frequent association of certain phage types of S. typhimurium with particular animals and birds and their food products (Anderson, 1962). Similarly, particular phage types of S. schottmuelleri have been associated with certain imported foods. It is tempting to suspect these foods when outbreaks due to the relevant phage types occur. Furthermore, sporadic cases of salmonellosis occurring simultaneously in several areas may be correlated by their phage types and associated with particular shipments or home-produced consignments of foodstuffs from which the same phage type has been isolated.

MEDIA, METHODS AND SAMPLING

The combination of media used for the isolation of salmonellae from foods varies from laboratory to laboratory in the United Kingdom but the liquid enrichment media selenite-F (SF) (Leifson, 1936; Hobbs, King and Allison, 1945), and tetrathionate broth (TT) (Rolfe, 1946) and the selective agar media, desoxycholate citrate agar (DC) (Leifson, 1935), sometimes containing 1 per cent sucrose, and Wilson and Blair agar (WB) (Difco) and the modification described by de Loureiro (1942) are commonly used.

The media selected for a particular foodstuff should depend on the usual bacterial flora of the food, some requiring a more inhibitory or selective liquid medium than others. For example, when it is unlikely that there are Gram-negative bacilli other than salmonellae present, nutrient broth (NB) may be more satisfactory for liquid enrichment than SF. While SF is excellent for the inhibition of coliform bacilli and pseudomonads, this medium may also discourage the growth of salmonellae, particularly when they are present in small numbers only and incapable of immediate active growth. It is recommended, therefore, that a full bacteriological examination be carried out on a selection of samples of the food to be investigated before the liquid media are chosen.

The use of two liquid enrichment media increased the number

of positive isolations, in some instances there was a 40-50 per cent increase; similarly the use of two agar media for subcultures increased the number of positives; also, individual observers sometimes found it easier to work with one medium than another. It is possible that duplicate quantities of one liquid medium of any kind would be as satisfactory as two quantities of different media although the results shown in Table 13.8 indicate that this may not be true unless the ideal medium is chosen. In general, SF and TT were used for foodstuffs such as liquid whole egg and meat which contain a wide variety of Gram-negative bacilli, two broths for foodstuffs such as spray and pan-dried egg albumen having a flora of Gram-positive cocci and salmonellae as the predominant Gram-negative bacilli, and SF and NB for material such as frozen albumen, dried whole egg and coconut with low counts and a mixed flora.

Samples of 50 to 60 grams of food were divided into two lots, each incubated in 100 ml. of liquid medium. Recommended incubation temperatures vary from 37^0 to 43^0C. but $37^0 \pm 1^0$C. was used in this work; subcultures were made after 1 day and again after 3 days incubation if the results of the first day's plating were negative when examined after incubation at 37^0C. for 2 days.

One or more characteristic colonies were picked, for fermentation reactions into Gillies' (1956) modification of Kohn's (1954) tubes and into NB, and also onto MacConkey agar for purity. Suspensions for slide agglutination were taken from the slope cultures of the Gillies tubes, and when necessary, Craigie tubes (Tulloch, 1939) were inoculated to change the phase; identification of the more usual serotypes could be carried out in 1 to 2 days. Obscure serological types were referred to the Salmonella Reference Laboratory, and S. schottmuelleri, S. typhimurium, S. thompson, S. enteritidis and S. pullorum were sent to the Enteric Reference Laboratory for phage typing.

The examination of imported foodstuffs known to be a source of salmonellae was planned by the Port Medical Officers to take place on the basis of 5 per cent of samples from all batches. Batches found to be positive were re-examined by sampling 10 per cent of the packs and the consignment was condemned if any of these further samples were positive. If negative results were obtained from the second sampling, only those packs found positive at the first sampling were condemned. Sixty to 80 grams or more of each sample, taken by the Port Health inspectors, arrived at the laboratory as soon as possible after being taken. A variety of receptacles were used including sterile $\frac{1}{2}$ lb. honey jars with metal screw caps for frozen egg samples, waxed cartons or aluminum tins with screw caps for dried products, and

polyethylene bags for frozen meat samples, which, because of
their irregular shapes, would not fit into other containers. The
honey jar caps were lined with cardboard washers which were
replaced each time the jars were washed and sterilized.

Frozen samples were examined immediately after thawing.
Separate rooms were used for putting up the samples of egg
products, meats and coconut. Plastic aprons, kept for the workers
in each room, were wiped with a solution of hypochlorite after
use. Whenever subcultures were made from the liquid enrichment to agar media, the workers were encouraged to dip their
hands into a solution of an iodophor detergent as often as practicable. All surfaces were swabbed down frequently with a solution of hypochlorite.

When, after the examination of many samples over a period
of time, a foodstuff was confirmed as a potential source of salmonellae, methods of treatment were considered. Such research
work often entailed collaboration with other organizations and
laboratories. Whether experiments were concerned with pasteurization by dry or steam heat, gamma irradiation or fumigation by ethylene oxide, naturally or artificially contaminated material was examined before and after treatment. A review of
some of the published work describing these procedures is given
under results.

When salmonella counts were required, they were carried out
by one of two methods. If sufficient numbers of salmonellae were
anticipated, a surface plate count on DC, WB or MacConkey agars
was used. When less than 50 salmonellae per gram were expected, a dilution technique was used similar to that for the estimation of coli-aerogenes in water (Report, 1956), SF or NB was
substituted for MacConkey broth and each tube was subcultured
onto DC and WB agars. The number of dilutions followed either
the technique for water analysis (5 x 10 ml., 5 x 1 ml. and
5 x 0.1 ml.) (Report, 1956), or a modified American technique
with three tubes of each dilution and an adjusted set of the most
probable number (MPN) tables (Hoskins, 1934).

RESULTS

The results selected for analysis were taken from 100 samples positive for salmonellae of any serotype of each of the foodstuffs, frozen whole egg, dried whole egg, frozen egg albumen,
pan-dried and spray-dried egg albumen, frozen boned meat and
coconut. All samples were received and examined during 1961 as
part of a general survey, and the results reported here were

Table 13.1. Frozen Whole Egg

Comparison between enrichment in Selenite-F (SF) and Tetrathionate (TT) and Desoxycholate Citrate (DC) and Wilson and Blair (WB) agars for the isolation of salmonellae from 100 positive samples

Liquid Enrichment		Subcultures on Agar Media			
Medium	No. and per cent	DC-total	WB-total	DC-only	WB-only
SF - total	72	68	49	23	4
TT - total	84	77	63	21	7
SF - only, but not TT	16	16	11	5	0
TT - only, but not SF	28	26	18	10	2
SF and TT - combined total	100	97	81	19	3
Positive 3rd day only SF and TT	12				

Table 13.2. Dried Whole Egg

Comparison between enrichment in Selenite-F (SF) and Nutrient Broth (NB) and Desoxycholate Citrate (DC) and Wilson and Blair (WB) agars for the isolation of salmonellae from 100 positive samples

Liquid Enrichment		Subcultures on Agar Media			
Medium	No. and per cent	DC-total	WB-total	DC-only	WB-only
SF - total	53	41	50	3	12
NB - total	89	69	85	4	20
SF - only, but not NB	11	9	9	2	2
NB - only, but not SF	47	39	44	3	8
SF and NB - combined total	100	84	95	5	16
Positive 3rd day only SF and NB	12				

Table 13.3. Frozen Egg White

Comparison between enrichment in Selenite-F (SF) and Nutrient Broth (NB) and Desoxycholate Citrate (DC) and Wilson and Blair (WB) agars for the isolation of salmonellae from 100 positive samples

Liquid Enrichment		Subcultures on Agar Media			
Medium	No. and per cent	DC-total	WB-total	DC-only	WB-only
SF - total	75	70	51	24	5
NB - total	72	65	46	26	7
SF - only, but not NB	28	26	16	12	2
NB - only, but not SF	25	20	18	7	5
SF and NB - combined total	100	92	75	25	8
Positive 3rd day only SF and NB	28				

Table 13.4. Dried Egg White — Flake

Comparison between duplicate Nutrient Broth (NB) cultures and
Desoxycholate Citrate (DC) and Wilson and Blair (WB) agars
for the isolation of salmonellae from 100 positive samples

Liquid Enrichment		Subcultures on Agar Media			
Medium	No. and per cent	DC-total	WB-total	DC-only	WB-only
NB (1) - total	76	60	71	5	16
NB (2) - total	84	69	79	5	15
NB (1) - only, but not NB (2)	16	12	13	3	4
NB (2) - only, but not NB (1)	24	19	22	2	5
NB (1) and NB (2) - combined total	100	85	94	6	15
Positive 3rd day only	25				

Table 13.5. Dried Egg White Powder

Comparison between duplicate Nutrient Broth (NB) cultures and
Desoxycholate Citrate (DC) and Wilson and Blair (WB) agars
for the isolation of salmonellae from 100 positive samples

Liquid Enrichment		Subcultures on Agar Media			
Medium	No. and per cent	DC-total	WB-total	DC-only	WB-only
NB (1) - total	77	66	74	3	11
NB (2) - total	85	68	81	4	17
NB (1) - only, but not NB (2)	15	14	13	2	1
NB (2) - only, but not NB (1)	23	19	21	2	4
NB (1) and NB (2) - combined total	100	87	96	4	13
Positive 3rd day only	20				

Table 13.6. Meat

Comparison between enrichment in Selenite-F (SF) and Tetrathionate (TT)
and Desoxycholate Citrate (DC) and Wilson and Blair (WB) agars
for the isolation of salmonellae from 100 positive samples

Liquid Enrichment		Subcultures on Agar Media			
Medium	No. and per cent	DC-total	WB-total	DC-only	WB-only
SF - total	79	70	61	18	9
TT - total	74	67	65	9	7
SF - only, but not TT	26	24	17	9	2
TT - only, but not SF	21	20	19	2	1
SF and TT - combined total	100	97	85	15	3
Positive 3rd day only SF and TT	4				

obtained during routine examinations. The comparative results obtained from the different media for each foodstuff are given in Tables 13.1 and 13.7. In Table 13.1 for example, the selenite-F and tetrathionate totals give the number of samples positive through those media, i.e., the "SF-total" includes positive samples from TT also, but not those which were positive through "TT-only" which are given under that heading. Similarly the "TT-total" includes positive samples from SF with the exception of those which were positive through "SF-only." Thus reading vertically in the second column, the figures for "SF-total" and "TT-only" add up to 100, and similarly the "TT-total" and "SF-only" figures added together cover the 100 samples.

The figures under the heading "Subcultures on Agar Media" should be read horizontally. They show the results of subcultures onto DC and WB agar plates from the "SF-total" and "TT-total" positives; for completeness the results of subcultures from the "SF-only" and "TT-only" positives have been given in the same way. So that the "DC-total" and "WB-only" figures add up to the corresponding figures in the second column, i.e., in Table 13.1 the figures 68 plus 4 and 49 plus 23 both equal the "SF-total" of 72, similarly 77 plus 7 and 63 plus 21 both equal the "TT-total" of 84; the same principle applies to the results of subcultures from the "SF-only" and "TT-only" cultures. The combined total and third-day plating results are self-explanatory.

Table 13.2 gives the results for dried whole egg and shows that NB was superior to SF as an enrichment medium; 47 per cent of positive results came through NB only. Of the 12 samples positive after 3 days incubation, 8 came through NB only and 3 through SF only. Somewhat better results were obtained by plating onto WB than onto DC, particularly when subcultures were made from NB. Twenty subcultures from NB were positive on WB only.

Table 13.3 indicates that SF and NB were similar in effect for the liquid enrichment of frozen egg white but that neither medium seemed to be entirely satisfactory; 25 to 28 per cent of positives would have been missed if one medium only had been used. Of the 28 samples which were positive after 3 days incubation, 14 came through SF only and 7 through NB only. Subcultures were positive more often on DC than on WB agar.

The results for flaked and powdered albumen were similar (Tables 13.4 and 13.5) and showed clearly that two NB cultures gave higher yields than one and that 20 to 25 per cent more positive results were obtained by third-day plating; subcultures from NB gave higher yields on the more inhibitory WB agar than on DC.

Meat was enriched in both SF and TT but there was little difference in the results obtained from these two media (Table 13.6);

Table 13.7. Coconut

Comparison between enrichment in Selenite-F (SF) and Nutrient Broth (NB) and Desoxycholate Citrate (DC) and Wilson and Blair (WB) agars for the isolation of salmonellae from 100 positive samples

Liquid Enrichment		Subcultures on Agar Media			
Medium	No. and per cent	DC-total	WB-total	DC-only	WB-only
SF - total	60	53	46	14	7
NB - total	70	34	63	7	36
SF - only, but not NB	30	29	24	6	1
NB - only, but not SF	40	21	35	5	19
SF and NB - combined total	100	77	87	13	23
Positive 3rd day only SF and NB	6				

the number of positive colonies picked from DC was slightly higher than from WB. A further 4 per cent of samples were found positive after the third-day subcultures.

Coconut (Table 13.7) was incubated overnight in SF and NB, but neither medium appeared to be entirely satisfactory as 40 per cent of positive samples would have been missed using SF only and 30 per cent if NB only had been used. Subcultures from SF gave a few more positives on DC than on WB, but for subcultures from broth the more inhibitory WB medium again gave higher yields and 36 per cent of positives would have been missed if subcultures had been made from NB to DC only. Of the 6 per cent of samples positive only after the third-day subculture, 4 came through NB only and 2 through SF only.

Table 13.8 summarizes the results for the seven different food products enriched in the two liquid media. Apparently, NB gives better results than a selective medium such as SF for a

TABLE 13.8.

Comparison of liquid enrichment media for the isolation of salmonellae from various foods (100 positive samples of each)

Food	Selenite-F Total	Tetra-thionate Total	Nutrient Broth Total	Selenite-F Only	Tetra-thionate Only	Nutrient Broth Only	3rd Day Plating
Egg							
Frozen whole	72	84		16	28		12
Dried whole	53		89	11		47	12
Frozen white	75		72	28		25	28
Dried white			A B			A B	
Flake			76 84			16 24	25
Powder			77 ^5			15 23	20
Meat	79	74		26	21		4
Coconut	60		70	30	40		6

substance with a low count of Gram-negative bacteria. For example, egg albumen dried in flakes may have a large bacterial population made up predominantly of streptococci used for fermentation, but small numbers only of salmonellae or other enterobacteriaceae.

The results suggest that a strongly inhibitory agar medium, such as WB, is necessary for subcultures from NB, and in most instances better yields were obtained from NB cultures subcultured on WB than on DC agar; the reverse was true when the liquid media SF and TT were used, when higher yields were obtained on DC agar.

It is necessary, therefore, to consider the bacterial flora of the food in relation to the most suitable liquid enrichment media, and the inhibitory properties of the liquid media in relation to the agar media used for subcultures. Whatever combination of media is used, there is no doubt that a large proportion of positive samples would be missed if one lot of liquid medium and one agar plate were used only; furthermore, from 4 to 28 per cent of positives were found only after plating on the third day.

Using these methods of examination, the numbers and percentages of food samples from which salmonellae were isolated in 1961 are given in Table 13.9. This table shows also the products from which S. schottmuelleri, S. typhimurium and S. thompson have been isolated; in the United Kingdom S. typhimurium is still the commonest cause of salmonellosis.

TABLE 13.9.

Contamination of imported foodstuffs with salmonellae with particular reference to Salmonella schottmuelleri, Salmonella typhimurium and Salmonella thompson (1961)

Food	No. Examined	Total No. Positive	Per cent	S. schottmuelleri		S. typhimurium		S. thompson	
				No.	Per cent	No.	Per cent	No.	Per cent
Egg									
Frozen whole	1042	165	15.8	8	4.8	35	21.2	24	14.5
Dried whole	898	107	11.9	10	9.3	1	0.9	17	15.9
Dried yolk	57	3	5.3	0		0		0	
Frozen white	675	41	6.1	0		13	31.7		
Dried white									
Flake	536	68	12.7	3	4.4	22	32.3	2	2.9
Powder	214	18	8.4	0		2	11.1		
Meat (Frozen and Boneless)									
Horsemeat for animal feeding	796	454	57.0	5	1.1	3	0.7		
For human consumption	795	41	5.1			15	36.6		
Coconut	683	28	4.1	6	21.4	2	7.1		

TABLE 13.10.

Salmonellae from egg products. Comparison between countries
(1961)

Country	Whole Egg Frozen			Whole Egg Dried			White Frozen			White Dried			Yolk Dried		
	No. Exd.	Salmonellae No.	Per cent	No. Exd.	Salmonellae No.	Per cent	No. Exd.	Salmonellae No.	Per cent	No. Exd.	Salmonellae No.	Per cent	No. Exd.	Salmonellae No.	Per cent
Argentina				10	7	70.0				57	34	59.6			
Australia	144	21	14.6				342	23	6.7				42	3	7.1
China	244	61	25.0	570	60	10.5				33	6	18.2			
Czechoslovakia				76	0	0	20	0	0				10	0	0
Israel	564	59	10.5				121	6	4.9						
Jugoslavia				40	20	50	100	4	4.0						
Netherlands				174	20	11.5	66	3	4.6						
Poland	58	1	1.7												
United Kingdom	454	34	7.5												
United States	32	23	71.9				20	5	25.0	658	44	6.6			

The analysis of the egg figures by countries, Table 13.10, indicates the proportion of contaminated material from each source; the predominant serotypes isolated from each food are shown in Table 13.11. S. schottmuelleri and S. typhimurium are included even when they were not the predominant types.

Each country contributed a particular pattern of serotypes. When the salmonellae from egg products are considered, for example, S. tennessee and S. cerro among other serotypes, came exclusively from the U.S.A., S. michigan from the Argentine, S. schottmuelleri, S. aberdeen, S. sundsvaal and S. potsdam from Chinese products, S. thompson from China and the U.S.A., S. give from the Netherlands and S. choleraesuis from Jugoslavia only. S. infantis was isolated from Israeli and U.S.A. samples, S. enteritidis came predominantly from Israel and Jugoslavia, while S. senftenberg was found in samples from China and Israel.

In addition to the importation of salmonellae in egg products, both desiccated coconut from Ceylon, and boneless meat for human and animal feeding from various countries, contributed salmonellae as shown in Table 13.12.

The number of positive samples of coconut declined from 9 per cent in 1959 (Galbraith et al., 1960) to 4.1 per cent in 1961 and 2.6 per cent in the first 3 months of 1962 through the efforts of manufacturers to improve the hygiene of production and shipping in Ceylon. Nevertheless, this commodity is still a source of the salmonella serotypes schottmuelleri, waycross, typhimurium, bareilly, litchfield and others.

Boneless meat for human feeding is also bringing S. schottmuelleri, S. typhimurium, S. bovis morbificans and S. newport into the United Kingdom (Hobbs and Wilson, 1959), and horsemeat for animal feeding is a potent source of S. minnesota, S. oranienburg, S. meleagridis, S. derby, S. anatum, S. give, S. bredeney and many other serotypes. Much of this horsemeat is consumed raw by household pets, a large amount is made up into heated, compressed open-pack pet food, which is frequently found to be recontaminated after processing (Galbraith et al., 1962). Wilson et al. (1961) give figures for the isolation of salmonellae from raw poultry and raw pork specimens.

There is no mention in this review of animal feedstuffs but the results of the investigation of their salmonella content have been published previously (Report, 1959, 1961b). The likelihood of contaminated feeds infecting animals so that they become symptomless excretors and reservoirs for carcass and offal contamination is suggested by a number of authors (Newell et al., 1959; Galbraith, Archer and Tee, 1961; Galton et al., 1954, and Galton et al., 1955). These authors conclude that contaminated

TABLE 13.11.

Foodstuffs and salmonella serotypes. Comparison between countries
(1961)

Country	Whole Egg	Egg White	Egg Yolk	Meat for Humans	Meat for Animals	Desiccated Coconut
Africa				typhimurium anatum montevideo		
Argentina	typhimurium michigan anatum	typhimurium newport montevideo		typhimurium minnesota newport	schottmuelleri typhimurium minnesota	
Australia	typhimurium pullorum potsdam	typhimurium pullorum hesserek				
China	schottmuelleri thompson senftenberg	schottmuelleri meleagridis thompson	enteritidis reading			
Ceylon						schottmuelleri typhimurium waycross
East Europe					minnesota meleagridis oranienburg	
Israel	typhimurium enteritidis infantis	typhimurium braenderup kentucky				
Jugo-slavia	enteritidis choleraesuis pullorum	typhimurium choleraesuis senftenberg				
Nether-lands	anatum bareilly give	typhimurium				
New Zealand				typhimurium		
United States	typhimurium cerro newport	typhimurium oranienburg tennessee				

TABLE 13.12.

Salmonellae from coconut and boneless meat. Comparison between countries
(1961)

Country	Coconut			Boneless Meat					
				Human feeding			Animal feeding		
	No. examined	No. positive	Per cent	No. examined	No. positive	Per cent	No. examined	No. positive	Per cent
Africa				675	28	4.1			
Argentina				40	7	17.5	697	426	61.1
Australia				9	4	44.4			
Ceylon	683	28	4.1						
East Europe							92	27	29.3
New Zealand				46	2	4.3			
Philippines	183	0							

feed may be the primary source of infection in outbreaks and sporadic cases of salmonellosis.

Neither S. heidelberg nor S. saint paul were isolated from egg products, meat or coconut in 1961. It seems likely that they are imported in feedstuffs and that the immediate vehicles of infection are home-killed meat or poultry products.

The results of various methods of treatment to kill salmonellae in foods indicate that pasteurization is the most practicable treatment for liquid whole egg prior to freezing and drying. Murdock et al. (1960) reviewed the situation with regard to pasteurization, and Heller et al. (1962), and Shrimpton et al. (1962) carried out field and laboratory trials to prove the efficiency of pasteurization at 64.4^0 C. for $2\frac{1}{2}$ min. as a means of removing salmonellae and leaving an acceptable bakery product. The work of Brooks (1962), compiled posthumously, has provided an enzyme test which indicates that liquid whole egg has been pasteurized at 64.4^0 C. for $2\frac{1}{2}$ min.

The pasteurization of liquid egg white is more difficult but not impossible, and the dry heat treatment of flaked and spray-dried albumen is a well-known process.

The use of gamma irradiation for frozen whole egg is also well known, and Ingram, Rhodes and Ley (1961) describe recent bacteriological and feeding tests with irradiated material.

Ethylene oxide may be used for dried products such as desiccated coconut, powdered whole egg and albumen, but there are difficulties with regard to the toxicity of residual ethylene glycol. At present, desiccated coconut may be steam treated and redried (Report, 1961c) until better methods of production eliminate the original sources of contamination.

DISCUSSION

The range of media used in this survey was limited by the excessive number of samples being examined in a comparatively small laboratory. Similar or better results may have been obtained using different media or different combinations of media.

Nevertheless, at least four factors were noteworthy: 1) the use of more than one liquid enrichment medium gave an increased number of positive results, not only because a greater bulk of each sample was examined but also because there were advantages in using different kinds of media. Two enrichment cultures gave 11 to 47 per cent more positives than would have been obtained with one liquid medium alone. 2) Two agar media, differing in selectivity, were also useful, partly to provide different conditions for the subcultures and partly for the benefit of the individual investigator who may have greater preference for one medium than another. 3) The choice of liquid enrichment should depend on the nature and bacterial content of the foodstuff to be examined. Nutrient broth, for example, was found to be superior to a selective medium designed to inhibit Gram-negative bacteria when the numbers of organisms were low, as with coconut, or when the predominant flora was Gram-positive and the few enterobacteriaceae present were principally salmonellae, as observed with egg white. Banwart and Ayres (1953) demonstrated that pure cultures of serotypes of salmonellae grew better on nutrient broth than in SF, TT or Ruys medium, but that the addition of whole egg minimized the differences. They concluded that consideration must be given to the material under examination when selecting a broth for the enrichment of salmonellae. It was worthwhile replating the liquid cultures after further incubation if the results from the first subcultures were negative. Four 28 per cent additional positives were obtained in this way. It is not suggested that 24 hr. and 3 days were necessarily the best intervals of time, and more salmonellae may have been isolated had further subcultures been made earlier or even later than 3 days. Similar results were obtained by Nottingham and Urselmann (1961) who carried out a survey of salmonella infection in calves and other animals. They found that the use of two liquid enrichment media and two plating media as well as plating after 24 and 72 hours increased the chances of detecting small numbers of salmonellae.

The effect of temperature on the incubation of the liquid medium was investigated in a small number of instances only, when duplicate samples were incubated at 37° and 43°C. The results favored the 37°C. incubation temperature and it was more convenient to use this temperature. However, Harvey and Thomson (1953) stated that 43° C. was the best temperature of incubation reliable for liquid enrichment media. Harvey (1956) chose a simple brilliant green MacConkey agar partly because of its selectivity and partly because it was consistent whereas more modern complex media have a higher batch variability. Taylor, Silliker and Andrews (1958) recommended brilliant green agar, and Gassner's medium (Gassner, 1917), water blue metachrome yellow agar, and brilliant green phenol red agar are still used by some workers (Report, 1960a). In this laboratory none of these media seemed to be sufficiently selective, nor was it easy to differentiate the salmonella colonies. It is probable, however, that incubation of the liquid cultures at 43°C. may suppress the growth of some of the Gram-negative flora so that the less selective agar media can be used effectively.

The dilution of food in the liquid enrichment medium has been debated and in this laboratory a dilution of approximately 1 in 5 was chosen. Pre-incubation of the food in lactose broth before sub-culture to more selective liquid was suggested by North (1961) and Taylor (1961); Silliker and Taylor (1958) said that centrifugation to remove soluble food substances improved the efficiency of liquid enrichment media. Kenner et al. (1953) suggested pre-incubation followed by membrane filtration for water. There are many variations in liquid media such as different formulae for TT and modifications of SF including the addition of brilliant green (Stokes and Osborne, 1955), cystine (North and Bartram, 1953) and sulfapyridine (Osborne and Stokes, 1955). Even the method of sterilization, whether by steam pressure or filtration, is impor-tant. Smith (1959) compared selenite media prepared with var-ious carbohydrates. There are variations in preparations of se-lective agar media; for example, of the different Wilson and Blair powders tested, Difco has always given the best results in this laboratory. Also the storage of this medium at 4^0C. overnight improves its efficiency.

For the fermentation of suspected salmonella colonies, we have chosen the Gillies two-tube method, but there are other combinations of sugars in single tubes (Taylor and Silliker, 1958; Papadakis, 1960) which may be as good or better.

The significance of the findings in terms of outbreaks and sporadic cases of salmonellosis is difficult to assess. The devel-opment of phage typing within salmonella serotypes is enabling the epidemiologist to associate sporadic cases or outbreaks oc-curring in widely separated areas with food sources known to be contaminated with particular phage types of a known serotype (Anderson, Galbraith and Taylor, 1961). A knowledge of the vari-ous serotypes and phage types occurring consistently in certain foods may lead investigators to a particular food source, so that an immediate examination of as many samples as possible may be undertaken.

The relevance of numbers of salmonellae in foodstuffs may be debated. Montford and Thatcher (1961) have published a compar-ative study on the isolation of salmonellae from egg products. Using six enrichment broths and five selective media, their rec-ommended method for the isolation of salmonellae from foods was shown to be effective for 0.15 salmonellae per gram, even in the presence of large numbers of coliform bacilli and other bac-terial contamination. We consider the isolation of any number of any serotype of salmonellae from a foodstuff to be significant and indicative of pollution, so that the amount of foodstuff examined is governed only by the limitation of laboratory facilities. When

50 grams of food are examined, the minimum number of salmonellae will be 0.02 per gram, assuming that the method used is
sufficiently sensitive to select a single salmonella per 50 gram
of sample. If it is desired to know the number of salmonellae
present in a foodstuff then separate investigations must be carried out. Taylor (1961) suggested that a modified MPN count
should be conducted routinely, but this seems impracticable when
large numbers of samples are being examined daily. We prefer
to regard even 0.02 salmonellae per gram as significant in relation to the danger of cross contamination in food establishments.
Furthermore the distribution of salmonellae within a pack may
be variable and counts from sample to sample may bear little
relation to each other.

Egg products, poultry, coconut and meat for humans and animals are all potential sources of contamination to other foods,
or they may act directly as vehicles of infection. Measures to
prevent the spread of contamination from these foodstuffs are
important not only in bakeries but in other establishments where
food is handled, but it is more important to rid the product of its
contamination before distribution to food manufacturers, stores
and kitchens.

Heat, irradiation, and fumigation are all effective methods of
treatment. At present, heat, applied to egg products and coconut,
is the only treatment accepted as entirely safe. The problem of
meat contaminated with salmonellae remains. The situation may
be improved by an increased knowledge of the sources and spread
of contamination on farms, by the elimination of salmonellae from
animal feeds and by market and abattoir hygiene. In the meanwhile, methods of treatment must be relied on to protect the consumer from contaminated food.

In the Annual Report for Food Poisoning in England and
Wales 1959 (Report, 1960b), the genus Salmonella is divided into
endogenous and exogenous types, and whereas the incidents due
to endogenous types has increased irregularly, those due to exogenous types have increased regularly from year to year. Increases in incidence are: S. newport from 35 in 1954 to 319 in
1959, S. thompson from 83 in 1954 to 207 in 1959, S. heidelberg
from 16 in 1954 to 308 in 1958, S. infantis from 1 in 1954 to 57 in
1959. Incidents of S. saint-paul increased from 25 in 1954 to 95 in
1958 and decreased to 83 in 1959. In 1959, all but 6 of the 39 different types from egg and egg products and all of the 23 serotypes from meat and meat products were isolated from humans.
Of the 63 types from animal feeds and fertilizers, 47 were isolated from human cases.

In 1960 (Report, 1961a), however, there was a general

reduction in the numbers of incidents due to the serotypes quoted, and it is hoped that the greater control exercised over foodstuffs is effective.

SUMMARY AND CONCLUSIONS

Results from 100 samples positive for organisms of the salmonella group of each of the following foods, frozen whole egg, dried whole egg, frozen egg white, dried egg white, desiccated coconut and frozen boneless meat, were analyzed according to the media used.

Twenty to 25 gram quantities of food were incubated in each of two liquid media, the choice between selenite F, tetrathionate and nutrient broth depending on the foodstuff and its usual bacteriological content. The use of two liquid media gave as many as 47 per cent more positive isolations than from a single enrichment medium. Two selective agar media, desoxycholate citrate and Wilson and Blair bismuth sulfite, were used consistently for subcultures after incubation for 1 day at 37^0 C. and again after 3 days incubation, if the results from the first-day subcultures were negative. The largest number of positives found only after subculture on the third day was 28 per cent.

The advantage of two agar media, particularly when one was more selective than the other, was shown both by subcultures from broth, which often yielded a higher percentage of positives on Wilson and Blair agar than on desoxycholate citrate agar, and by subcultures from an inhibitory liquid medium such as SF, when the reverse occurred. The preference of individual workers for a particular agar medium with regard to picking colonies was also important.

Of samples examined in 1961 the proportion positive for salmonellae varied according to the foodstuff and the country of origin, as follows: 2 to 72 per cent for frozen whole egg, 0 to 70 per cent for dried whole egg, 0 to 25 per cent for frozen egg white, 7 to 60 per cent for dried egg white, 0 to 4 per cent for desiccated coconut and 4 to 61 per cent for frozen boneless meat.

A study of the predominant salmonella serotypes indicated that S. schottmuelleri came not only from China in egg products, but also from Ceylon in coconut and the Argentine in frozen boneless horsemeat. S. typhimurium was imported from a variety of countries in egg products, meat and coconut; certain serotypes were found more often in foods from one country than another. For example, S. bareilly was associated with Holland and S. choleraesuis with Yugoslavia. Similarly, particular phage types of

S. schottmuelleri and S. typhimurium were associated with different countries and foods.

The importance of obtaining detailed information about serotypes and phage types is stressed in relation to epidemiological investigations to trace the source of infection in outbreaks and sporadic cases of salmonellosis. Sporadic cases caused by a particular foodstuff distributed over a large area may be linked together when the serotype, or better still the more detailed phage type, is known.

Methods of heat, irradiation and fumigation treatments to rid foodstuffs of salmonellae, either during production or immediately before distribution, are reviewed briefly.

ACKNOWLEDGMENTS

I am grateful to Dr. Joan Taylor for the identification of salmonella serotypes, Dr. E. S. Anderson for phage typing, Miss M. E. Smith for the practical organization of food examinations on a large scale, and Miss N. Cockman for the analysis of much data.

LITERATURE CITED

Anderson, E. S. 1960. The occurrence of Salmonella paratyphi B in desiccated coconut from Ceylon. Mon. Bul. Minist. Hlth. Lab. Serv. 19, 172-75.

_____. 1962. Salmonella food poisoning. In: Food Poisoning: Symposium. London (Royal Society of Health).

_____, Galbraith, N. S., and Taylor, C. E. D. 1961. An outbreak of human infection due to Salmonella typhimurium phage-type 20a associated with infection in calves. Lancet, i, 854-58.

Banwart, G. J. and Ayres, J. C. 1953. Effect of various enrichment broths and selective agars upon the growth of several species of Salmonella. Appl. Microbiol., 1, 296-301.

Brooks, J. 1962. α-Amylase in whole egg and its sensitivity to pasteurization temperatures. J. Hyg., Camb., 60, 145-51.

Buxton, A. 1957. Public health aspects of salmonellosis in animals. Vet. Rec. 69, 105-9.

_____. 1958. Salmonellosis in animals. Vet. Rec. 70, 1044-49.

Callow, B. R. 1959. A new phage-typing scheme for Salmonella typhi-murium. J. Hyg., Camb., 57, 346-59.

Cameron, A. D. C. S. 1959. A localised outbreak of paratyphoid B fever (phage-type Taunton). Med. Offr. 102, 330-34.

Dixon, J. M. S. and Pooley, F. E. 1961. Salmonellae in a poultry-processing plant. Mon. Bul. Minist. Hlth. Lab. Serv. 20, 30-33.

_____. 1962. Salmonellae in two turkey-processing factories. Mon. Bul. Minist. Hlth. Lab. Serv. 21, 138-41.

Galbraith, N. S., Archer, J. F., and Tee, G. H. 1961. Salmonella saint-paul infection in England and Wales in 1959. J. Hyg., Camb., 59, 133-44.

_____, Hobbs, B. C., Smith, M. E., and Tomlinson, A. J. H. 1960. Salmonellae in desiccated coconut. An Interim Report. Mon. Bul. Minist. Hlth. Lab. Serv. 19, 99-106.

_____, Taylor, C. E. D., Cavanagh, P., Hagan, J. G., and Patton, J. L. 1962. Pet foods and garden fertilizers as source of human salmonellosis. Lancet, i, 372-74.

Galton, M. M., Smith, W. V., McElrath, H. B., and Hardy, A. B. 1954. Salmonella in swine, cattle and the environment of abattoirs. J. Infect. Dis. 95, 236-45.

_____, Mackel, D. C., Lewis, A. L., Haire, W. C., and Hardy, A. V. 1955. Salmonellosis in poultry and poultry-processing plants in Florida. Am. J. Vet. Res. 16, 132-37.

Garside, J. S., Gordon, R. F., and Tucker, J. F. 1960. The emergence of resistant strains of Salmonella typhimurium in the tissues and alimentary tracts of chickens following the feeding of an antibiotic. Rec. Vet. Sci. 1, 184-99.

Gassner, G. 1917. Ein neuer Dreifarbennäkrböden zur Typhus-Ruhr-Diagnose. Zbl. Bakt. (I Abr. Orig.) 80, 219-22.

Gillies, R. R. 1956. An evaluation of two composite media for preliminary identification of shigella and salmonella. J. Clin. Path. 9, 368-71.

Gordon, R. F. 1959. Broiler diseases. Vet. Rec. 71, 994-1003.

Harvey, R. W. S. 1956. Choice of a selective medium for the routine isolation of members of the salmonella group. Mon. Bul. Minist. Hlth. Lab. Serv. 15, 118-24.

_____, and Thomson, S. 1953. Optimum temperature of incubation for isolation of salmonellae. Mon. Bul. Minist. Hlth. Lab. Serv. 12, 149-50.

Heller, C. L., Roberts, B. C., Amos, A. J., Smith, M. E., and Hobbs, B. C. 1962. The pasteurization of liquid whole egg and the evaluation of the baking properties of frozen whole egg. J. Hyg., Camb., 60, 135-43.

Hobbs, B. C. 1961. Public health significance of salmonella carriers in livestock and birds. J. Appl. Bact. 24, 340-52.

_____, King, G. J. G., and Allison, V. D. 1945. Studies on

the isolation of Bact. typhosum and Bact. paratyphosum B. Mon. Bul. Minist. Hlth. Lab. Serv. 4, 40-46.

_____ and Wilson, J. G. 1959. Contamination of wholesale meat supplies with salmonellae and heat-resistant Clostridium welchii. Mon. Bul. Minist. Hlth. Lab. Serv. 18, 198-206.

Hoskin, J. K. 1934. Most probable numbers for evaluation of coli-aerogenes tests by fermentation tube method. Publ. Hlth. Rep. 49, 393-405.

Ingram, M., Rhodes, D. N., and Ley, F. J. 1961. The use of ionizing radiation for the elimination of salmonellae from frozen whole egg. Low Temperature Research Station, Cambridge. Record Memo. No. 365.

Kenner, B. A., Rockwood, S. W., and Kabler, P. W. 1957. Isolation of members of the genus Salmonella by membrane filter procedures. Appl. Microbiol., 5, 305-7.

Kohn, J. 1954. A two-tube technique for the identification of organisms of the enterobacteriaceae group. J. Path. Bact. 67, 286-88.

Leifson, E. 1935. New culture media based on sodium deoxycholate for the isolation of intestinal pathogens and for the enumeration of colon bacilli in milk and water. J. Path. Bact. 40, 581-99.

_____. 1936. New selenite enrichment media for the isolation of typhoid and paratyphoid (salmonella) bacilli. Am. J. Hyg., 24, 423-32.

Leistner, L., Johantges, J., Deibel, R. H., and Niven, C. F., Jr. 1961. The occurrence and significance of salmonellae in meat animals and animal by-product feeds. Proc. XIII res Confr., Am. Meat Inst. Fdn., Chicago. P. 9.

de Loureiro, J. A. 1942. A modification of Wilson and Blair's bismuth sulphite agar (stabilized stock solutions). J. Hyg., Camb., 42, 224-26.

Montford, J. and Thatcher, F. S. 1961. Comparison of four methods of isolating salmonellae from foods, and elaboration of a preferred procedure. J. Food Sci. 26, 510-17.

Murdock, C. R., Crossley, E. L., Robb, J., Smith, M. E., and Hobbs, B. C. 1960. The pasteurization of liquid whole egg. Mon. Bul. Minist. Hlth. Lab. Serv. 19, 134-52.

Newell, K. W., Hobbs, B. C., and Wallace, E. J. G. 1955. Paratyphoid fever associated with Chinese frozen whole egg. Brit. Med. J. ii, 1296-98.

_____, McClarin, R., Murdock, C. R., MacDonald, W. N., and Hutchinson, H. L. 1959. Salmonellosis in Northern Ireland, with special reference to pigs and salmonella-contaminated pig meal. J. Hyg., Camb., 57, 92-105.

North, W. R., Jr. 1961. Lactose-pre-enrichment method for isolation of salmonella from dried egg albumen. Appl. Microbiol. 9, 188-95.

North, W. R. and Bartram, M. T. 1953. The efficiency of selenite broth of different compositions in the isolation of salmonella. Appl. Microbiol. 1, 130-34.

Nottingham, P. M., and Urselmann, A. J. 1961. Salmonella infection in calves and other animals. N.Z. J. Agr. Res. 4, 449-60.

Osborne, W. W. and Stokes, J. L. 1955. A modified selenite brilliant-green medium for the isolation of salmonella from egg products. Appl. Microbiol. 3, 295-99.

Papadakis, J. A. 1960. Dulcitol-sucrose-salicin-iron-urea agar (DSSIU) — a new medium for differential diagnosis of salmonellae. J. Hyg., Camb., 58, 331-36.

Report. 1947. The bacteriology of spray-dried egg with particular reference to food poisoning. Med. Res. Coun. Lond. Spec. Rep. Serv., No. 260.

Report. 1956. The bacteriological examination of water supplies. H.M.S.O. London. No. 71.

Report. 1958. The contamination of egg products with salmonellae, with particular reference to Salm. paratyphi B. Mon. Bul. Minist. Hlth. Lab. Serv. 17, 36-51.

Report. 1959. Salmonella organisms in animal feeding stuffs and fertilizers. Mon. Bul. Minist. Hlth. Lab. Serv. 18, 26-35.

Report. 1960a. Die mikrobiologischen Methoden fur die Untersuchung von Milch und Milcherzeugnissen. Milchwissenschaft, 15, 120-29.

Report. 1960b. Food poisoning in England and Wales, 1959: a report of the Public Health Laboratory Service. Mon. Bul. Minist. Hlth. Lab. Serv. 19, 224-37.

Report. 1961a. Food Poisoning in England and Wales, 1960: a report of the Public Health Laboratory Service. Mon. Bul. Minist. Hlth. Lab. Serv. 20, 160-71.

Report. 1961b. Salmonella organisms in animal feeding stuffs. A Report of a Working Party of the Public Health Laboratory Service. Mon. Bul. Minist. Hlth. Lab. Serv. 20, 73-85.

Report. 1961c. The destruction of salmonella in desiccated coconut by means of steam treatment. The British Food Manufacturing Industries Research Association, Leatherhead. Tec. Cir. No. 193.

Rolfe, V. 1946. A note on the preparation of tetrathionate broth. Mon. Bul. Minist. Hlth. Lab. Serv. 5, 158-59.

Sadler, W. W., Yamamoto, R., Adler, H. E., and Stewart, G. F. 1961. Survey of market poultry for salmonella infection. Appl. Microbiol. 9, 72-76.

Savage, W. G. 1932. Some problems of salmonella food poisoning. J. Prev. Med. 6, 425-51.

_____. 1956. Problems of salmonella food poisoning. Brit. Med. J. ii, 317-23.

Shrimpton, D. H., Monsey, J. B., Hobbs, B. C., and Smith, M. E. 1962. A laboratory determination of the destruction of α-amylase and salmonellae in whole egg by heat pasteurization. J. Hyg., Camb., 60, 153-62.

Silliker, J. H. and Taylor, W. I. 1958. Isolation of salmonellae from food samples. II. The effect of added food samples upon the performance of enrichment broths. Appl. Microbiol. 6, 228-32.

Smith, H. G. 1959. Observations on the isolation of salmonellae from selenite broth. J. Appl. Bact. 22, 116-24.

Stokes, J. L. and Osborne, W. W. 1955. A selenite brilliant green medium for the isolation of salmonella. Appl. Microbiol. 3, 217-20.

Taylor, W. I. 1961. Isolation of salmonellae from food samples. V. Determination of the method of choice for enumeration of salmonella. Appl. Microbiol. 9, 487-90.

_____ and Silliker, J. H. 1958. Isolation of salmonellae from food samples. III. Dulcitol lactose iron agar, a new differential tube medium for confirmation of microorganisms of the genus Salmonella. Appl. Microbiol. 6, 335-38.

_____, and Andrews, H. P. 1958. Isolation of salmonellae from food samples. I. Factors affecting the choice of media for the detection and enumeration of Salmonella. Appl. Microbiol. 6, 189-93.

Thatcher, F. S. and Montford, J. 1962. Egg-products as a source of salmonellae in processed foods. Canad. J. Publ. Hlth. 53, 61-69.

Tulloch, W. J. 1939. Observations concerning bacillary food infection in Dundee during the period 1923-38. J. Hyg., Camb., 39, 324.

U.S. Livestock Sanitary Association, 1961. Proceedings of 65th Annual Meeting.

Wilson, J. G. 1962. Imported foods and the nation. Roy. Soc. Hlth. J., 82, 4-11.

Wilson, E., Paffenbarger, R. S., Foter, M. J., and Lewis, K. H. 1961. Prevalence of salmonellae in meat and poultry products. J. Infect. Dis. 109, 166-71.

14

Parasites in Food

LEON JACOBS
NATIONAL INSTITUTES OF HEALTH

THE VERSATILITY of animal parasites is so great that it is not surprising that many of them have developed life cycles involving their residence in animals or on plants that serve as food for their definitive hosts. Every large class of parasites has some representatives which use such means of development and transport. Some human parasites are cited as examples from each class, regardless of the locales in which they are important while the final part of this discussion is spent focussing on two forms. One of these is well known and has been studied for over 300 years. The other is much younger from the standpoint of human knowledge, although its adaptations indicate it is most likely an ancient resident of many hosts.

CESTODES IN FOOD

The common tapeworms of man are adult forms of the genus Taenia, found as encysted larvae, called bladderworms or cysticerci, in the flesh of cattle and swine. The life cycle of these worms is maintained in areas where there is unsanitary disposal of human feces and access of the domestic animals to feed or water that has become contaminated with the excreta of a human being who harbors the tapeworm.

The Beef Tapeworm

Taenia saginata, the beef tapeworm, lives as an adult in the human intestine. It consists mainly of a broad ribbon of segments, as long as 30 to 40 feet. Each segment, or proglottid, has

male and female reproductive organs. The proglottids are pro-
liferated from an anterior neck section just behind the scolex, or
head, which is equipped with 4 suckers and serves as an attach-
ment organ.

The most posterior segments become gravid sequentially.
Each produces many thousands of eggs. These proglottids become detached and are passed out with the feces. The eggs contain small 6-toothed embryos, or oncospheres, which are already
infective when passed in the feces. They can remain viable in the
ground or in water for as long as 6 months. When the eggs are
ingested by bovines, the oncospheres hatch out under the influence of digestive juices and penetrate the intestinal wall.
Through the lymphatics and blood vessels, they wander to intramuscular connective tissue, where they develop into cysticerci.
A cysticercus consists of a small fluid-filled sac ca 5 x 8 mm
into which a miniature scolex is invaginated. The cysticerci are
widely distributed in the flesh; various authors have reported different groups of muscles as the most heavily infected. They
occur in the muscles of the tongue, neck, jaws, heart, diaphragm,
and esophagus, and within all the large muscles. They remain
viable in the flesh for as long as a year.

When meat containing a cysticercus is ingested by a human
being, the scolex evaginates in the intestine and growth begins.
The development to an adult worm takes 8 to 10 weeks; gravid
proglottids begin to appear in the feces by that time.

The Pork Tapeworm

Taenia solium, the pork tapeworm, has a life cycle similar to
that of T. saginata. The adult worm differs morphologically from
T. saginata in that its scolex is armed with hooks. Its cysticercus also occurs principally in striated muscles throughout the
body, but in addition, the liver, lungs, and other organs of its intermediate host may be invaded. Taenia solium is a much more
dangerous parasite for man, because man can serve as its intermediate host, i.e. as the site for development of the cysticerci.
It sometimes happens that auto-infection with eggs from the adult
worm occurs in man, due to intestinal disturbances that cause
eggs to reach the upper intestine where they can be hatched by
digestive enzymes. Under these circumstances, the larvae penetrate the intestine and develop into cysticerci in subcutaneous
tissues, brain, eye, heart, liver, or lung, as well as in muscle.
Severe symptoms result if the cysticerci localize in vital organs.

Taenia saginata and T. solium are world-wide in distribution.

The importance of the problem in different countries varies, of course, with the extent to which sanitary disposal of human feces is practiced, and with the meat-eating habits of the people. Even within a generally well-sanitated country, poor local conditions can result in considerable infection. Schwartz (1956) stated that in the United States about 16,000 to 27,000 infected beef carcasses were found annually in abattoirs under Federal inspection, where a total slaughter of 12-15 million head of cattle and 5.5-7.5 million calves was performed. However, to point up the importance of local conditions, Schwartz (1938) cited three examples of outbreaks of bovine cysticercosis with high rates of infection due to contamination of the feed and water with human excreta. In other countries, such as Abyssinia and Syria where beef is eaten raw, the prevalence rates of human infection are very high indeed. Stoll (1947) estimated that about 39 million people throughout the world, mostly in Africa, Asia, and the U.S.S.R., were infected with Taenia saginata. T. solium is most prevalent in countries in central and eastern Europe where pork is enjoyed raw. It is also common in China, India, South America, and Africa. In parts of South Africa it is found in 10-15 per cent of pigs. Within the United States, it is rarer than the beef tapeworm. Schwartz (1956) reported only 11 infected swine carcasses in more than 57 million inspected in 1953, and only 4 in 1954.

The cystic stage of these tapeworms is large enough to be discerned easily in meat inspection. The practice in this country is to condemn as unfit for human consumption all carcasses that have numerous bladderworms, i.e. those in which the cysticerci are readily found. If only an occasional cyst is seen, or a dead or degenerated cyst, the carcass may be passed for human food, after cutting away the infected portion and additional treatment. Such carcasses must be thoroughly cooked at a temperature of 60°C. Refrigeration of carcasses at -10°C or below for 10-15 days will destroy the viability of bladderworms. Chilling at 0°C for one to two months is not destructive. According to Ershov (1956) cysticerci die rapidly in concentrated salt solutions. In Russia, light infections in carcasses are handled by soaking the meat in brine for 3 weeks. However, other authors do not regard brine pickling as adequate (LaPage, 1956).

The Broad Fish Tapeworm

One other tapeworm, which man acquires from food, also merits mention here. This is the broad fish tapeworm, Diphyllobothrium latum. Like its taenioid relatives, D. latum is widely distributed throughout the world. It has principal foci in the Baltic region and in French Switzerland in Europe, from which areas it has spread to many other countries on that continent and to the New World. It has been reported from Siberia, Manchuria, Japan, and the Philippine Islands, from Palestine, and from various parts of Africa. It is particularly prevalent in Finland, Turkestan, and Japan.

The adult tapeworm resides in the intestinal tract of man, where it may attain a length up to 10 meters. The proglottids have a uterine pore through which the eggs are shed. The small ciliated embryo within the egg is liberated from the shell within about 10 days after the egg reaches fresh water, and is ingested by small copepods of the genera Cyclops and Diaptomus. In the copepod the coracidium loses its cilia, penetrates the intestinal wall, and grows within the body cavity into a procercoid larva about 0.5 mm long. This procercoid, in turn, is capable of further development within the flesh of fish that consume the copepods. Here it grows to a length of up to 20 mm. When the fish is eaten by certain mammals, the plerocercoid develops into the adult tapeworm.

A number of species of fish serve as the source of infection for man. In the United States, these are the wall-eyed pike (Stizostedeon vitreum) the sandpike (S. canadense-griseum), the pickerel (Esox lucius) and burbot (Lota maculosa). In Europe, similar fish, such as the pike, river perch, burbot, trout, grayling, and whitefish, have been incriminated, and in Japan the Pacific salmon.

In addition to the gastrointestinal complaints arising from the residence of this tapeworm within the gut, human beings — especially in Finland — sometimes suffer a severe pernicious anemia which is initiated by the ability of the tapeworm to absorb large quantities of vitamin B_{12} (Nyberg, 1958). This subject has been reviewed extensively by Birkeland (1932) and von Bonsdorff (1956).

Because of the fact that fish-eating mammals other than man, such as the dog, bear, and other wild carnivores, may be infected with this tapeworm (Vergeer, 1930) proper sanitation by itself cannot be expected to effect control of diphyllobothriasis. The best preventive measure is the proper preparation of fish. Thorough cooking at 50^0-$55\,^0$C for 10 minutes is adequate to kill plerocercoids. Freezing at -10^0 C is stated to destroy the larvae. Cold-smoking or light salting are not adequate treatments. Inspection of various delicacies such as caviar, prepared from fish, should be performed, because plerocercoids are frequently encountered in the roe of infected fish.

TREMATODES IN FOOD

Trematodes, or flukes, have a wide variety of life cycles. They are characterized by enormous proliferation of larvae within some of their intermediate hosts, so that the final larval stages, destined for development to maturity in man or other animals, can under the proper circumstances abound in situations favorable for this transference.

Clonorchis sinensis

As an example of the complicated trematode life cycle, we can select one of the important flukes of man, Clonorchis sinensis. This parasite, which is a most important pathogen in Asia, resides in the biliary ducts and produces extensive pathology in the liver leading to cirrhosis. Its eggs are deposited in the biliary passages, are passed into the intestine, and leave the body in the feces. The eggs contain ciliated larvae, or miracidia, which are liberated when the eggs are ingested by susceptible snails in water. The miracidia develop into sac-like sporocysts within the lymph spaces of the snail, and within these sporocysts numerous new larvae, called rediae, develop. The rediae again wander in the snail, then settle down and produce another generation of larvae called cercariae. These cercariae break out of the tissues and leave the body of the snail. They actively invade fish, penetrating under the scales, and forming ovoid cysts in the flesh. The encysted metacercariae then await the ingestion of the fish by an appropriate mammalian host. The action of digestive enzymes liberates the larva from the cyst and it rapidly enters the biliary tract to develop to maturity.

Some 40 species of fresh-water fish serve as the secondary intermediate hosts. Among these are some species of cyprinoid fishes that are customarily consumed raw in Japan and China. Here again, the parasite has a number of mammalian hosts; so elimination of stream contamination with human feces is not adequate to prevent its dissemination. The dog, cat, hog, and other animals can harbor the adult worms and serve as reservoirs of infection.

Heating at 50°C for 15 minutes is adequate to destroy encysted metacercariae. However, neither drying, salting or marinating, nor refrigeration is sufficient to render fish non-infective. The disease will probably exist for as long as the custom of eating raw fish is maintained.

Other Trematodes

Relatives of <u>Clonorchis</u>, <u>Opisthorchis felineus</u> and <u>O. viverrini</u>, have similar habitats and life histories, and involve similar types of fish. Both species are common in southeast Asia (Sadun, 1955). <u>O. felineus</u> is endemic in central and eastern Europe and in Siberia. The cat is an important reservoir host in East Prussia, (Vogel, 1957) and probably in other localities. Dogs, wild carnivores, and hogs are also hosts. The fact that man is not commonly infected in these areas is due to the differences in food habits as compared with Asia. Nevertheless, human cases do occur in central Europe.

Still other species of flukes, of the family Heterophyidae, arrive in man, where they occupy intestinal locations, by means of infected fish. <u>Heterophyes heterophyes</u> is a small intestinal fluke naturally parasitic in man, cats, dogs, foxes, and probably other fish eaters. Its first intermediate hosts are brackish water snails. Its fish hosts are species of <u>Mugel</u> (millet), <u>Tiliapa</u>, and <u>Acanthogobius</u>. The importance of this fluke was enhanced by the findings of Africa, de Leon, and Garcia (1940) that, although they cause only slight pathology in the gut, they do liberate eggs in sites from which they are carried by the blood to the heart, brain, or other organs where serious inflammatory reactions result.

Other species of flukes have crustacea as their secondary intermediate hosts. Most important among these is <u>Paragonimus westermani</u>, the lung fluke of man. After development of the first larval stages in snails, the cercariae shed from the snails penetrate through the articulations in the skeleton of various crustacea and encyst in the muscles. Edible fresh-water crabs, such as <u>Eirocheir japonicus</u> and <u>Potamon</u> species, and various crayfish of the genera <u>Astacus</u> and <u>Cambarus</u> serve as the secondary hosts.

When the crustacean is eaten by a mammalian host, the metacercariae excyst, penetrate through the intestinal wall into the abdominal cavity, and then make their way through the diaphragm into the pleural cavity, and thence to the lungs. The oriental custom of eating crustaceans raw in brine or wine favors the acquisition of the parasite by human beings. While the crabs succumb to this treatment, the metacercariae can survive in salted crabs for up to 24 hours, and in vinegar for over 4 hours.

<u>P. westermani</u> has a wide distribution, in Asia, Africa, some South Pacific areas, and North and South America. It occurs in man, the cat, dog, goat, hog, muskrat, opossum, weasel, and wolf. Some doubt exists as to whether or not these hosts all harbor the same species of <u>Paragonimus</u>. Recently, in Japan, some new species of this genus were described, and the careful distinction

of species may help in gaining a better appreciation of the epidemiology of this helminthiasis (Miyazaki, 1959).

Before leaving the flukes, it is worth while to mention two
other parasites which are important in some areas. These are
Fasciola hepatica, the liver fluke, and Fasciolopsis buski, the
large intestinal fluke. The cercariae of these species, instead of
encysting within the flesh of secondary animal hosts, attach themselves to the integument of water plants. Fasciola hepatica is
principally a parasite of sheep and cattle, but many cases have
occurred in human beings, in such widely separated places as
Cuba, France, Chile, and Russia, due to the consumption of water
cress from contaminated areas. The chief plants from which
fasciolopsiasis is derived are the water caltrop, water chestnut,
and water hyacinth. In the Orient, the pods of the caltrop and the
bulbs of the water chestnut are eaten in the fresh state during
certain seasons and are then dangerous. The dried plants are innocuous because the metacercariae do not withstand desiccation.

While it may be argued that these are not truly food-borne infections, but are merely happenstance contaminants on water
plants, their habit of encystation on edible vegetation places them
in a different category from the usual contaminants found in polluted waters, and which no attempt has been made to describe.
The metacercariae may be found detached from plants, but their
adaptation to plants is a fundamentally important factor in the
epidemiology of the human disease.

NEMATODES IN FOODS

The class Nematoda contains some food-borne representatives that are occasional parasites of man and one that is of considerable public health importance. Among the less well-known
forms are two that have only recently been described as pathogenic for human beings, Eustoma rotundatum and Angiostrongylus
cantonensis. The other worm we shall discuss is the one already
mentioned in my introduction as known for 300 years, Trichinella
spiralis.

Eustoma rotundatum

Van Thiel et al. (1960) have described, in a fascinating report,
the discovery of the relation of Eustoma rotundatum to a syndrome of very severe abdominal pain and fever in otherwise
healthy individuals. The attack was so severe that in 9 of 10

cases laparotomy had to be performed and solid, infiltrated portions of the small intestine resected. Within these intestinal sections, a nematode larva was found which was eventually identified as <u>Eustoma rotundatum</u>. The adult form of this worm lives in the intestine of sharks and rays; the larva occurs in cod, haddock, herring, and mackeral. The exact life cycle is not known.

All of the cases studied in Holland occurred in the Schiedam-Rotterdam district, where the custom of eating raw or "green" herring is well established, and all could be related to herring because of the time of the year in which they occurred. In the herring, <u>Eustoma</u> larvae lie coiled in the peritoneal cavity. It was found that the worms have a tendency, after the fish are killed, to bore into the abdominal musculature. Since the practice of icing the herring catch and deferring evisceration and salting until docking, is recent, the observation of human cases of infection only since 1955 coincides with this change in the method of handling the fish. When the fish were eviscerated and salted immediately after catching, the worms, which are numbed by salt, had little chance of becoming embedded in the flesh.

While heating is rapidly lethal to <u>Eustoma</u> larvae, 55°C for 10 seconds being adequate to kill them, the worms can withstand saturated salt solution for 1 3/4 hours, and the low concentrations of salt used in the very slightly salted Dutch herring have no harmful effect on them. The larvae are also resistant to acids and can stand acetic acid-salt mixtures usually used in pickled herring in Holland. Also, cold-smoked herring, during which process the temperature does not exceed 40°C, can serve as a source of infection.

As is true with many parasitic worms, there is a great tendency for migration to occur in an abnormal host. When ingested larvae arrive in the human intestinal tract, therefore, they bore into the tissues and thus produce the severe symptomatology of this disease. The exact methods of pathogenesis are still little understood.

The disease caused by <u>Eustoma</u> is certainly not of great statistical importance at present. Nevertheless, it is extremely interesting that its occurrence in man seems to coincide with an altered practice in fish-handling. It is likely that more cases will be observed in the Netherlands and in other countries where raw or lightly salted or pickled herring is popular. Herring is such an important food fish that it is well to point out this particular hazard connected with it.

Angiostrongylus cantonensis

The history of <u>Angiostrongylus cantonensis</u> is not so clear on a number of points. This worm was found by Rosen <u>et al</u>. (1962), in the brain of a patient with eosinophilic meningo-encephalitis who died in Hawaii in 1960. In another case of similar disease that occurred at the same time in Hawaii, the pathologic lesions and foreign material seen in the brain at autopsy suggested that this was another case of the same type.

The worms found in the brain of the first case were young adult specimens. <u>Angiostrongylus cantonensis</u> is a parasite normally found in the rat, where it dwells in the pulmonary arteries after maturing in the brain and meninges. Thus the forms found in Rosen's case could have been derived from the same source or a source similar to that from which the rat acquires infection. The intermediate hosts of <u>A. cantonensis</u> are terrestrial molluscs, which pick up larvae passed in the feces of the rat. It is impossible to say definitely how the two cases discovered by Rosen acquired their infections, because the poor mental state of the victims precluded questioning. These cases stand by themselves, except for a report by Horio and Alicata (1961). These authors reported on a Japanese in Hawaii who was found to have an eosinophilic meningitis after he reportedly ate two live slugs. The incubation period between the ingestion of the slugs and the appearance of his complaints is shorter than might be expected. If it were not for a more general problem of eosinophilic meningitis in other areas, which appear to be associated with the consumption of raw fish (Rosen <u>et al</u>., 1962) this would appear to be too freakish a situation to merit much attention.

Sporadic cases of eosinophilic meningitis of unknown etiology have been reported from many parts of the world. However, it is only in four areas of the Pacific, Ponape, New Caledonia, Saipan, and Tahiti, that outbreaks of the disease, involving many people, have occurred. Rosen's extensive epidemiologic data on the outbreak in Tahiti pointed to the consumption of fresh pelagic fish, as the most likely source of infection. As a result of his diligent searching for a chance to identify the cause, he was able, in Hawaii, to discover the two old cases of a similar disease, resting on the pathologists' shelves, from one of which <u>Angiostrongylus</u> was identified. The correlation with eating raw fish does not correspond with the known life cycle of <u>A. cantonensis</u>, and the cause of the Tahitian outbreak is still unknown. It is possible, however, since the worm larvae may also exist in fresh-water shrimp, that a juice from shrimp which is used as a sauce, commonly eaten with raw fish in Tahiti, is a source of human infection.

The entire story is certainly not clear. Eosinophilia is associated with many helminthic infections and it is because of the
wandering within the brain that characterizes A. cantonensis larvae that suspicion points to this particular worm. If the nematode is able to develop or survive in many different intermediate
hosts, it may indeed prove to be a relatively important parasite
in food. It is not possible to relate sporadic cases of eosinophilic
meningitis outside the Pacific area to this parasite. Considering
the wide variety of nematode parasites related to A. cantonensis,
other forms from fish or other food may be involved.

Trichinella spiralis

The causative agent of trichinosis is a most important parasite in food, responsible every year for considerable morbidity
and mortality. T. spiralis exists in all areas of the world except
Asia, Puerto Rico, the South Pacific islands and Australia. It is
especially common in Europe and North America, less frequently
encountered in Africa and South America. It is associated most
frequently with populations that eat raw or insufficiently cooked
pork.

The life cycle of Trichinella spiralis is relatively simple.
Both the larval and adult stages are passed in the same host.
The larvae encyst in muscle and when the flesh is eaten by another host, adults develop in the intestine, discharge larvae into
the blood, and these larvae again settle in striated muscle and
become encysted as infective forms. This cycle is maintained in
carnivorous and omnivorous animals and, in cold regions where
scavenging by ordinarily herbivorous animals occurs, also in
some species, like the hare, not usually found infected in temperate areas (Rausch et al., 1956). A considerable number of
hosts are involved, including man, swine, dog, cat, rodents, bear,
and other carnivora. The cycle of most importance to man is
maintained by swine-to-swine transfer through infected pork
scraps in garbage.

The wisespread occurrence of infection with T. spiralis in
human beings in the United States was reported by Wright, Kerr
and Jacobs (1943) summarizing the results of necropsy examinations, done by various workers, of almost 12,000 individuals who
died of many causes. Of all these people, 16.2 per cent harbored
Trichinella larvae in their muscles. While the infection rate was
fairly uniform throughout the country, two general areas of
higher incidence of clinical trichinosis have been noted: these
are the northeastern and the Pacific states (Wright, 1939a).

The areas of highest incidence are those in which the feeding of uncooked garbage to swine has been, at least until recently, extensively practiced by many municipalities. Garbage-fed swine raised in northeastern states have been found, in various surveys conducted between 1933 and 1952, to have prevalence rates of from 6 to 11 per cent (Schwartz, 1956). It is also likely that these particular areas have larger numbers of population groups who enjoy raw pork in sausages and other delicacies.

Grain-fed hogs in the Midwest, hogs fed on cooked garbage, and swine reared in the South on pasture, have prevalence rates of trichinosis ranging from 1.5 to 0.5 per cent (Wright, 1939b). It is apparent that so-called grain-fed swine have probably received some garbage scraps or swill from the farm kitchen. This is the practice in many countries, where a variety of vegetable feeds is supplemented by garbage. In Poland, for instance, cooked potatoes are the main feed for swine, but the utilization of garbage supplements is widely done in accordance with the "waste not - want not" philosophy of frugal farmers.

The feeding of uncooked garbage to swine has been reduced in most of the United States by laws enacted during 1953-1955. [It is noteworthy that this was accomplished not as the result of the repeated recommendations of public health workers (Wright, 1939a, b) to effect a diminution in cases of human trichinosis, but primarily to eradicate a nationwide outbreak of vesicular exanthema.] Nevertheless, there is still danger, although diminished, in raw pork. A study by Zimmerman, Schwarte and Biester (1961) conducted at Ames between 1953 and 1960 revealed Trichinella in 1 per cent of about 8400 fresh bulk sausage samples, 2.4 per cent of 1400 fresh link samples, and 0.2 per cent of 861 processed link samples. This was a marked decrease from their previous findings (1956) of up to 12.4 per cent in sausage samples. They attributed the decrease to the garbage feeding regulations, better swine management, widespread use of home freezers, and decrease in home processing of meat.

The lowered prevalence figures in pork found by Zimmerman, Schwarte and Biester are certainly encouraging. However, trichinosis remains a threat as long as some pork samples are parasitized. It is necessary not only to pass laws on garbage-cooking but to enforce them, and this is not always accomplished. There are still plants that feed uncooked or improperly cooked garbage, and swine in some northeastern areas still show high prevalence rates of trichinosis. Furthermore, the worm exists in nature in many feral hosts, from which it can on occasion reach our food animals. Vigilance in meat inspection and processing is therefore necessary. Trichinosis is a serious, painful, and sometimes fatal disease in man, and merits such vigilance.

Meat inspection for trichinosis is not practiced in the United States. Microscopic examination of all pork for T. spiralis is practiced in Germany, Poland, Czechoslovakia, and in various other countries in central Europe. However, such inspection does not detect all cases of infection and is economically unfeasible in our rapid packing-house operations. Indeed, reports of epidemics in Poland (Kozar, 1961) following the consumption of microscopically inspected pork, and of technical difficulties in inspection and identification of carcasses (Cironeanu, 1960; Merkushew, 1960) indicate that microscopic inspection allows breakthroughs and presents problems even in the areas where it is relied on and where the meat-processing techniques are less rapid.

Processing of pork products that are prepared for the use of the consumer without cooking is required by the Federal meat inspection service in the United States. Such foods as cooked hams, frankfurters, salami and smoked sausages, luncheon meats, etc. must be processed by special heating, curing, or refrigeration. Heating must be done in such a manner as to obtain a temperature of 137^0F (58.3^0C) in the center of the product; in actual practice this is exceeded for some products such as frankfurters. Freezing times and temperatures depend on the thickness of the pork or pork products, ranging from 20, 10, or 6 days at 5^0F (-15^0C), -10^0F (-23.3^0C), or -20^0F (-28.9^0C) respectively for pieces 6 inches or less in thickness, to 30, 20, and 12 days at the same respective temperatures for pieces 6 to 27 inches thick (Wright, 1954). The efficacy of these procedures has been attested to by a number of surveys of processed products (Harrington, Spindler, and Hill, 1950). However, at least 30 per cent of ready-to-eat pork products consumed in the United States come from plants that are not federally inspected, where the processing requirements may neither be as rigid nor as rigidly enforced. And, of course, there is no assurance whatsoever that fresh pork passed as safe for human consumption by any State or Federal inspection service is free of Trichinella spiralis.

There have been a number of studies on the effect of irradiation on Trichinella spiralis. Alicata and Burr (1949) found that doses of 12,000 r sterilized from 60 to 100 per cent of the adult females developing from larvae in meat irradiated with cobalt 60. Alicata (1951) reported that X-irradiation of small meat samples at a dosage of about 10,000 r rendered 100 per cent of larvae sterile. Only a few larvae irradiated at this and higher dosages of 15,000 or 20,000 r were able to reach maturity in the intestine when fed to rats, and they were all sterile. Gomberg and Gould (1953) found that a dose of 12,800 r applied to encysted larvae in rat muscle, using a cobalt 60 source, was adequate to sterilize

99 per cent of adult females maturing from the larvae when the flesh was fed to rats. Complete sterilization was reached by a dose of 15,000 r. Maturation of larvae was reduced to less than 1 per cent when the trichinous muscle was given a radiation dose of 18,000 r. Thus, the irradiation necessary to remove the danger of trichinosis in pork is very small in relation to dosages necessary to preserve foods.

In addition to the domestic pig, the bear and wild boar are relatively common sources of human trichinosis. Meat from these animals is eaten in many locales, and in a few places especially in the Orient, dog meat which may be heavily infected is also eaten. It is patently impossible to eliminate such sources of human disease, except by health education measures.

PROTOZOA IN FOOD

An organism that is relatively new to human knowledge appears to be gaining importance as a pathogen that may be acquired in food. This is Toxoplasma gondii, an intracellular protozoan of very cosmopolitan distribution, both geographically and in regard to its hosts. Since the time of its original discovery, in 1908, in a small North African rodent, T. gondii has been found in a wide range of hosts, including representatives of every order of mammals, some birds, and possibly reptiles. In the last 25 years, it has been found to cause a variety of severe forms of human disease, such as hydrocephalus and blindness of newborn due to intra-uterine infection, lymphadenopathy and encephalitis in juveniles and adults, and chronic chorioretinitis in adults.

The biology of this organism will not be covered in detail. It is sufficient for this discussion to mention the two forms in which it is now known. In early acute infections, Toxoplasma propagates within many different types of cells as a vegetative or trophozoite form. This is very delicate and it is not efficient in oral transmission of the infection. In later stages and in chronic infections, Toxoplasma is found encysted in the brain and other tissues, notably skeletal muscle. The mature cysts consist of many parasites packed together within a resilient cyst wall. The enclosed parasites show an important difference from the trophozoites; they are able to withstand at least 2 hours of digestion by pepsin-HCl and over 6 hours of exposure to trypsin. This makes them capable of being transmitted through the digestive tract. Indeed, experiments on the feeding of chronically infected tissues are much more successful than those using tissues containing only the vegetative forms.

Various workers on toxoplasmosis have hypothesized that infected meat is a source of human toxoplasmosis. The assumulation of reports of this infection in swine, cattle, and sheep has lent support to these hypotheses (e.g., Farrell et al., 1952; Sanger et al., 1953; Roever-Bonnet, 1957). Weinman and Chandler (1954, 1956) suggested that toxoplasmosis was maintained in swine in a manner similar to trichinosis, and reported finding evidence of the presence of the parasite more frequently in swine fed uncooked garbage than in swine fed cooked garbage. Jacobs (1957) pointed out the importance of demonstrating the encysted stage of the parasite in the flesh of domestic animals; this has been done in several surveys in the writer's laboratory by means of mouse inoculation following a digestion technique which allows sampling larger amounts of meat than would otherwise be possible (Jacobs and Melton, 1957). A survey of 50 pork diaphragms from a Baltimore slaughter house revealed 24% positive for Toxoplasma (Jacobs, Remington and Melton, 1960). It is not believed that toxoplasmosis in swine and human beings is as intimately associated with trichinosis as Weinman and Chandler hypothesize, because the epidemiology of the two infections is considerably different. This arises from the fact that toxoplasmosis occurs in sheep as well as in swine, and possibly in cattle.

Most of the studies on the occurrence of toxoplasmosis in animals have been serological surveys accompanied by examination of the brains of animals. The validity of these observations is attested to by a detailed study of toxoplasmosis in sheep in New Zealand. This study showed that when the serological test for toxoplasmosis was positive at a moderate titer, 1:256, or higher, the existence of Toxoplasma in the animals could be confirmed in almost 90 per cent. When the dye test was positive at somewhat lower titers, parasites were demonstrable in 50 per cent (Jacobs, 1961). It is noteworthy, also, that diaphragm and psoas muscle were found positive more frequently than brain. All in all, then, the demonstration of antibodies for toxoplasmosis and organisms in the brain of meat animals in many countries indicates the extent to which pork and mutton are infected.

The encysted parasites in muscle or brain are scattered and too small, less than 100 microns, to be seen with the unaided eye. Serological or skin tests of animals for inspection purposes are impracticable. The processor must rely on heating, freezing, or other methods, to render meat safe for human consumption. Data on isolated cysts show that Toxoplasma has a thermal death point similar to that for Trichinella larvae. Toxoplasma is also more sensitive to freezing, which destroys the cysts almost immediately (Jacobs et al., 1960). As yet, no data are available on the effects of brine pickling on cysts. Some irradiation studies being done in the writer's laboratory, by Dr. Akio Kobayashi who is a guest worker from Japan, indicate that radiation of Toxoplasma destroys the organisms at about 15,000 r. This is similar to what is required to sterilize Trichinella larvae.

While pork and mutton show a high rate of Toxoplasma infection, thus far the parasite has not been found in beef except in one equivocal instance. This is somewhat surprising, at least in the New Zealand survey (Jacobs, 1961), because cattle and sheep commonly graze on the same paddocks in many areas of that country. Furthermore, there are reports of clinical toxoplasmosis in cattle, and serological surveys have shown antibodies in herds in the United States and other countries (Feldman and Miller, 1956). It may be that the distribution of the parasite is different in beef.

Jacobs and Melton (1957) have demonstrated the potential of some meats as sources of human toxoplasmosis, but this is not adequate to explain the epidemiology of the infection as revealed by serological surveys. Vegetarians show Toxoplasma antibodies also (Jacobs, 1957), and the highest prevalence rates of toxoplasmosis, in serological surveys, occur in places like Tahiti where meat consumption is very low. There must be other sources. At present Jacobs, Melton, and Stanley (1962) are studying the occurrence of Toxoplasma in the eggs of chronically infected hens, because they were successful in isolating the parasite from oviducts and ovaries of healthy birds processed at a Baltimore packing plant. Thus far no infected eggs have been found, but enough samples have not as yet been examined to rule this source out. Chicken meat is rarely positive. Even these additional sources, however, are not sufficient to explain the entire epidemiological picture. It is probable that in addition to meat, sources similar to those from which herbivores may acquire the infection serve also for human beings.

At any rate, a very definite danger exists in certain meats and this danger is not subject to inspection and control by meat inspection agencies. At the present time, the best advice for those who like raw or rare meat is that they freeze and thaw the meat before processing it further. This will take care of toxoplasmosis, but of course it will not protect against all other agents.

CONCLUSION

No attempt has been made to include in this discussion any mention of the parasites which may exist on food as contaminants, such as the eggs or cysts of parasitic worms and protozoa. It would be legitimate to mention these, but they are not quite comparable to the bacterial contaminants in that they do not increase in numbers during food storage, as bacteria may do. The multitude of parasites that are spread by contaminative means would make a discussion of these forms too voluminous. Therefore, this chapter is restricted to those organisms that have stages of their life cycle within the flesh of animals or fish or encysted on edible plants. These are parasites which may be acquired by human beings even though cleanliness is practiced in the distribution and preparation of the products.

This survey of animal parasites in food demonstrates many hazards that exist for large segments of the human population throughout the world. The customs of people are hard to change, and economic factors impinge on many food problems. Therefore, it is likely that, despite present knowledge of these organisms, the infections and diseases produced by parasites in food will be continual problems for years to come. Nevertheless, work to remove these threats to man's well-being should continue, by health education, improvements in sanitation, and improvements in inspection and control of processed meats. Disease due to food is not inevitable.

LITERATURE CITED

Africa, C. M., de Leon, W., and Garcia, E. Y. 1940. Visceral complications in intestinal heterophyidiasis in man. Acta Med. Philippina, Monographic Series 1.

Alicata, J. E. 1951. Effects of Roentgen radiation on Trichinella spiralis. J. Parasitol. 491-501.

_____, and Burr, G. O. 1949. Preliminary observations on the biological effects of radiation on the life cycle of Trichinella spiralis. Science 109:595-96.

Birkeland, I. W. 1932. Bothriocephalus anemia. Diphyllobothrium latum and pernicious anemia. Medicine 11:1-139.

Cironeanu, I. 1960. Considerati asupra trichinellozei si examenului trichinoscopic. Ind. Alimentara 11:338-42. [Abstract in Wiadomosci Parazytologiczne 8(special no.): 149-50.]

Ershov, V. S. 1956. Parasitology and Parasitic Diseases of Livestock. State Publishing House for Agr. Literature, Moscow. Translation 1960 by the Israel Program for Scientific Translations, available from Office of Technical Services, U.S. Dept. of Commerce, Washington, D. C.

Farrell, R. L., Docton, F. L., Chamberlain, D. M., and Cole,

C. R. 1952. Toxoplasmosis I. Toxoplasma isolated from swine. Am. J. Vet. Res. 13:181-85.

Feldman, H. A. and Miller, L. T. 1956. Serological study of toxoplasmosis prevalence. Am. J. Hyg. 64:320-35.

Gomberg, H. J. and Gould, S. E. 1953. Effect of irradiation with cobalt-60 on trichina larvae. Science 118:75-77.

Harrington, R. F., Spindler, L. A., and Hill, C. H. 1950. Freedom from viable trichinae of pork products, prepared to be eaten without cooking under federal meat inspection. Proc. Helm. Soc. Wash. 17:90-91.

Horio, S. R. and Alicata, J. E. 1961. Parasitic meningoencephelitis in Hawaii. Hawaii Med. J. 21:139-40.

Jacobs, L. 1957. The interrelation of toxoplasmosis in swine, cattle, dogs, and man. Public Health Repts. 72:872-82.

_____. 1961. Toxoplasmosis in man and animals. N.Z. Vet. J. 9:85-91.

_____, and Melton, M. L. 1957. A procedure for testing meat samples for Toxoplasma, with preliminary results of a survey of pork and beef samples. J. Parasitol. 43 (Suppl.): 38-39.

_____, Melton, M. L., and Stanley, A. M. 1962. The isolation of Toxoplasma gondii from the ovaries and oviducts of naturally infected hens. J. Parasitol. 48 (Suppl.) (to be published).

_____, Remington, J. S., and Melton, M. L. 1960. The resistance of the encysted form of Toxoplasma gondii. J. Parasitol. 46:11-21.

_____. 1960. A survey of meat samples from swine, cattle, and sheep for the presence of encysted Toxoplasma. J. Parasitol. 46:23-28.

Kozar, Z. 1961. Wolne od wlośni. Med. Weterynar. 17:332-36. [Abstract in Wiadomosci Parazytologiczne 8 (special no.): Proc. Intern. Comm. Trichinellosis) 153-54.]

LaPage, G. 1956. Veterinary Parasitology. Chas. C. Thomas, Springfield, Ill. 964.

Merkushev, A. V. 1960. On trichinelloscopy in pigs. Veterinariya 37:69. [Abstract in Wiadomosci Parazytologiczne 8(special no.): 153.]

Miyazaki, Q. 1959. Four species of Paragonimus occurring in Japan. J. Parasitol. 45 (Suppl.):20.

Nyberg, W. 1958. Uptake and distribution of Co^{60}-labeled vitamin B_{12} by the fish tapeworm, Diphyoolbothrium latum. Exptl. Parasitol. 7:178-90.

Rausch, R., Babero, B. B., Rausch, R. V., and Schiller, E. L. 1956. Studies on the helminth fauna of Alaska XXVII. The occurrence of larvae of Trichinella spiralis in Alaskan mammals. J. Parasital. 42:259-271.

Roever-Bonnet, de H. 1957. The epidemiology of toxoplasmosis. Document. Med. Geograph. 9:17-26.

Rosen, L., Chappell, R., Laqueur, G. L., Wallace G. D., and Weinstein, P. P. 1962. Eosinophilic meningoencephalitis caused by a metastrongylid lung-worm of rats. J. Am. Med. Assoc. 179:620-24.

Sadun, E. 1955. Studies on Opisthorchis viverrini in Thailand. Am. J. Hyg. 62:81-115.

Sanger, V. L., Chamberlain, D. M., Chamberlain, K. W., Cole, C. R., and Farrell, R. L. 1953. Toxoplasmosis V. Isolation of Toxoplasma from cattle. J. Am. Vet. Med. Assoc. 123: 87-91.

Schwartz, B. 1956. Parasites that attack animals and man. In: Animal Diseases, Yearbook of Agriculture, Washington, D. C. pp. 21-28.

_____. 1938. Animal parasites transmissible to man. Sci. Monthly 47:400-10.

Stoll, N. R. 1947. This wormy world. J. Parasitol. 33:1-18.

Van Thiel, P. H., Kuipers, F. C., and Roskam, P. Th. 1960. A nematode parasitic to herring, causing acute abdominal syndromes in man. Acta Leidensis 30:143-62.

Vergeer, T. 1930. Causes underlying increased incidence of broad tapeworm in man in North America. J. Am. Med. Assoc. 95:1579-81.

Vogel, H. 1937. Beobachtungen uber die Lebens-geschichte von Opisthorchis felineus in Ostpreussen. Zentrl. Bakt. I. Abt. orig. 138:250-54.

von Bonsdorff, B. 1956. Diphyllobothrium latum as a cause of pernicious anemia. Exptl. Parasitol. 5:207-30.

Weinman, D. and Chandler, A. H. 1956. Toxoplasmosis in man and swine. An investigation of the possible relationship. J. Am. Med. Assoc. 161:229-32.

_____. 1954. Toxoplasmosis in swine and rodents. Reciprocal oral infection and potential human hazard. Proc. Soc. Exptl. Biol. Med. 87:211-16.

Wright, W. H. 1954. Control of trichinosis by refrigeration of pork. J. Am. Med. Assoc. 155:1394-95.

_____. 1939a. Studies on trichinosis. XI. The epidemiology of Trichinella spiralis infestation and measures indicated for the control of trichinosis. Am. J. Public Health 29:119-27.

_____. 1939b. Studies on trichinosis. IX. The part of the

veterinary profession in the control of human trichinosis. J.
Am. Vet. Med. Assoc. 94 (n. s. 47):601-8.

Wright, W. H., Kerr, K. B., and Jacobs, L. 1943. Studies on trich
inosis XV. Summary of the findings of Trichinella spiralis
in a random sampling and other samplings of the population
of the United States. Public Health Repts. 58:1293-1313.

Zimmerman, W. J., Schwarte, L. H., and Biester, H. E. 1961.
On the occurrence of Trichinella spiralis in pork sausage
available in Iowa (1953-60). J. Parasitol. 47:429-32.

Zimmerman, W. J., Schwarte, L. H., and Biester, H. E. 1956.
Incidence of trichiniasis in swine, pork products, and wild-
life in Iowa. Am. J. Public Health 46:313-19.

15

Harmful and/or Pathogenic Organisms:
Commentary and Discussion

M. T. BARTRAM
FOOD AND DRUG ADMINISTRATION

COMMENTARY

PAPERS PRESENTED in this section have dealt with a different type of food additive from that covered in the preceding portions of the book. These additives may be referred to in the broad sense as microorganisms. We may break these down to those which are intentionally added and which are essential in the production of the specific food and to those which are accidental to its preparation. This latter group can be further reduced to those microorganisms which are unavoidable and to those which are avoidable under sanitary production conditions.

The first group, the intentional additives, need not greatly concern us for these usually have been in use for many years and by their very nature offer protection for, and characterization of, the food. This is not to say that they should be accepted without scrutiny and without regard to the proper safeguards necessary to insure their correct use and continuous safety.

It is the unintentional additives, and most particularly those which are avoidable, which constitute the greatest problem and the greatest hazard and to which we must devote the major attention. This is especially true when we consider the rapid changes occurring in our food technology.

These changes have introduced entirely new problems and created additional responsibility for those concerned with the safety of the food supply. They have resulted, as has been said many times, in food preparation being transferred from the kitchen to the processing area of the manufacturer, removed from the control and supervision of the housewife who now must accept in faith the integrity of the producer and the awareness of the regulatory agency. With this has been introduced a time lag between production and consumption, during which interval the food passes through many hands in transportation, storage, and sale. Here only an alert regulatory authority can offer protection.

The means by which the food control official can most adequately discharge his obligation and fulfill the trust placed upon him has created much discussion and study. It is natural that local authorities in particular should emphasize the need for control in the form of bacterial standards which would hopefully serve to insure adequate sanitary precautions in food production, proper handling during distribution, and eliminate hazards of food poisoning microorganisms.

It is obvious that the selection and adoption of these standards is not an easy matter. First, there is a need to select a group or groups of organisms most useful for the particular product and to decide upon acceptable methods for their detection. These problems are not too difficult of solution but require considerable research and much willingness to compromise differences.

By far the most difficult of the problems is the selection of the actual limits of microorganisms in the many and varied products making up our food supply. It is natural that the control official, on the one hand, wishes to establish low numbers which will offer the greatest safeguard for the consumer and possibly permit latitude in application. On the other hand, the producer desires more lenient levels which can be consistently met and which will furnish him some latitude above that which can normally be achieved. These divergent viewpoints, while obviously regarded with the same goal in mind, seem for the present almost impossible to reconcile.

To this difficulty must also be added the variations introduced by the type of product, the method of preparation (for example the type and degree of cooking and the stage at which it is applied), variation in ingredients, extent of processing steps and handling, and the times and temperatures of holding associated with production. All of these can markedly influence the bacteriological spectrum of the finished product and, lacking specific knowledge of the details involved, all too often lead to erroneous interpretation. For this reason, there are many who feel that a more realistic approach is through a program of control based on thorough observation of the factors of production, supported by examination of samples from each step in the process, and coupled with analysis of the finished product. With this approach the producer and the control agency should more readily find a basis for mutual agreement. Often, when industry, by laboratory control, discovers and eliminates trouble spots in production, it finds it can offer a product with bacterial levels previously believed to be unattainable.

By the same procedure, the control agency gains a better understanding of sanitation problems associated with the various production steps and can evaluate potential hazards with greater assurance and develop bacteriological guide lines offering maximal consumer protection.

Unfortunately these bacterial levels, attainable with good commercial production practices, may not offer complete protection from the many types of potential food poisoning microorganisms. To this end we must develop not only better, easier, and more accurate means of identifying these microorganisms but we must develop more complete understanding of the means by which they gain access to our food products and of procedures for their elimination or control.

In this country we have been acutely aware of the problems associated with staphyloccoccal food poisoning and recognized the possibilities, now a reality, that situations could occur where the processing or other conditions resulted in a product free of viable members of the group but with the enterotoxin remaining. This subject will be more fully explored in the following chapters.

A review of yearly summaries of food-borne outbreaks of salmonellosis will convince us that we not only have a problem in this area, but that it is becoming progressively more important. It is obvious that we do not have a complete understanding of all of the factors responsible for the apparent increase in the incidence of outbreaks involving Salmonella but it is doubtful that we can ignore the possibility that the more widespread use of some prepared foods plays a substantial role. Other countries, both abroad and on this continent, have taken steps to eliminate from the market many food products containing Salmonella and we have similar programs contemplated or under way.

Investigations both here and abroad have convinced us that we have much to learn about methods for the isolation of this group from specific food products. Indeed, it is apparent that the test procedures virtually must be tailored to the specific products under investigation.

While it may seem that the multiplicity of foods implicated as agents in Salmonella outbreaks poses an almost insurmountable problem in methodology, it is noteworthy that with some exceptions the greatest incidence involves animals or animal products. At this stage we are doubtlessly lacking in a complete understanding of all control measures that may be indicated. However we should give careful consideration to any steps necessary to eliminate this potential animal reservoir of Salmonella and in so doing, we may hope to reduce the problem in other areas.

As has been stated earlier, we are poorly informed concerning the food-borne potentialities of viral agents. Certainly the occurrences of the past year must convince us that hepatitis should be so regarded. We cannot at present rule out others as having similar possibilities. Without satisfactory methods of isolation and identification we can only hope that application of known principles of sanitation will be effective in minimizing the hazard.

Another and important problem introduced by our advancing technology has been the advent of new processes for food preservation, specifically, the use of radiation, gases, and antibiotics. These demand our most careful consideration and evaluation since they may give rise to new and unforeseen difficulties. Probably the greatest cause for concern, common to some degree to all of these agents, is that they possess selective action. This varies from inherent selectivity in the case of antibiotics to a calculated application, in the case of radiation, designed to destroy specific spoilage producing forms. Such action then requires exact methods of examination for the particular organisms of interest. With the gases, such as ethylene oxide, selective action is not presently apparent and our concern here is primarily with its effect on specific food components.

It has been pointed out earlier that authorization for the use of antibiotics in this country has been granted in only two instances, for poultry and whole fish. In our consideration of these and other applications, there has been concern that the use of antibiotics could give rise to bacterial flora not normally encountered in the food product or that resistant forms of the organisms, for which control was sought, would develop. There is some evidence for and against these changes in poultry. In any event, this use of antibiotics has not gained wide acceptance and it is also clear that its use in ice for preservation of fish is very limited at present.

In some work conducted in our laboratories and soon to be published it was found that when chlortetracycline was applied as a dip to fish fillets the growth of bacteria and onset of odors of decomposition were retarded. However the chemical indices of decomposition, the volatile acids and bases, increased as rapidly, and in a few instances more rapidly than in the untreated fillets. These findings appear to introduce a new problem into this field. Recent reports of the increased incidence of antibiotic resistant strains of Salmonella raise the question as to whether a situation is developing analogous to that involving Staphylococcus.

The presence of parasites in the flesh of meat animals and fish is undoubtedly a problem of considerable magnitude and one that with some exceptions has not received deserved attention as to significance and control. This may be due in part to the fact that in most cases there appears to be little if any health hazard and means of control are presently unknown.

Intestinal parasites are to be numbered among the exceptions to this statement and, as has been said in this section, they merit greater attention because of the increase in distribution of uncooked foods from areas known to be rather heavily infected. We may have been fortunate in that cases attributed to fruits and vegetables, consumed without cooking, have not occurred. However, this possibility is being considered by health-minded individuals who have traveled in countries where infection occurs and where sanitation is poor. Research is needed to evaluate the safeguards offered by washing and to determine more effective means of eliminating these agents.

It also appears that the consumer is becoming more conscious and observant of the "worms" in many species of fish. In some instances these may be eliminated or reduced by avoiding areas of high infestation; in others the more obvious parasites can be removed by inspection and trimming. With fish, means of control at the source are not known at present, but it is reasonable to expect that, as is true of trichinosis in pork, control in animal flesh may be possible although not easily accomplished.

I can only repeat for emphasis that, in all too many cases, our knowledge of the potential hazards and means of control of microorganisms in foods is presently far too limited.

DISCUSSION

Question: Has any work been done on propylene oxide for gaseous sterilization?

Bartram: Work has been done, but there has been no approval.

Question: Does the problem of residual compounds extend to glycols?

Bartram: No, we don't think there is a problem with glycols.

Question: Dr. Hobbs, how many organisms did you encounter in your cases with Clostridium perfringens and Salmonella; how heavy was the contamination?

Hobbs: Counts are done on foods from outbreaks, but only on occasional samples sent for routine investigation for salmonellae. Results from coconut, whole egg and egg white vary greatly from batch to batch and from country to country.

Question: With Cl. perfringens — you had outbreaks, how many spores were necessary to cause an outbreak?

Hobbs: Up in the millions for vegetative forms — not spores. We cannot count spores in cooked foods because they have germinated; in cooked meat it is unusual to find any spores at all.

Question: With respect to technique, have you compared the two-tube Gillies' method with dulcitol lysine?

Hobbs: We have tried several single-tube fermentation methods but feel happier with the Gillies' technique.

Mossel: In lieu of Gillies' medium e.c. we use four tubes, lactose, Kligler's, urea, and lysine, but it's hard to get a group to change its technique. The French recently have done some good work on short methods for detection of Salmonella (Buttiaux and Catsaras, 1961).

El-Bisi: Dr. Mossel, in studying population dynamics, that is growth and metabolism of mixed cultures, what is the status of available methodology and available knowledge?

Mossel: What happens with food is not the same as with pure culture. To study a given food, we need a pre-sterilized substrate and then inoculate with representative known mixed culture.

El-Bisi: Can you give examples and describe the techniques you use for enumeration and identification?

Mossel: We generally use three classical groups — Salmonella (e.g., S. typhimurium), Group D streptococci and staphylococci. Say we have, for example, a total count of 100,000, with 10 salmonellae, 100 streptococci, and 1000 staphylococci. To make up the rest, we choose some nonfermentative Gram-negative rods as an inoculum. We can do this with food or a model system. We apply a given temperature and storage time and then study the changes.
      What do we do for Salmonella? What serotype? Let's take S. typhimurium. For the enumeration of Salmonella we use pour plates of sulfite agar and confirm black colonies.

Hobbs: Do you count surface colonies? You may have other sulfite reducers.

Mossel: True, but we confirm biochemically and serologically (Mossel, 1956a). Obviously, one can use a suitable MPN method.
      For Lancefield Group D streps we use Packer's medium (Packer, 1943) and confirm suspect colonies.
      For the enumeration of S. aureus — we definitely prefer

mannitol salt agar. For <u>Clostridium perfringens</u>, we like to use sulfite agar, but I know that Dr. Hobbs suggested that we should use blood agar in a Fildes jar.

<u>Hobbs:</u> Otherwise you can't tell if you have a food poisoning strain or not.

<u>Mossel:</u> No, we can't differentiate between food poisoning and nonfood poisoning strains of <u>Cl. perfringens</u> just by judging the colonies found in sulfite agar, but subsequent biochemical and thermobacteriological confirmation is not too difficult.

<u>Lewis:</u> We don't use blood agar; our <u>perfringens</u> aren't the same types as yours.

<u>Hobbs:</u> They must be similar! Have you tracked down stool samples? You're not getting sufficient tie-up between stool samples and food in your outbreaks.

<u>Mossel:</u> I think Dr. Angelotti, at the Robert A. Taft Sanitary Engineering Center, confirmed the suitability of our sulfite medium. For bacilli, it is very difficult. For the enumeration of non-fermentative Gram-negative rods we use the method published by Dr. Harold Olsen in the Journal of Dairy Science (1961). This uses Standard Plate Count Agar plus 1 ppm. crystal violet. It is generally a satisfactory method, but does not allow the growth of all <u>Achromobacter</u>.

<u>Question:</u> How about by difference?

<u>Mossel:</u> We do not consider this a reliable method.

<u>Question:</u> We had examined two samples for which the results were reported as having less than 10 salmonellae per gram, one negative. Is less than 10 safe?

<u>Hobbs:</u> There is a tremendous variation between samples and reporting methods by different laboratories. One lab gets positive results and one negative; it's just hit or miss. Even less than 10 is significant, when salmonellae are isolated through enrichment cultures.

<u>Question:</u> Does less than 10 mean zero?

<u>Question:</u> If you had only 4% positive, but some had very high counts, would you reject them?

<u>Hobbs:</u> I would.

<u>Loy:</u> What's wrong with tellurite glycine for staphylococci?

<u>Mossel:</u> I'll refer that question to Dr. Niven.

Niven: We don't know why it works or doesn't work in some peoples' hands. With cured meats, we find it best. I'll defend it to that extent. The medium was designed for use with cured meats; and I'm not familiar with its use for other foods.

Mossel: At the A.S.M. last year, there were two objections to tellurite glycine for cheese and milk: 1) Inhibition of Staphylococcus aureus in primary isolation from food and 2) Black colonies are not invariably S. aureus; the authors found only 14 out of 140 black colonies to be S. aureus.

Niven: In contrast, we found less than 1% of black colonies were coagulase-negative. Our biggest worry is inhibition. Why did Mr. Loy ask the question?

Loy: I asked the question because the medium is speedy and useful; why the objection? I'll subscribe to what Dr. Niven said.

Mossel: Dr. Hobbs has asked if anyone has tested tellurite egg agar.

Niven: One person we know got a bottle of tellurite glycine that was mislabeled. It should be tried on known cultures.

Mossel: We have used tellurite glycine agar prepared in the United States and observed that it was inhibitory to many S. aureus on primary isolation from foods, while it also allowed growth of quite a few S. saprophyticus and micrococci as black colonies (Mossel, 1956b).

Question: Would you clarify the situation regarding fungal toxins?

Hobbs: The story goes that turkeys died from eating ground nut meal traced to a toxin from Aspergillus flavus. The turkeys had lesions in the liver. Toxicity was verified by feeding the toxin back with symptoms showing up. Mutation of molds creates difficulties when isolating metabolic products. Dr. Raymond at the Tropical Products Laboratory in London is working on this.

Question: What kind of lesions were produced?

Hobbs: I don't know — I only recall that the work was reported (Sargeant et al., 1961a, 1961b).

Bartram: We have started work on this in this country.

Comment: There was a paper given at the A.S.M. meeting this year with reference to Aspergillus flavus. In lesions developed in cattle and swine, hepatitis type, the liver was yellow. The difficulty came from moldy corn. Dr. Lindstrom of Davis College, Nashville, Tennessee, described this. Also Fusarium and other molds were mentioned.

Raj: Concerning the problem of staphylococci in frozen foods, we used blood agar, mannitol salt agar, Chapman Stone and tellurite glycine agar. Our medium is called sorbic acid mannitol salt agar. Tellurite glycine was very inhibitory; we got only 30 colonies compared to 100 on sorbic acid mannitol salt agar. We started with pure cultures and mixed coagulase-positive and coagulase-negative on the same medium and got only coagulase-positives on Staph 110 medium with egg yolk.

Oser: The hearings on fish flour have made news — are there any fecal indicators for the intestinal contents of fish in fish flour? Is there a parasite problem?

Jacobs: I would think that there might be some parasites as fecal indicators — this is just a guess.

Tarr: This is very complicated — the flour is finely ground, and it would be hard to identify eggs and bodies.

Comment: Isn't fish flour made by extraction? How could you find fragments?

Tarr: The methods give a finely ground material. Both dry and wet extractions are used.

Comment: But this is all basic protein. How could you know where it came from?

Jacobs: Some structures are chitinous; one may be able to identify these microscopically. They're not all reduced to fish protein.

Oser: There's no objection to parasites found in the flesh. The Food and Drug Administration objects to viscera and visceral contents, so the question is if any specific parasites for intestinal contents could be detected.

Jacobs: I don't know.

Question: What is eosinophile meningitis?

Jacobs: Actually, this is not meningitis, but meningo encephalitis, with a variety of encephalitic syndromes. Symptoms include rigidity of the neck, headache, mental disturbance. You find a large number of leucocytes in the spinal fluid. It has been found in various Pacific islands in epidemic form and has also been reported in other places. I can't give you a complete description of all symptoms and signs.

Question: Will you recover from it?

Jacobs: Yes, in Tahiti they did. With the deaths in Hawaii, there were a number of aspects that were not clear. People had encephalitis, not diagnosed as meningeal, until death. In Tahiti, it was associated with eating raw pelagic fish. This doesn't fit with other cases. In epidemiological studies, A. cantonensis had a life history involving rats, slugs, and fresh water shrimp. It is possible that pelagic fish in Tahiti ingest shrimp containing slugs and pass on the parasites.

Question: What is the period of onset?

Jacobs: The incubation period for larvae to wander through the brain is about 14 to 16 days. There was one case of a Japanese who ate 2 raw slugs on advice of a faddist. He chewed these very carefully. He had a gastrointestinal upset, but ate another. Two days later, he had encephalitis with eosinophiles in spinal fluid. The incubation period was short, but larvae were found in slugs in the garden.

LITERATURE CITED IN DISCUSSION

Buttiaux, R. and Catsaras, M. 1961. Le milieu au tétrathionate additionné de novobiocyne pour l'enrichissement des salmonella des matières fécales. Ann. Inst. Pasteur Lille 12:13-18.

Mossel, D. A. A. 1956. Aufgaben und Durchführung der modernen hygienischbakteriologischen Lebensmittelüberwachung. Wien. tierarztl. Monatsschr. 43:321-44; 596-610.

_____, de Bruin, A. S., van Diepen, H. M. J., Vendrig, C. M. A., and Zoutewelle, G. 1956. The enumeration of anaerobic bacteria, and of Clostridium species in particular, in foods. J. Appl. Bact. 19:142-54.

Olson, H. C. 1961. Persistence of contamination in lactic cultures. J. Dairy Sci. 44:970 (abstract).

Packer, R. A. 1943. The use of sodium azide (NaN₃) and crystal violet in a selective medium for streptococci and Erysipelothrix rhusiopathiae. J. Bact. 46:343-49.

Sargeant, K., O'Kelly, J., Carnaghan, R. B. A., and Allcroft, R. 1961a. The assay of a toxic principle in certain groundnut meals. Vet. Rec. 73:1219-23.

_____, Sheridan, A., O'Kelly, J., and Carnaghan, R. B. A. 1961b. Toxicity associated with certain samples of groundnuts. Nature 192:1096-97.

PART FIVE

Microbial Toxins

16

Anaerobe Toxins

HANS RIEMANN
DANISH MEAT RESEARCH INSTITUTE

BOTULISM

BOTULISM IS CAUSED by exotoxins formed during the growth of <u>Clostridium botulinum</u> in foods. The number of cases of human botulism per annum has been fluctuating but is small compared to many other diseases. However, the mortality rate is high, and strict precautions are therefore taken to prevent botulinum food poisoning. Botulism among animals is much more common and causes severe losses every year.

Geographical Distribution of <u>C. botulinum</u>

<u>Human Botulism</u>. This is most often caused by types A, B, and E of <u>C. botulinum.</u> The first case of botulism seems to have been reported in 1735 (Geiger, 1941), and the organism responsible for botulism was isolated in 1895 (van Ermengen, 1896). Several reviews on botulinus food poisoning have been published, one of the more recent by Meyer (1956). About 1830 outbreaks of human botulism with approximately 5640 cases have been reported. The average mortality per cent is 30, somewhat higher in the U.S.A. than in Europe.

Botulism in the U.S.A. has never exceeded 65 cases with 48 deaths in a single year. The average annual incidence in Germany was constant for several years with 9-15 outbreaks and 30-40 victims, but now seems to be declining. The mortality per cent being 17-30. Botulism was seldom reported in France before World War II, but more than 1000 cases were reported during the war. The incidence in France is similar to that of Germany. Over the past 50 years isolated cases or small group intoxications with a fatality rate of 2-15 per cent have been reported in various countries in Europe and South America. A few outbreaks have also been reported in Japan and India (Meyer, 1956). These reports indicate that C. botulinum has an almost world-wide distribution.

Animal Botulism. Botulism among animals has been given various names: "lamziekte," "spinal paralysis," "limberneck," etc. The various names reflect the difficulty involved in the diagnosis. Animal botulism is most frequently caused by types C and D, although types A and B have also been implicated.

C. botulinum type C toxin is toxic for: horses, cattle, sheep, mink, chickens and ducks. Most cases of "lamziekte" in South Africa are caused by type D but 10-20 per cent of the outbreaks are due to type C. About 100,000 cattle die yearly from "lamziekte." The animals get the toxin by eating decomposing carcasses in which C. botulinum multiply. Similar large outbreaks have occurred among sheep in Australia (Meyer, 1956; Müller, 1961). Large outbreaks of botulism in aquatic wild birds living in certain areas of lakes and mud flats in America, Australia, and South Africa are almost invariably caused by type C.

Large numbers of mink on farms have occasionally been killed by botulinus toxin formed in spoiled animal or fish food. Type A or more frequently type C has been responsible. Spoiled botulinogenic food has caused the death of barnyard fowl in the U.S.A. and Canada. C. botulinum type A and less frequently type B have been responsible. Botulism in animals kept in stables has been reported from various countries. Types B and C have been involved, but the type of food which caused the intoxication has not always been traced (Meyer, 1956; Müller, 1961). The mortality of animal botulism equals that of human botulism.

Spontaneous outbreaks of botulism are very rare among cats, dogs, and pigs, but all but one vertebrate — the American turkey vulture — seem to be affected by the botulism toxin. Massive growth of C. botulinum occurs in organs, e.g. the liver of animals which die from botulism. This may lead to dissemination of large numbers of the organism and it is fortunate that humans are relatively resistant to types C and D (Meyer, 1956) which are the types that most frequently cause animal botulism. The increased risk of human botulism in areas where outbreaks of animal botulism caused by type A or B occur should not be overlooked.

In Soils. By examination of soil, manure, etc. from various countries, C. botulinum type A was found in 18 per cent and type B in 7 per cent of samples of virgin soils. Of cultivated soil samples, 7 per cent contained type A and 6 per cent type B. Six per cent of pasture samples contained type A and 21 per cent type B (Dewberry, 1959). C. botulinum types A and B were detected in

3 per cent of Scottish soil samples (Dewberry, 1959). C. botulinum type A and/or type B was detected in 5-14 per cent of English soil samples (Haines, 1942). C. botulinum type C was found in about 8 per cent of examined soil samples but is presumably more frequent in areas where outbreaks of type C botulism take place (Meyer, 1956).

C. botulinum type E was not recovered in the earlier examinations of soil samples, etc. The reason for this is apparently that the samples in most cases were heated at 80° C to kill contaminating vegetative cells before incubation in nutrient media. Such a heat treatment will kill most C. botulinum E spores present (Prevot and Sillioc, 1958). Furthermore an incubation temperature of about 37°C was used in most investigations, and type E requires a lower temperature, about 30°C, to form appreciable amounts of toxin. Special techniques must therefore be used for the isolation of C. botulinum type E. Mouse passage has been found successful (Pederson, 1955). A preliminary drying of the samples at 37° C is also useful. The drying kills a large number of vegetative cells but apparently does not harm type E spores (Johanssen, 1962).

By applying an appropriate technique it has been possible to detect C. botulinum type E in marine mud samples, fish, etc., in various countries. Some of the most recent investigations in Scandinavian countries have indicated a rather high incidence of C. botulinum type E. Eighty-four per cent of mud samples from the harbor of Copenhagen and 26 per cent of soil samples taken in a city park contained type E organisms (Pederson, 1955). The organism is found in 50 per cent or more of marine mud samples taken close to the coast of Southern Sweden. Soil samples from the area show an equal or even higher incidence of C. botulinum type E. It has been suggested that the presence of the organism in marine samples is due to the fairly large quantities of soil, which are washed out into the sea by rainfall and other precipitation (Johanssen, 1962).

C. botulinum has been isolated from the intestinal tract or liver of diseased or healthy cattle, dogs and fish (Meyer, 1956). No systematic studies seem to have been made to determine the extent of a carrier stage.

The available information indicates that C. botulinum is distributed in soils and marine mud over large areas of the world. Type A appears more frequently in the western states in the U.S.A. while type B is found more frequently in the eastern states and in Europe. Type E may be more common than hitherto suspected. The numbers of C. botulinum present in soils, etc., is presumably small, i.e. less than one per gram, except in areas with outbreaks of animal botulism.

Cultural Characteristics of C. botulinum

C. botulinum occurs as Gram-positive anaerobic rods with subterminal oval spores. It ferments glucose, fructose, maltose and glycerol and liquifies gelatin. There are 6 types — A, B, C, D, E, and F — on the basis of the serological specificity of the neurotoxin produced. Strains vary in ability to hydrolyze proteins more complex than gelatin and also in their ability to ferment carbohydrates (Johanssen, 1961; Smith, 1954).

The Toxins of C. botulinum

Formation of C. botulinum Toxin.   Glucose or maltose has been found essential for optimum toxin formation in trypticase yeast extract cultures of C. botulinum type A. Lysis of the cells which is a mechanism in toxin liberation was not complete in glucose-free media. However, sonic disintegration of cells in non-lysing cultures resulted in only negligible increase in toxicity (Bonventre and Kempe, 1959). Maximum toxin formation of type A toxin in trypticase yeast extract took place between pH 5.5 and 8.0. Autolysis and liberation of the toxin was most rapid between pH 6.5 and 7.0. Alkaline environment resulted in inactivation of the toxin, but did not affect its synthesis. It was noticed in these experiments as well as in others that maximum toxicity was reached only when autolysis was complete (Ohye and Scott, 1957). C. botulinum A produced toxin in a defined medium consisting of 12 amino acids, vitamins, phosphate and magnesium. Most of the toxin was formed after the cessation of growth.

EDTA in concentrations that had no influence on growth suppressed toxin formation in resting cells. Streptomycin, chloramphenicol and chlortetracycline suppressed toxin formation. The requirements for amino acids, glucose, organic phosphorus and magnesium suggested a de novo synthesis in resting suspensions (Kindler, Mager and Grossowicz, 1956a, 1956b). In other experiments (Bonventre and Kempe, 1960) there was no de novo synthesis of toxin in resting cells in a stationary phase, but type A and B toxins may be synthesized during the first 16 hours of growth as large molecules with a comparatively low biological activity due to masking of active chemical groupings. These toxin precursors must be partially degraded, probably by proteolytic enzymes of the organism, before manifesting their full toxic potentialities; maximum toxicity is, therefore, reached only after maximum growth has been obtained.

The Nature of C. botulinum Toxins. The botulinum toxins are heat labile, simple proteins. They are distinguishable from each other by being specifically neutralized by the homologous antibodies. Upon treatment with formaldehyde they lose their toxicity, but retain their antigenicity. They are insoluble in alcohol, ether or chloroform.

Most research work has been done with type A toxin which in crystalline form consists of 19 amino acids. Possible toxiphoric groups are unknown, but splitting of hydrogen bonds and reactions which involve free amino and carboxyl groups cause detoxification (Lamanna, 1959). Purified type A toxin has a molecular weight of 900,000 to 1,000,000 (Heyningen, 1955). Irradiation experiments with crude dilute toxin solutions have indicated a molecular weight of 1.6 million (Skullberg, 1951). Purified preparations of type A toxin agglutinate red blood cells but the agglutinin is separate from the toxic molecule.

The crystalline toxin dissociates into smaller units at pH above 6.5. This takes place without the intervention of proteolytic enzymes. The suggestion has been made that the toxin molecule with a molecular weight of 1,000,000 consists of 12 smaller fragments each with a molecular weight of 70,000. Only one out of every three of the fragments seems to be toxic. Attached to this complex is a small hemagglutinin residue. Crystalline or any highly concentrated or purified toxin has an unusual ability to form relatively insoluble compounds or complexes with normal serum proteins. These serum proteins have no toxin neutralizing effect. The toxicity of type D toxin can be increased 20,000-fold by 0.2 per cent gelatin solution; the mechanism is unknown (Heyningen, 1955; Lamanna, 1959). Type B toxin has the same toxicity as type A but its molecular weight is only 70,000; it may be a mixture of toxic and inert fragments of a larger molecule (Lamanna, 1959).

The Stability of Botulinum Toxins. Type A toxin is not hydrolyzed by papain at pH 4.9. Pepsin causes approximately 40 per cent breakdown and 25 per cent loss in toxicity in 17-18 hours. Trypsin at pH 6.6 causes 90 per cent loss in toxicity in 3 hours, but there is only a slight degradation and this takes place after the toxicity has been diminished (Heyningen, 1955; Lamanna, 1959). During treatment with proteases the loss of toxicity proceeds without initial lag which indicates that the maintenance of structural integrity of the protein molecule is necessary for toxicity.

The toxin is very resistant to acid and not destroyed by 1-3 per cent hydrochloric acid during 24-36 hours at 35^0 C. Alkali exerts a powerful destroying effect upon the toxin (Dewberry, 1959). Incubation of toxin at neutral pH and 37°C for 24 hours decreases its potency by 90-99 per cent.

The heat resistance of botulinum toxin is low. Eighty degrees C inactivates type A toxin in 6-10 minutes, type B toxin in 15 minutes, C toxin in less than 30 minutes (Dewberry, 1959). Type E toxin is destroyed at 60°C in 5 minutes at pH 7.5 and 3.5, 40 minutes being required at pH 4.6 and 4.9 (Bonventre and Kempe, 1959b; Dewberry, 1959). There is no decrease in the potency of toxin during storage at freezing temperatures for several months, but the toxin is gradually destroyed by direct sunlight, diffused daylight and air.

The toxin is fairly resistant to ionizing radiation and is partially protected by the presence of food materials. It has been found that 0.9 megarad is required to inactivate 99.99 per cent of the toxin (Zeissler et al., 1949). Other experiments have indicated that the destruction of types A and B toxins by beta rays is an exponential function of the dose. 0.65 to 1.25 megarads were required for 90 per cent destruction of the toxin at 15°C and pH values of 3.8 to 6.2 (Skullberg, 1951). Ultrasonic treatment has been found to cause a 90 per cent reduction in the toxicity of a comparatively pure type A toxin, but had no influence on type C toxin, which was present in liver broth (Scheibner, 1955).

Various microorganisms, especially protease-producing bacilli, seem to be able to destroy the toxin of C. botulinum (Hall and Peterson, 1923; Ingram and Robinson, 1951; Legroux, Jeramec and Second, 1947).

Mode of Action of Botulinum Toxins.   The death of a mouse can be caused by 10^{-4} microgram botulinal toxin. Only tetanus toxin and shigella neurotoxin have equivalent toxicity (Lamanna, 1959).

Trypsin and chymotrypsin in the small intestine have detoxification capabilities. In spite of this it seems that the major site for systemic absorption of the toxin is the small intestine. The explanation is probably that detoxification is so slow that toxin may be present in feces for several days after ingestion. The toxin molecule which is absorbed from the intestinal tract is too big to be dialyzable. It is not fragmented before absorption neither does the enzymatic lysis of toxin make it diffusible. The toxin is probably passed into the lymph in small amounts from the intestinal tract (Lamanna, 1959). Absorption of small quantities of toxin has been demonstrated in obstructed segments of the small intestine of dogs (Haerem, Dack and Dragstedt, 1938). The slow and incomplete absorption is in agreement with the observation that type A toxin is between 50,000 and 250,000 times more toxic to mice when administered intraperitoneally than when administered orally (Lamanna and Meyers, 1960).

Botulism is a paralysis of the efferent autonomic nervous system, and the cause of death is asphyxia. The central nervous system is not affected and sensory nerves are not harmed. The actual site of action is at the synapses of efferent parasympathetic and somatic motor nerves (the endplates). Only the cholinergic system and not the adrenergic system is affected by the toxin. Apparently the production of acetylcholin is not interfered with. The reaction of toxin with an unknown receptor at the site of action is rapid and not reversible, but it takes at least half an hour before the action of the toxin becomes manifest. In contrast to curare the effect of botulinus toxin is not reversed by choline esterase inhibitors. This indicates that the specific effect of botulinus toxin is interference with the release of acetylcholine, but not with its action (Lamanna, 1959).

Symptoms and Treatment of Botulism

The symptoms produced by all the toxins are similar. In some individuals the symptoms involving the central nervous system which are characteristic of botulism are preceded by an acute digestive disturbance and vomiting. The typical symptoms of botulism usually appear within 12-36 hours. Such disturbances as double vision occur early. Blepharoptosis, mydriasis and loss of reflex to light stimulation are present with the double vision. Nystagmus and vertigo are occasionally recorded. Difficulty in swallowing occurs early and difficulty in speech is observed later. Death is usually due to respiratory failure. The duration of illness in fatal cases is usually from 3 to 6 days.

The treatment is unsatisfactory at best. Antitoxin therapy is of less value once symptoms have appeared, but since botulinum antitoxin is the only known specific therapeutic agent for botulism it should be used even though the disease is advanced. Botulinum toxin is not formed in toxic food after ingestion but damage may continue as the result of a gradual absorption of toxin as long as the latter remains in the intestinal tract.

Toxoid

Toxoids can easily be prepared from toxin containing C. botulinum cultures. The toxoid appears very stable and gives protection lasting 6 months or more (Barnes and Ingram, 1956; Barton

and Reed, 1954; Lamanna, 1959; Prevot and Sillioc, 1958). A
polyvalent (ABCDE) highly purified toxoid for use in human beings is available. The toxoid is generally used only under special
circumstances which involve an increased botulism hazard.

Types of Food Involved in Outbreaks of Botulism

A great variety of low-acid (pH above 4.5) foods have been
found to permit growth and toxin production of C. botulinum;
blood pudding, liver sausage, smoked pork and ham, meatballs
liver paste, salmon, sturgeon, cured herring, smoked salmon,
sardines, mackerel, canned vegetables, olives, spinach.

In Europe most cases of botulism have followed the eating
of preserved meat and fish. In the U.S.A. home canned vegetables
have been responsible for the greatest number of cases. It is a
characteristic that foods which have caused botulism are such
items which have been given a preserving treatment, stored for
some time and then eaten without appropriate heating during
the preparation of the meal. The growth of C. Botulinum in
foods frequently, but not always, gives rise to a foul rancid
odor, which serves as a warning for the consumer. In cured
foods as hams and herrings, in canned string beans and canned
bread as well as in the more acid foods the foul odor may not
be sufficiently pronounced to serve as a warning (Dack, 1956).
Toxin formation without appreciable spoilage may take place in
cured meats with salt contents between 7 and 10% (g NaCl/100
ml H_2O).

Control of C. botulinum Food Poisoning

The Resistance of C. botulinum Spores. The heat resistance
of C. botulinum spores varies, but in general 300 minutes at
100°C, 120 minutes at 105°C and 10 minutes at 120°C can be depended upon to kill most suspensions of C. botulinum spores
(Smith, 1954).

The resistance expressed as the number of minutes required
to kill 90 per cent of the spores present, the decimal reduction
time, is of the order 0.96 minutes at 240°F (151°C) and 3.3 minutes at 230°F (145°C). For C. botulinum B decimal reduction
times of 1.07 minutes at 240°F (151°C) and 2.93 minutes at 230°F
(145° C) have been reported (Knock and Lambrechts, 1956). This
heat resistance was found for spores produced in heat sterilized
laboratory media and heated in phosphate buffer. There are

indications that spores which are produced in media containing raw animal tissue have a lower heat resistance (Vinton, Martin and Gross, 1947). The addition of native serum albumin to heat sterilized sporulation media may also lower the heat resistance (Sugiyama, 1951). There are few published data on the heat resistance of type C spores but they seem to be less resistant than types A and B (Müller, 1961). C. botulinum type E spores have a very low heat resistance compared to the other types. Decimal reduction time at 100° C is of the order of 2 seconds (Ohye and Scott, 1957), and heat treatment of naturally contaminated mud samples have indicated that only 4-5 per cent of the spores survived 5 minutes at 80°C (Pederson, 1961). The pH influences the heat resistance but it has to be reduced well below 4 before there is any appreciable reduction in the thermal death time.

C. botulinum A and B spores are fairly resistant to ionizing irradiation. A dose of about 0.5 megarads is required to kill 90 per cent of the spores. Preirradiation with 1 megarad reduces the heating time required to kill C. botulinum spores approximately 80 per cent (Kempe, Graikoski and Bonventre, 1957; 1958). The radiation resistance of C. botulinum type E is roughly one third of the resistance of types A and B (Schmidt, 1962).

The resistance of spores of C. botulinum to chemical agents is marked. Five per cent phenol or lysol does not kill in exposure up to 7 days. Ten per cent hydrochloric acid destroys the spores in 1 hour. Commercial formalin diluted with an equal volume of hot water and acting over a period of at least 24 hours has been recommended for the disinfection of material which cannot be treated with hydrochloric acid or heated. Quaternary ammonium compounds are probably rather ineffective (Smith, 1954).

Factors Affecting Growth and Toxin Formation

Temperature. C. botulinum is a mesophilic organism and growth and toxin production is inhibited at low temperatures. Spore inocula of C. botulinum types A and B could initiate growth at 15-42.5°C but not at 12.5 or 45°C. Inocula of actively growing cells could initiate growth at 12.5 or 47.5°C but not 10 or 50°C (Ohye and Scott, 1953).

Vegetative cells, but not spores die out rapidly at temperatures above 50°C (Ohye and Scott, 1953). C. botulinum type A did not produce toxin in defrosted chicken à la king during 5 days' incubation at 10° C, but some samples which were incubated at 30° C became toxic after 2 days incubation (Saleh and Ordal, 1955). In experiments with inoculated meat C. botulinum type A did not produce toxin at 3^0 C during 30 days incubation. At 21^0C toxin was present in pork after 16 days incubation but not in beef after 30 days incubation (Greenberg, et al., 1958).

C. botulinum type E strains differ from types A and B in their temperature requirements. Type E strains have minimum temperatures for growth $8-10^0$ C below the minimum temperatures for types A and B, and the maximum temperature is about 5^0 C less. Three out of 9 type E cultures failed to grow at 5^0C, but the strains which did grow formed toxin. Six out of 9 strains failed to grow at 40^0C. The inocula used in these experiments consisted of spores, and the lag period varied between 9 and 33 hours at $10-15^0$C (Ohye and Scott, 1957). Some type E strains can form toxin slowly at 3^0C (Schmidt, 1962). The toxin production by type E strains is higher at 30^0C than at 37^0C although the multiplication is equally good at the two temperatures. The toxin is unstable at 37^0C, and a considerable decrease in toxicity is found after a few days (Beerens, 1955; Pederson, 1955).

The influence of temperatures on the germination of clostridial spores has not been extensively investigated. It is generally believed that the temperature requirements for germination are close to those for growth, but it has been observed that clostridial spores can germinate at temperatures as low as 4.4^0C in spleen infusion (Mundt, Mayhew and Stewart, 1954).

pH.   Growth of C. botulinum in laboratory media seems to be inhibited around pH 5. The growth is quite rapid between pH 5.5 and 7.0 but is somewhat inhibited at pH 8.5. Maximum toxin formation seems to take place between pH 5.5 and 8.0 (Bonventre and Kempe, 1959b). The pH which limits growth is dependent on other factors present and differs for various foods — from pH 5.7 in banana purée to 4.8 in pineapple rice pudding (Townsend, Yee and Mercer, 1954). The limiting pH for growth of C. botulinum A and B in bread seems to be between 4.8 and 5.0 somewhat dependent on the amount of water present (Ingram and Robinson, 1951a and 1951b; Ingram and Handford, 1957; Kadavy and Dack, 1951; Wagenaar and Dack, 1954). Toxin production has been found to take place in canned foods with pH values as low as 4.0, when yeast was present (Meyer and Gunnison, 1929). This suggests that growth and toxin formation at pH values below 4.5 may require the presence of essential factor(s) which may be provided by the growth of other organisms. It appears safe however to conclude that no growth of C. botulinum will take place at pH values below 4.5 if no other organisms will grow which might interfere with the inhibitory effect of pH.

Germination of botulinum spores seems to be affected by pH in much the same way as vegetative growth. Germination in complex

laboratory media takes place at pH 6.3-7.9 but no germination
could be observed at pH 4.0 (Treadwell, Jann and Salle, 1958). In
experiments with bread inoculated with spores or vegetative cells
no consistent difference between the two cell types with regard
to resistance to pH was demonstrated (Ingram and Robinson,
1951b).

The Effect of the Oxidation-Reduction Potential. C. botulinum does not grow on the surface of media exposed to air. The
atmospheric pressure has to be reduced to 4-8 per cent before
growth will take place in liquid laboratory media (Meyer, 1929).
In most foods it is the oxidation-reduction potential which will determine whether growth takes place or not. The ability of clostridia to grow at a certain oxidation-reduction potential depends,
among other factors, on the size of the inoculum, but in general a
potential close to zero or a negative potential is required. Such
potentials are established in many foods by reducing systems
present, e.g. SH groups in meat, ascorbic acid and enzyme
systems in plants, other microorganisms, etc. Compact masses
of foods will therefore frequently have potentials which are low
enough for C. botulinum to grow (Mossel and Ingram, 1955), and
it seems to make little difference whether the food is vacuum
packed or not (Johanssen, 1961).

The Effect of Salt. The salt concentrations required for
complete inhibition of growth from large inocula of C. botulinum
in laboratory media vary between 5 and 10 per cent, and it has
been recommended that brined foods should contain not less than
10 per cent of common salt (Dewberry, 1943). When heating is
applied and $NaNO_2$ is present salt seems to be more inhibiting towards clostridial spores. It has thus been found that inoculated
cured meat which had been processed to F_0 values as low as 0.3[1]
did not become toxigenic if the NaCl concentration was 5.3 or
above (gram NaCl/100 ml H_2O) (Halvorson, 1955; Jensen, 1954).

The inhibitory effect on C. botulinum spores in such items is
due to the combined effect of heat, NaCl and $NaNO_2$; and a change
in any of these factors may have an influence on the inhibitory effect of the others. A situation similar to that in canned cured
meat seems to exist in cheese. No toxin production in cheese inoculated with C. botulinum types A and B was detected in cheese
spread stored at 90°F. The pH of the cheese spread varied between 5.1 and 6.1 and the NaCl concentration varied from 3.4
to 7.7 (gram NaCl/100 ml H_2O). The spores seemed to germinate during storage as indicated by a drop in spore counts (Wagenaar and Dack, 1955). Other experiments have demonstrated

---

[1] An F_0 value of 0.3 equals heating at 250°F for 0.3 minutes.

that the inhibitory effect of NaCl in surface ripened cheese was in addition to the effect of free fatty acids. The highest NaCl concentration which permitted growth in surface-ripened cheese was around 8% (gram NaCl/100 ml H_2O) (Grecz, Wagenaar and Dack, 1959b).

The germination of C. botulinum spores in nutrient media was inhibited by NaCl, KCl and $MgCl_2$ in concentrations from 4 to 9 (gram/100 ml H_2O). Fifteen per cent Na_2SO_4 and $MgCO_3$ did not inhibit germination (Halvorson, 1955; Hitzman, Halvorson and Utika, 1957). The available data regarding the effect of salt indicate that the concentration of salt should be considered in relation to the amount of water in the food, and it is likely that the effect of salt can largely be expressed in terms of water activity.

Water Activity (a_w). Almost the same a_w is found in cooked hams whether calculated from the H_2O and NaCl content or determined by freezing point determinations (Scott, 1955). Only limited information exists with regard to the relation between a_w and NaCl content in other food materials. The available data indicate that the germination of C. botulinum spores is inhibited at water activities below 0.94 (Halvorson, 1958; Ingram and Handford, 1957; Kodavy and Dack, 1951; Wagenaar and Dack, 1954; Williams and Purnell, 1953).

The Effect of Nitrate, Nitrite and Hydroxylamin. Nitrate has been used to inhibit the growth of clostridia in cheese. The effect of nitrate is believed to be that it increases the oxidation-reduction potential. It has been reported that 0.15 per cent nitrate has a considerable effect on C. botulinum spores in cured, canned meat (Jensen, 1944; Jensen, 1954). A pronounced effect of nitrate has not been found however in other experiments (Hansen and Appleman, 1955; Silliker, Greenberg and Schack, 1958).

Nitrite has a considerable inhibitory influence on spores of C. botulinum and other clostridia (Gross, Vinton and Martin, 1946; Jensen, 1944; Jensen, 1954; Silliker, Greenberg and Schack, 1958). The effect of 78 ppm nitrite has been found to be comparable to that of 3.5 gram NaCl/100 ml H_2O in canned cured meat. Hydroxylamin which may be produced by reduction of nitrite has a considerable inhibitory effect. 25-500 ppm may inhibit C. botulinum type A for more than a month at optimum temperatures and at pH ranges of 5.9 to 7.6 (Tarr, 1953).

The Effect of Antibiotics. The effect of a large number of antibiotics on C. botulinum has been tested. Many of these antibiotics have a destroying effect on germinated spores, but none has so far been put into practical use. One of the reasons being that it has not yet been sufficiently proved that any of these antibiotics may justify a relaxation of the other precautions against C. botulinum.

Other Inhibitory Compounds. No food preservative whether antioxidant, antimycotic agent or others should be able to stimulate the growth of C. botulinum. It has been demonstrated that sorbic acid, caproic acid and propionic acid did not inhibit or stimulate the growth of C. botulinum types A and B (Hansen and Appleman, 1955). Oxidizing agents as persulfate, chlorate, perchlorate and bromate have been used to inhibit clostridia in cheese. Rancid fatty acids have an inhibitory effect on the germination of clostridial spores (Roth and Halvorson, 1952) and this effect seems to be significant in certain kinds of cheeses. Other inhibitory compounds naturally present in foods as lysozyme and lactenin in milk may have an influence on the growth of C. botulinum (Johanssen, 1961), but the practical significance of these compounds has not been evaluated.

Effect of Other Microorganisms. A number of other microorganisms may interfere with the production of botulinum toxin or may destroy the toxin. This is probably the explanation of why toxin formation very often cannot be demonstrated in inoculated foods which contain many other organisms (Johanssen, 1961). Some microorganisms are inhibitory; others have a synergistic effect on clostridia.

The inhibitory effect of other organisms has been demonstrated in experiments designed to evaluate the botulinum hazard in foods prepacked in plastic bags (Johanssen, 1961). Toxin formation took place in inoculated cured herring provided it had been eluted with tap water before it was inoculated. Sealing of the plastic bag under vacuum had no influence on toxin formation in the herring. No toxin production could be detected in a variety of inoculated meat products packed in plastic bags and incubated for up to 32 days at 25°C. In many cases spores could be recovered from the samples at the end of the incubation period. The pH values in the samples varied from 4.6 to 6.6. The contents of water, salt, nitrate and nitrite were not stated. Most of the samples spoiled before the incubation was terminated.

The conclusions of these experiments were that vacuum packaging had no important influence on the ability of C botulinum to form toxin in these products but factors inherent in the food material or produced by other organisms present determined whether or not C. botulinum would grow and form toxin.

A synergistic effect of lactic acid bacteria on clostridia has been observed. This effect may be due to a lowering of the oxidation-reduction potential, to a destruction of nisin or to production of growth-stimulating factors (Benjamin, Wheather and Shepherd, 1956). The synergistic effect of yeast on the growth of C. botulinum in canned fruit has already been mentioned (Meyer and

Gunnison, 1929). A stimulating effect of <u>Bacillus mesentericus</u> on the growth of <u>C. botulinum</u> in bread has been observed; the nature of the effect is unknown (Kadavy and Dack, 1951). Several specific organisms seem to have an inhibitory effect on the growth of the toxin formation by <u>C. botulinum</u>. It has thus been found that <u>C. sporogenes</u> may interfere with the development of botulinus toxin in laboratory media, diminish the amount that is produced or cause an early disappearance of the toxin (Jordan and Dack, 1924). Lactic acid starters have been found to inhibit toxin formation in precooked frozen food (chicken à la king) inoculated with <u>C. botulinum</u> type A. The germination of the spores was inhibited due to lactic acid formed by the starter cultures.

The inhibitory effect of mixed cultures which are able to produce a low pH in carbohydrate media is also known from attempts to isolate <u>C. botulinum</u> from soil samples. Some of the acid-producing organisms from soil prevented toxin formation (Hall and Peterson, 1923). Peroxides formed by lactic acid bacteria may also have an inhibitory effect. The inhibitory effect of other organisms is also evident in the case of cheese where the production of free fatty acids and other compounds during maturation may prevent growth of <u>C. botulinum</u> (Grecz and Wagenaar, 1959a, 1959b, 1959c). Some cheese organisms, e.g. <u>Bacterium linens</u> may even form a specific stable antibiotic against <u>C. botulinum</u> (Jensen, 1944). Nisin-producing <u>Streptococcus lactis</u> strains are well-known examples of specific inhibitors.

CONCLUSION

A number of factors can influence the growth and toxin production of <u>C. botulinum</u> in foods. Some of these factors are well known and applied on a rational basis, others are not well understood. It is generally agreed that foods are safe with regard to botulism if they:

1. are heated to $90-100\,^{\circ}C$ immediately before consumption.
2. have a pH which has not been higher than 4.5.
3. contain 10 grams or more of common salt per 100 grams of water, or have been dried to an equivalent of a_w.
4. have been kept at temperatures below $3\,^{\circ}C$.
5. have been heat processed in a tight container to an F_0 value of 3.
6. have been irradiated, at a time when no toxin production could have occurred, with approximately 5 megarad.

It is certainly true that these treatments will give a high insurance against botulism. The fact is, however, that the named requirements are not always fulfilled in practice. Canned cured meat is one example. This product has an extremely good public health record in spite of the fact that it cannot be heat processed at F_0 values much above one tenth of the values recommended as "botulinum cook." It is true that the salts present in canned cured meats are inhibitory to the heat-damaged spores but it is of equal importance that the contamination with spores is low.

The "botulinum cook," recommended as a minimum for low-acid canned foods is expected to reduce the number of C. botulinum spores approximately 12 log cycles. In canning cured meat not much more than a 3 log cycle reduction can be obtained on a comparable basis. It thus seems that the safety factor is lower for canned cured meats than for other low-acid canned foods. However the public health record of canned cured meat is as good as that of any other canned food, possibly the requirement of a 12 log cycle reduction; which is arbitrary, is excessive for other products. The safety factor built into canned cured meat may actually be greater than it seems to be. Not only is the natural contamination and clostridia very low in meat — and the safety factor should be considered in relation to the likely initial contamination — but it is also known that spores in contact with raw meat may lose heat resistance. The spores may actually initiate germination and there are indications that such changes may take place during commercial curing of meat trimmings. The consequence of this is that changes in curing methods as well as other changes in canning technology should be evaluated with regard to public health hazard. Another consequence is that inoculated pack studies in which the spores frequently have been in contact with raw meat for a short time or not at all will tend to show little inhibition of the spores.

When evaluating the botulism hazard in foods the variance in resistance of the individual spores should be considered so the degree of inhibition can be put on a quantitative basis. The result of such a variation is that the frequency with which the spores can germinate and produce growth is diminished with increasing effect of the preserving agent used and increased with the number of cells present. This is considered in relation to heat- and radio-resistance but rarely with regard to other preserving factors. What has been said above concerning canned cured meats applies in principle to many other foods. The importance of the microflora in nonsterilized foods and foods which do not contain significantly inhibitory compounds should be studied. The natural flora frequently can prevent toxin formation and in the future it might be desirable to inoculate foods in which the natural flora is greatly diminished, e.g. by aseptic techniques or by irradiation,

with organisms which will either prevent toxin formation or
serve as an indicator if the food is mishandled. The possibility
of using lactic acid starters in frozen precooked foods has been
demonstrated. The use of starters, e.g. <u>Pediococcus cerevisiae</u>,
in sausage manufacture where the temperature conditions are
favorable for <u>C. botulinum</u> may serve as another example.

A continued study of botulism toxin will probably give very
useful results. Areas for such studies are, e.g. 1) the mechanism
of toxin production, 2) the ability of the toxin to pass through
the intestinal wall, 3) mechanism of the synaptic poisoning and
4) evaluation of the molecular structure of the toxin that accounts
for its effect. If such research should lead to an efficient method
for eliminating botulism the processing of foods could be revolu-
tionized.

CLOSTRIDIUM PERFRINGENS FOOD POISONING

<u>C. perfringens</u> seems to be gaining increasing importance as
a food poisoning organism. The number of annual outbreaks in
England and Wales has increased from 25 in 1953 to 93 in 1957
(Dewberry, 1959).

The Occurrence of <u>C. perfringens</u>

<u>C. perfringens</u> is present in soil, water, milk, dust, sewage,
and the intestinal tract of man and animals. This very common
distribution is probably the explanation for the slow recognition
of the role of <u>C. perfringens</u> in outbreaks of food poisoning (Dack,
1956). The fact that <u>C. perfringens</u> in contrast to <u>C. botulinum</u> is
a normal inhabitant of the digestive tract probably has the effect
that intravital or agonal invasion of this organism can take place
in slaughtered animals, and <u>C. perfringens</u> is not uncommonly foun
by bacteriological meat inspection (Jepsen, 1960). <u>C. perfringens</u>
has been found present in small numbers in horse muscle where
no other anaerobes are found. The organisms did not start mul-
tiplying until four hours after death, and this was attributed to the
high oxidation-reduction potential in prerigor muscle (Barnes and
Ingram, 1956). It is also found sometimes in boiled canned cured
meat although it is much less frequent than the putrefactive anaerobes (Larsen, 1956). From a survey on the occurrence of <u>C.
perfringens</u> in hospital foods (meat, fish, and poultry) it can be
calculated that the maximum number which was found in uncooked
food was approximately 1.3 per gram and the minimum number in

cooked food was below 0.009 (McKillop, 1959). C. perfringens
has been found in 56 per cent of swabs from sewage, 2-18 per
cent of samples of feces from pigs, rats, mice and cattle and in
14-20 per cent of samples of raw pork, beef and veal obtained in
retail shops. The organisms have also been isolated from blow
flies (Hobbs, et al., 1953).

Outbreaks of C. perfringens Food Poisoning

Most outbreaks of C. perfringens food poisoning are caused
by meat dishes or poultry dishes that have been cooked and allowed to cool slowly overnight. This type of poisoning has therefore become more common with increased canteen feeding.

The symptoms are the following: 8-20 hours after the food
has been eaten symptoms appear in a large fraction of the persons who have eaten of the food. A mild brief type of illness is
observed in almost all outbreaks, characterized by colic and di-
arrhoea without vomiting. The duration of the illness is generally not more than 24 hours (McClung, 1945; Hobbs, et al., 1953;
Mikkelsen, Petersen and Skovgaard, 1962; Osterling, 1952).
In outbreaks which have been properly examined, heat resistant strains of C. perfringens dominated in the food; organisms with the same characteristics were isolated from the feces
of about 90 per cent of the people who had eaten the food, in contrast to only 5 per cent of normal persons (Hobbs, et al., 1953).
The number of C. perfringens present in the food causing food
poisoning is generally very high while the aerobic counts might
be low (Hobbs, 1955).

Food Poisoning C. perfringens Strains

A number of cases of severe gastrointestinal illness some of
which were caused by home-canned rabbit were reported in Ham-
burg in 1946. C. perfringens was isolated, and it was found that
its spores resisted boiling from 1 to 4 hours. It differed morphologically and culturally from all types of C. perfringens so
far described and the name C. perfringens type F was proposed
(Osterling, 1952). Eight cultures from these outbreaks were
studied thoroughly (Zeissler, et al., 1949), and it was found that
the organisms most closely resembled type B but it did not produce eta-toxin and the spores were much more heat resistant.

The strains isolated in connection with outbreaks of C. perfringens food poisoning in England have different but distinctive

characters. They form nonhaemolytic colonies which may later
show a faint haemolytic zone. They are heat resistant and will
stand steaming for one hour, some for several hours. They produce Nagler reaction which is inhibited by type A antitoxin. In
the toxicological reactions they resemble type A. No new toxins
were found in the food poisoning strains. Eight serological types
were recognized among the strains; they cross-reacted serologically with some type A strains but no relation to type B, C, D,
E, or F was shown (Hobbs, et al., 1953). It is not yet known how
uniform the food poisoning strains of C. perfringens are (Jepsen,
1960). In most of the English outbreaks the strains have been
nonhaemolytic but outbreaks involving haemolytic strains have
been reported (Hessen and Riemann, 1959; Mikkelsen, Petersen
and Skovgaard, 1962). More work is apparently needed to characterize the strains which can give food poisoning.

Toxicity

Attempts to demonstrate enterotoxin formation in cultures of
C. perfringens involved in food poisoning by feeding experiments
with mice and guinea pigs have failed (McClung, 1945), and sterile filtrates of toxigenic cultures of C. perfringens gave no symptoms in human beings (Osterling, 1962). Also veal infusion broth
cultures or chicken broth cultures had failed to cause illness in
human volunteers (Dack, 1956). However, cooked meat cultures
swallowed by volunteers produced symptoms similar to those observed in spontaneous outbreaks (Hobbs, et al., 1953). The varying results obtained by feeding cultures to human volunteers may
be a reflection of the difference in the medium. The presence of
whole meat in the culture may be of importance in the ability of
the culture to cause illness. Further work with human volunteers
is required to elucidate the mechanism of the mode of action of
C. perfringens in causing illness.

Growth and Growth Inhibition

The optimum pH for growth of C. perfringens is approximately
pH 7. No growth was observed with pH below 5 or above 9 and the
temperature range of growth is between 24° C to 48°C (Fuchs and
Bonde, 1957). C. perfringens is less sensitive to oxygen than C.
botulinum (Meyer, 1929). The growth of C. perfringens strains
was studied in a meat digest medium at pH 6.0 poised to different
oxidation-reduction potentials. The effect of a high potential was

on the lag phase. At E_h = -45 mV there was hardly any lag in growth but the lag phase increased with increasing E_h until about +250 mV where the organisms slowly died (Barnes and Ingram, 1956). The growth rate of C. perfringens is very high under optimum conditions. In 20 hours 50 spores can produce 50 x 10^6 cells at 37°C and 15 x 10^3 at 22°C. The germination time of the spores varies. Addition of starch to a cooked meat medium speeds up the germination and/or growth. Visible growth was present after 5-24 hours with starch but 48 hours were required without starch (Hobbs, et al., 1953). With regard to other inhibitory factors C. perfringens seems to behave very much like C. botulinum. Strains of C. perfringens were inhibited in laboratory media at pH 5.6 and 7 by 5 per cent NaCl which would give a water activity of 0.97. Two strains were inhibited by 2.5 per cent $NaNO_3$, one was not inhibited (Beerens, 1955).

Heat Resistance

The resistance of C. perfringens spores varies considerably The following values have been recorded: 1 minute at 98°C, $1\frac{1}{2}$ hours at 100°C, 1-4 hours at 100°C, $\frac{1}{2}$ hour at 90°C (Dewberry, 1959). The most resistant forms are type F and the food poisoning strains (Dewberry, 1959; Hobbs, et al., 1953; Zeissler, et al., 1949).

Prevention of C. perfringens Food Poisoning

The precautions which must be taken to prevent C. perfringens food poisoning are much the same as those applied with regard to C. botulinum. There is, however, one outstanding feature in C. perfringens food poisoning which should be mentioned. This type of poisoning is almost always caused by meat or poultry dishes which have been cooked and allowed to cool slowly. There might be several reasons for this. The cooking will drive out most of the oxygen present in the food and at the same time heat activates the spores with the result that they germinate fast and give rise to growth. The most important preventive measure is the establishment of adequate conditions for rapid cooling of cooked foods. It is recommended that the cooling of such foods from 50°C to approximately 10°C should not take more than 3 hours.

LITERATURE CITED

Barnes, E. M. and Ingram, M. 1956. The effect of redox potentia. on the growth of Clostridium welchii strains isolated from horse muscle. J. Applied Bacteriol. 19:117-28.

Barton, A. L. and Reed, G. B. 1954. Clostridium botulinum type E toxin and toxoid. Canadian J. Microbiol. 1:108-17.

Beerens, M. H. 1955. Facteurs influençant le développement et le métabolisme des Clostridium mesophiles et thermophiles autres que les toxigènes dans les semi-conserves de viande. Ann. L'Institute Pasteur, Lille 7:75-82.

Benjamin, M. J. W., Wheather, D. M., and Shepherd, P. A. 1956. Inhibition and stimulation of growth and gas production by Clostridia. J. Applied Bacteriol. 19:159-63.

Bonventre, P. F. and Kempe, L. L. 1959a. Physiology of toxin production by Clostridium botulinum types A and B. II. Effect of carbohydrate source on growth, autolysis and toxin production. Applied Microbiol. 7:372-74.

_____. 1959b. Physiology of toxin production by Clostridium botulinum types A and B. III. Effect of pH and temperature during incubation on growth, autolysis and toxin production. Applied Microbiol. 7:374-77.

_____. 1960. Physiology of toxin production by Clostridium botulinum types A and B. IV. Activation of the toxin. J. Bacteriol. 79:24-32.

Dack, G. M. 1956. Food Poisoning. Third ed. University of Chicago Press, Chicago.

Dewberry, E. B. 1943. Food Poisoning. Leonard Hill, Ltd., London.

_____. 1959. Food Poisoning. Food-Borne Infection and Intoxication. Fourth ed. Leonard Hill, Ltd., London.

Ermengen, E. van. 1896. Recherches sur des cas d'accidents alimentaires produits par des saucissons. Revue d'Hygiene 18:761-819.

Fuchs, A. R. and Bonde, G. J. 1957. The nutritional requirements of Clostridium perfringens. J. Gen. Microbiol. 16:317.

Geiger, J. C. 1941. An outbreak of botulism. J. Am. Med. Assoc. 117:22.

Grecz, N., Wagenaar, R. O., and Dack, G. M. 1959a. Inhibition of Clostridium botulinum by culture filtrates of Brevibacterium linens. J. Bact. 78:506-10.

_____. 1959b. Inhibition of Clostridium botulinum and molds in aged surface ripened cheese. Applied Microbiol. 7:33-38.

_____. 1959c. Relation of fatty acids to the inhibition of

Clostridium botulinum in aged surface ripened cheese. Applied Microbiol. 7:228-34.

Greenberg, R. A., Silliker, J. H., Nank, W. K., and Schmidt, C. F. 1958. Toxin production and organoleptic breakdown in vacuum packaged fresh meats inoculated with Clostridium botulinum. Food Research 23:656-61.

Gross, C. E., Vinton, C., and Martin, S. 1946. Bacteriological studies relating to thermal processing of canned meats. IV. Viability of spores of a putrefactive anaerobic bacterium in canned meat after prolonged incubation. Food Research 11: 399-404.

Haerem, S., Dack, G. M., and Dragstedt, L. R. 1938. Acute intestinal obstruction. II. The permeability of obstructed bowel segments of dogs to Clostridium botulinum toxin. Surgery 3:339-50.

Haines, R. B. 1942. The occurrence of toxigenic anaerobes, especially Clostridium botulinum, in some English soils. J. Hygiene 42:323-27.

Hall, I. C. and Peterson, E. 1923. The effect of certain bacteria upon the toxin production of Bacillus botulinus. J. Bacteriol. 8: 319-41.

Halvorson, H. O. 1955. Factors determining survival and development of Clostridium botulinum in semi-preserved meats. Ann. L'Institut Pasteur, Lille 7:53-67.

_____. 1958. The physiology of the bacterial spore. Technical University of Norway, Trondheim.

Hansen, J. D. and Appleman, M. D. 1955. The effect of sorbic, propionic, and caproic acids on the growth of certain clostridia. Food Research 20:92-96.

Hessen, I. and Riemann, H. 1959. Arsskrift den kgl. Veterinaer- og-Landbohøjskole, Copenhagen. 50-69.

Heyningen, W. E. van. 1955. Recent developments in the field of bacterial toxins. Schweiz. z. allgem. Pathol. u. Bakteriol. 18:1018-35.

Hitzman, D. O., Halvorson, H. O., and Utika, T. 1957. Requirements for production and germination of spores of anaerobic bacteria. J. Bacteriol. 74:1-7.

Hobbs, B. C. 1955. The laboratory investigation of non-sterile canned hams. Ann. L'Institut Pasteur, Lille 7:190-202.

_____, Smith, M. E., Oakley, C. L., Warrack, G. H., and Cruickshank, J. C. 1953. Clostridium welchii food poisoning. J. Hygiene 51:75-101.

Ingram, M. and Handford, M. 1957. The influence of moisture and temperature on the destruction of Clostridium botulinum in acid bread. J. Applied Bacteriol. 20:442-53.

Ingram, M. and Robinson, R. H. M. 1951a. The growth of Clostridium botulinum in acid bread media. Proc. Soc. Appl. Bacteriol. 14:62-72.

_____. 1951b. A discussion of the literature on botulism in relation to acid foods. Proc. Soc. Appl. Bact. 14:73-84.

Jensen, L. B. 1944. Microbiological problems in the preservation of meats. Bacteriol. Revs. 8:161.

_____. 1954. Microbiology of Meats. Third ed. Garrard Press, Champaign, Ill.

Jepsen, A. 1960. Diagnostik bakteriologi og Levnedsmiddel-bakteriologi, København, Denmark.

Johanssen, A. 1961. S. I. K. Rapport No. 100. Svenska Institutet för Konserveringsforskning, Göteberg, Sweden.

_____. 1962. Personal communication.

Jordan, E. O. and Dack, G. M. 1924. The effect of Cl. sporogenes on Cl. botulinum. J. Inf. Dis. 35:576-80.

Kadavy, J. L. and Dack, G. M. 1951. Clostridium botulinum in canned bread. The effect of experimentally inoculating canned bread with spores of Clostridium botulinum and Bacillus mesentericus. Food Research 16:328-37.

Kempe, L. L., Graikoski, J. F., and Bonventre, P. F. 1957. Combined irradiation-heat processing of canned foods. I. Cooked ground beef inoculated with Clostridium botulinum spores. Applied Microbiol. 5:292-95.

_____. 1958. Combined irradiation-heat processing of canned foods. II. Raw ground beef inoculated with spores of Clostridium botulinum. Applied Microbiol. 6:261-63.

Kindler, S. H., Mager, J., and Grossowicz, N. 1956a. Nutritional studies with the Clostridium botulinum group. J. Gen. Microbiol. 15:386-93.

_____. 1956b. Toxin production by Clostridium parabotulinum. J. Gen. Microbiol. 15:394-403.

Knock, G. G. and Lambrechts, S. J. 1956. A note on the heat resistance of a South African strain of Clostridium botulinum type B. J. Sci. Food Agric. 7:244-48.

Lamanna, C. 1959. The most poisonous poison. Science 130: 763-72.

_____, and Meyers, C. E. 1960. Influence of ingested foods on the oral toxicity in mice of crystalline botulinal type A toxin. J. Bacteriol. 79:406-10.

Larsen, A. E. 1956. Found: a prevention against botulism. Am. Fur Breeder 29:12-13; 54-56.

Legroux, R., Jeramec, C., and Second, L. 1947. Destruction des toxines bacteriennes par les proteases microbiennes. Ann. L'Institut Pasteur 73:828-30.

McClung, L. S. 1945. Human food poisoning due to growth of Clostridium perfringens (C. welchii) in freshly cooked chicken: preliminary note. J. Bacteriol. 50:229-31.

McKillop, E. J. 1959. Bacterial contamination of hospital food with special reference to Clostridium welchii food poisoning. J. Hygiene 57:31-46.

Meyer, K. F. 1929. Maximum oxygen tolerance of Cl. botulinum A, B, and C, of Cl. sporogenes and Cl. welchii. J. Inf. Dis. 44:408-11.

_____. 1956. The status of botulism as a world health problem. Bul. World Health Organization 15:281-98.

_____, and Gunnison, J. B. 1929. Botulism due to home canned bartlett pears. J. Inf. Dis. 45:135-47.

Mikkelsen, H. D., Petersen, P. J., and Skovgaard, N. 1962. Tre tilfaelde af Clostridium perfringens − levnedsmiddelforgift-ning. [Three outbreaks of Clostridium perfringens food poi-soning.] Nordisk Veterinaermedicin 14:200-11.

Mossel, D. A. A. and Ingram, M. 1955. The physiology of the microbial spoilage of foods. J. Applied Bact. 18:232-68.

Müller, J. 1961. S. I. K. Rapport No. 100. Svenska Institutet för Konserveringsforskning, Göteberg, Sweden.

Mundt, J. O., Mayhew, C. J. and Stewart, G. 1954. Germination of spores in meats during cure. Food Technol. 8:435-36.

Ohye, D. F. and Scott, W. J. 1953. The temperature relations of Clos-tridium botulinum types A and B. Australian J. Biol. Sci. 6:178-89.

_____. 1957. Studies in the physiology of Clostridium botuli-num type E. Australian J. Biol. Sci. 10:85-94.

Osterling, S. 1952. Matförgiftninger orsakade av Clostridium perfringens (welchii). [Food poisoning caused by C. welchii.] Nordisk Hyg. Tidsskrift. 173-79.

Pederson, H. O. 1955. On type E bolulism. J. Appl. Bacteriol. 18:619-29.

_____. 1961. S. I. K. Rapport No. 100. Svenska Institutet för Konserveringsforskning, Göteberg, Sweden.

Prevot, A. R. and Sillioc, R. 1958. Recherches sur l'immunisation anti-botulique C B due lapin par injection unique d'anatoxine concentrée adsorbée. Ann. L'Institut Pasteur 95:208-10.

Roth, N. G. and Halvorson, H. O. 1952. The effect of oxidative rancidity in unsaturated fatty acids on the germination of bacterial spores. J. Bacteriol. 63:429-35.

Saleh, M. A. and Ordal, Z. J. 1955a. Studies on the growth and toxin production of Clostridium botulinum in a precooked frozen food. I. Some factors affecting growth and toxin pro-duction. Food Research 20:332-39.

_____. 1955b. Studies on growth and toxin production of

<u>Clostridium botulinum</u> in a precooked frozen food. Food Research 20:340-50.

Scheibner, G. 1955. Der Einfluss von Ultraschallwellen auf das Botulinustoxin. Tierarzliche Umschau 10:364-66.

Schmidt, C. F. 1962. Personal communication.

Scott, W. J. 1955. Factors in canned ham controlling <u>Clostridium botulinum</u> and <u>Staphylococcus aureus</u>. Ann. L' Institut Pasteur, Lille 7:68-73.

Silliker, J. H., Greenberg, R. A., and Schack, W. R. 1958. Effect of individual curing ingredients on the shelf stability of canned comminuted meats. Food Technol. 12:551-54.

Skullberg, A. 1961. S. I. K. Rapport No. 100. Svenska Institutet för Konserveringsforskning, Göteberg, Sweden.

Smith, L. D. S. 1954. Introduction to the Pathogenic Anaerobes. Univ. of Chicago Press, Chicago.

Sugiyama, H. 1951. Studies on factors affecting the heat resistance of spores of <u>Clostridium botulinum</u>. J. Bact. 62:81-96.

Tarr, H. L. A. 1953. The action of hydroxylamine on bacteria. J. Fish. Res. Bd. Canada 10:69-75.

Townsend, C. T., Yee, L., and Mercer, W. E. 1954. Inhibition of the growth of <u>Clostridium botulinum</u> by acidification. Food Research 19:536-42.

Treadwell, P. E., Jann, G. J., and Salle, A. J. 1958. Studies on factors affecting the rapid germination of spores of <u>Clostridium botulinum</u>. J. Bacteriol. 76:549-56.

Vinton, C., Martin, S., and Gross, C. E. 1947. Bacteriological studies relating to thermal processing of canned meats. VII. Effect of substrate upon thermal resistance of spores. Food Research 12:173-83.

Wagenaar, R. O. and Dack, G. M. 1954. Further studies on the effect of experimentally inoculating canned bread with spores of <u>Clostridium botulinum</u>. Food Research 19:521-29.

_____. 1955. Studies on canned cheese spread inoculated with spores of <u>Clostridium botulinum</u>. Food Research 20: 144-48.

_____, and Murrell, C. B. 1959. Studies on purified type A <u>Clostridium botulinum</u> toxin subjected to ultracentrifugation and irradiation. Food Research 24:57-61.

Williams, O. B. and Purnell, H. G. 1953. Spore germination, growth and spore formation by <u>Clostridium botulinum</u> in relation to the water content of the substrate. Food Research 18:35-39.

Zeissler, J., Rassfeld-Sternberg, L., Oakley, C. L., Dieckmann, C., and Hain, E. 1949. Enteritis necroticans due to <u>Clostridium welchii</u> type F. Brit. Med. J. 1:267-71.

17

Enterococci

P. M. FRANCES SHATTOCK
UNIVERSITY OF READING, ENGLAND

THE SUBJECT of this chapter must first be defined because the terms "enterococcus" and "enterococci" have been used in various ways. They have been used loosely to include all streptococci of fecal origin, and also more precisely to denote only those streptococci which fulfill the criteria used by Sherman (1938) to characterize his "enterococcus group." During the last decade there has been renewed interest in the taxonomy of the fecal streptococci and use of these studies in various fields of applied microbiology. These studies have been largely concerned with investigations of the alimentary flora of various animal species in relation to digestion and nutrition, with attempts to find a useful indicator of fecal pollution of water and foodstuffs and with the possible pathogenicity of these organisms.

Because not all streptococci of fecal origin fulfill the criteria used by Sherman (1938) to define his "enterococcus group" all streptococci which possess the Group D antigen (Lancefield, 1933) are considered here. This serological group includes, in addition to the "enterococcus group" as defined by Sherman, two other species of fecal origin, namely Streptococcus bovis (Shattock, 1949a) and Streptococcus equinus (Smith and Shattock, 1962).

IDENTIFICATION OF GROUP D STREPTOCOCCI

Group Serology

To identify these fecal streptococci the serological grouping is of cardinal importance. In this respect Group D streptococci have presented some difficulties not encountered in other groups. The production of potent grouping sera may be troublesome and some commercial grouping sera have often proved unreliable. Recent investigations on the location of the group antigen in the streptococcal cell have helped to elucidate some of these difficulties. Unlike Group A streptococci, in which the group antigen is located in the cell wall (McCarty, 1952a, b) and therefore readily accessible to the antibody-forming cells of the animal, in Group D streptococci the group antigen is not an integral part of the cell wall, but is in the cell contents which remain when the cell wall has been removed (Elliott, 1960; Jones and Shattock, 1960). This helps to explain the fact that Group D antisera are more readily produced in rabbits by using as vaccine disintegrated cocci rather than whole organisms. The method of preparing antigens for precipitin ring tests is also important; with some strains it may even be necessary to concentrate the group antigen in HCl extracts (Shattock, 1949). The growth medium should contain a fermentable carbohydrate (Medrek and Barnes, 1962a).

Physiology

As we shall see the Group D streptococci may be divided into three divisions on the basis of physiological characters. These divisions are:

1. Streptococcus faecalis and its varieties liquifaciens and zymogenes.
2. Streptococcus faecium and Streptococcus durans.
   The organisms in divisions 1 and 2 fulfill the criteria for the "enterococcus group" of Sherman (1938).
3. Streptococcus bovis and Streptococcus equinus.
   S. equinus has recently been shown to possess the Group D antigen (Smith and Shattock, 1962).

Shattock (1955) summarized the position about the differentiation of species within Group D. Several papers have since been published on the identification of fecal streptococci (e.g. Barnes, 1956a; Lake, Deibel and Niven, 1957; Jones, 1958; Dunican and Seeley, 1962; Smith and Shattock, 1962) and their distribution among various animal species (e.g. Cooper and Ramadan, 1955; Kjellander, 1960; Bartley and Slanetz, 1960; Raibaud et al., 1961). Taking these more recent papers into account Group D streptococci may be differentiated by the selection of physiological characters given in Table 17.1.

It will be seen that all Group D streptococci are able to grow at 45 °C and to grow in the presence of 40 per cent ox bile.

Table 17.1. Group D Streptococci: Selected Differential Physiological Characters

	Division 1	Division 2		Division 3	
	S. faecalis and varieties	S. faecium	S. durans	S. bovis	S. equinus
β Hemolysis	-/+	-	+/-	-	-
Growth 10°	+	+	+	-	-
45°	+	+	+	+	+
50°	+	+ *	-	-	-
pH 9.6	+	+	+/-	-	-
6.5% NaCl	+/-	+/-	+/-	-	-
40% bile	+	+	+	+	+
Resists 60°C for 30 min.	+	+	+/-	-	- *
NH₃ from arginine	+	+	+	-	-
Gelatin liquefied	-/+	-	-	-	-
Tolerates 0.04% Pot. tellurite	+	-	-	-	-
Acid from:					
Glycerol (anaerobic)	+ *	-	-	-	-
Mannitol	+	+	- *	-/+	-
Sorbitol	+ *	- *	-	-/+	-
L-arabinose	-	+ *	-	+/-	-
Lactose	+	+	+	+	-
Sucrose	+ *	+/-	-	+	+ *
Raffinose	- *	- *	-	+	- *
Melibiose	-	+ *	- *	+	- *
Melezitose	+ *	-	-	-	-
Starch hydrolyzed	-	-	-	+ *	- *
Tetrazolium reduced at pH 6.0	+	-	-	+/-	-

+ = positive result.

- = negative result.

+/- = variation between strains, majority positive.

-/+ = variation between strains, majority negative.

* occasional strains atypical.

Division 1 and Division 2 fulfill the criteria for the "enterococcus group" of Sherman (1938).

The ability to grow at 10°C and 45°C, to grow in the presence of 6.5 per cent NaCl at pH 9.6, to produce ammonia from arginine and to resist heating at 60°C for 30 minutes serve to define the "enterococcus group" of Sherman (S. faecalis and varieties; S. faecium; S. durans), although growth in broth containing 6.5 per cent NaCl and resistance to heating at 60°C for 30 minutes are somewhat variable characters.

S. faecalis is well defined; its varieties liquifaciens and zymogenes differ from it only in their proteolytic properties and zymogenes also in being β-hemolytic in horse blood agar. The following properties serve to distinguish S. faecalis from S. faecium; S. faecalis is tolerant to the presence of potassium tellurite 0.04 per cent (Skadhauge, 1950) and will grow as luxuriant black colonies on glucose nutrient agar containing this concentration of potassium tellurite. S. faecium and all other members of Group D are greatly inhibited by 0.04 per cent potassium tellurite. Those strains of S. faecium which are able partially to tolerate this concentration of potassium tellurite grow as tiny dusty grey colonies.

The anaerobic fermentation of glycerol is characteristic of S. faecalis. In this test however it is essential to ensure that the growth medium provides a suitable hydrogen acceptor such as yeast extract (Gunsalus, 1947). The anaerobic fermentation of glycerol may be conveniently tested for in a medium containing peptone (1 per cent), yeast extract (1 per cent), glycerol (0.5 per cent) and should be heated in a boiling water bath for 15-20 minutes immediately before inoculation, the tubes layered with medicinal paraffin oil and incubated aerobically. The results should be read within seven days as some strains of S. faecium are able to produce acid oxidatively from glycerol and such strains may produce a late acidity due to oxygen entering the culture medium through the oil layer.

S. faecalis will ferment sorbitol and melezitose (with occasional exceptions) but is unable to ferment L-arabinose or melibiose; these tests also serve to differentiate S. faecalis from S. faecium. S. faecium has a higher maximum temperature (most strains will grow at 50° C) than S. faecalis. The ability of S. faecalis to reduce tetrazolium (2, 3, 5-triphenyltetrazolium chloride) to a formazan in a glucose-containing medium (initially at pH 6.0) also helps to distinguish S. faecalis from S. faecium (Barnes, 1956 and from all other members of Group D. Most strains of S. bovis from bovine feces will reduce tetrazolium (Barnes, 1959) but S. bovis is readily distinguished by other characters. This physiological separation of S. faecalis from S. faecium is substantiated by serological typing (Sharpe and Shattock, 1952).

S. durans attacks fewer carbohydrates and the inability of this species to produce acid from mannitol, sucrose and melibiose helps to differentiate it from S. faecium. However the distinction between these two species is not always clear and Lake, Deibel and Niven (1957) suggested that these should be regarded as varieties of one species, S. faecium (Orla-Jensen, 1919), on grounds priority. There is some serological evidence to support this view. S. faecalis is distinguished from these two species on the basis of type antigens but the serological distinction between S. faecium and S. durans is not so clear. Whether or not it is eventually decided to retain these as separate species, there is no doubt that they together form a physiological division within Group D.

S. bovis and S. equinus do not conform to the criteria for the "enterococcus group" of Sherman. They are able to grow at 45°C but not at 10°C, they do not produce ammonia from arginine, they do not grow in 6.5 per cent NaCl or at pH 9.6, they do not tolerate heating at 60°C for 30 minutes although most strains will survive 60°C for 15 minutes. Raffinose fermentation and hydrolysis of starch are characteristic of S. bovis strains of bovine origin but strains of human origin may fail to hydrolyze starch. S. equinus is characteristically unable to ferment lactose and this, together with the inability of the majority of its strains to ferment raffinose or to hydrolyze starch, serve to separate it from S. bovis. But cultures having physiological properties intermediate between these two species are not uncommonly encountered (Raibaud et al., 1961).

Type Serology

The broad divisions on physiological grounds are supported by serological studies.

1. S. faecalis (and its varieties zymogenes and liquifaciens) are at present divided into a relatively small number of serological types (Skadhauge, 1950; Sharpe and Shattock, 1952); subsequent unpublished observations by Dr. Elisabeth Sharpe suggest that probably these will not exceed 20. These type antigens are located in the cell wall (Elliott, 1960) and are distinct from the type antigens of other members of Group D streptococci (Sharpe, 1962).

2. S. faecium and S. durans are divided into more numerous antigenic types on the basis of cell wall antigens, some of which are shared by these two species but are distinct from the type antigens of S. faecalis (Sharpe and Fewins, 1960; Sharpe, 1962).

3. S. bovis may be divided into numerous serological types mainly on the basis of their capsular antigens (Medrek and Barnes, 1962b) which are not shared by other species within Group D. The type antigens of S. equinus have not yet been studied.

Recently, Hartsell and Caldwell (1961) studied the action of lysozyme on a comprehensive collection of Group D streptococci and devised a scheme for differentiating species on the basis of their reaction to lysozyme, alone or in conjunction with trypsin. In this scheme also three clear-cut divisions were indicated: S. faecalis and its varieties were lysed by a combination of lysozyme and trypsin but not by lysozyme alone; S. faecium and S. durans were lysed by lysozyme alone; S. bovis was completely resistant to lysis. These results again emphasize the differences in the composition of the cell walls of the various species of Group D streptococci already indicated by serological typing. The divisions obtained are in accordance with the broad divisions based on both serological and physiological studies.

SOURCES OF GROUP D STREPTOCOCCI

The natural habitat of Group D streptococci appears to be the alimentary tract of man and animals and because these organisms are being widely considered as an indicator of fecal pollution it is appropriate to discuss their distribution.

During the last decade the incidence of streptococci in human and animal feces has been investigated by several workers. Cooper and Ramadan (1955) studied in Great Britain the streptococci in human, bovine and sheep feces and found Streptococcus faecalis mainly in human feces. Comparative studies in U.S.A. have yielded similar results, e.g. Bartley and Slanetz (1960) made a survey of streptococci from the feces of humans, cows, pigs and chickens and found that, although there was considerable variation in the predominant types of streptococci in the feces of different individuals, typical S. faecalis was the dominant species in human feces and in the feces of chickens but was seldom found in the feces of most domestic animals. This is also borne out by the investigations of Kenner, Clark and Kabler (1960). In contrast, some European workers have reported different distributions of streptococcus species in human feces. Buttiaux (1958) in France found both S. faecalis and S. faecium in human feces with the latter species predominating. In Denmark, Kjellander (1960) also found a slight predominance of S. faecium over S. faecalis. Guthof (1957) in Germany found that although S. faecalis was most prevalent in fecal specimens from children, the incidence of S. faecium rose with increasing age and became dominant in the adult. Thus there appears to be some geographical differences in the predominant streptococci of the human intestine which might possibly be associated with diet. This is an interesting point and may well explain why for many years in the U.S.A. and Great Britain S. faecium (Orla-Jensen, 1919) and S. faecalis (Andrewes and Horder, 1906) were erroneously assumed to be synonymous (Sherman, 1937). Although S. faecalis may occur in the feces of domestic animals, this organism is usually present in relatively small numbers. In cows, pigs and sheep the dominant streptococci are S. bovis, S. equinus and S. faecium (Kjellander, 1960; Raibaud et al., 1961; Kenner, Clark and Kabler, 1960).

Group D Streptococci as Indicators of Fecal Pollution

In recent years the significance of Group D streptococci in food of various kinds has attracted a considerable amount of attention. Because Group D streptococci are invariably found in the intestinal contents of man and animals they have been widely advocated as indicators of fecal pollution (e.g. in frozen fruits, vegetables and fruit juice concentrates, Larkin, Litsky and Fuller, 1955a, b, c; in meat pies, Kereluk and Gunderson, 1959; Kereluk, 1959; in frozen sea foods, Raj, Wiebe and Liston, 1961). Buttiaux (1959) in discussing the value of the association of Escherichieae and Group D streptococci as indicators of contamination regarded Group D streptococci as affording a more sensitive test of fecal contamination of food than the coli-aerogenes bacteria because the Group D streptococci are able to withstand a wider variety of environmental conditions. Their relatively high resistance to unfavorable conditions might, however, be an objection to their use as an indicator of pollution. This objection was admitted and discussed by Buttiaux and Mossel (1961) who pointed out that, provided a quantitative approach be made to the problem, this objection is readily overcome. Fecal contamination of foods is, of course, highly undesirable whatever its origin because, apart from aesthetic consideration, domestic animals and birds as well as human beings are very frequent carriers of Salmonella (Hobbs, 1961).

Because of the many recent surveys of the incidence of Group D species in human and animal feces we are now in a position to apply species identification, and more precisely serological typing, to trace the source of Group D streptococci in some commercially processed foodstuffs. To illustrate this point I will quote two examples. Barnes, Ingram and Ingram (1956) studied the relative numbers and distribution of different species of Group D streptococci in bacon factories and were able to relate the presence of S. faecalis, which is normally absent from the gut of the pig, to human contamination beginning at a particular point in the factory process. Sharpe and Fewins (1960) serologically typed strains of S. faecium and certain unclassified Group D streptococci isolated from canned hams in order to establish the serological types associated with these sources and to provide a precise means of identifying and comparing isolates for subsequent investigations.

Methods of Isolating Group D Streptococci

One cannot leave this topic without some mention of methods for detecting Group D streptococci. In recent years much work has been done to develop selective techniques for the detection and enumeration of these organisms in water and foods (see reviews by: Barnes, 1959; Kjellander, 1960).

The incorporation of sodium azide as a selective agent alone or in conjunction with other inhibitory substances (e.g. crystal violet) have been extensively used and well reported for the detection of Group D streptococci in foods (e.g. Mossel, van Diepen and de Bruin, 1957; Zaborowski, Huber and Rayman, 1958; Raj, Wiebe and Liston, 1961).

Thallous acetate has also been used as a selective agent for the isolation of Group D streptococci (e.g. Mattick and Shattock, 1943; Sharpe, 1952). Barnes (1956b) used thallous acetate as the selective agent in a medium, containing tetrazolium, which was devised for the isolation and presumptive identification of Group D streptococci under conditions where fecal streptococci are likely to be heavily outnumbered by other organisms.

The choice of selective techniques must, of course, be governed by the objective and the nature of the material to be examined. The fact that a method is efficient under one set of conditions does not mean that it will be efficient under different conditions. Most selective media for fecal streptococci have been devised for the isolation of those which fall into the "enterococcus group" of Sherman. However, both S. bovis and S. equinus are less resistant to inhibitory agents and may be appreciably suppressed in media which allow good recovery of S. faecalis, S. faecium and S. durans. Recently Raibaud et al. (1961), in a study of the alimentary flora of pigs, described a medium containing sodium azide, sodium glutamate, acridine orange and tetrazolium chloride which, they claim, allows the selective enumeration of streptococci, including S. bovis and S. equinus, in the presence of a superior number of lactobacilli.

GROUP D STREPTOCOCCI AS POTENTIAL SOURCES OF FOOD POISONING

In any assessment of the significance of fecal streptococci in food the question of their pathogenic potentialities must be considered. There have been reports of food poisoning in which fecal streptococci have been implicated.

The clinical symptoms reported have been relatively mild with incubation times which varied in different outbreaks and between different individuals from 2 hours to 18 hours; in cases with short incubation times, vomiting was the dominant symptom and in cases with longer incubation times the dominant symptom was diarrhea. In such outbreaks streptococci were invariably reported to be present in very large numbers in the food, of the order of several millions per gram. Various kinds of food have been implicated: meat, particularly dishes prepared from meat previously cooked, gravy or stock previously prepared; dishes containing processed milk or synthetic cream; cheese. Moore (1955) reviewed the literature on streptococci and food poisoning and Dack (1956) discussed Streptococcus faecalis in relation to it. Table 17.2 briefly summarizes the main outbreaks which have been reported in some detail.

The first report of food poisoning attributed to streptococci was by Linden, Turner and Thom (1926). They isolated streptococci from cheeses implicated in two outbreaks of food poisoning and observed symptoms of food poisoning in cats fed living cultures of these streptococci. The streptococci were characterized but not named by the authors; some years later Sherman, Smiley and Niven (1943) identified them as S. faecalis. This species was further incriminated by Sherman, Gunsalus and Bellamy (1944) in a report on streptococci isolated by other workers from six outbreaks and they expressed the opinion that S. faecalis was the only species of streptococcus capable of causing food poisoning. However from the outbreaks summarized in Table 17.2 it may be seen that not all implicated strains can be identified with certainty.

The fundamental cause of illness has remained a mystery. The success obtained in demonstrating toxin production by staphylococci and the etiological role of the toxin in cases of staphylococcal food poisoning obviously influenced ideas about the basic cause of food poisoning attributed to fecal streptococci. Various attempts were made to demonstrate an enterotoxin both in strains of streptococci which had been associated with outbreaks of food poisoning and in cultures from other sources. The results of such work have been equivocal. Jordan and Burrows (1934) carried out feeding tests on monkeys with a streptococcus isolated from a coconut cream pie implicated in an outbreak of food poisoning. Sterile filtrates from semi-solid agar cultures repeatedly produced symptoms in some monkeys but not in others of a group and therefore no definite conclusions could be drawn about the toxicity of the cultures.

Cary, Dack and Davison (1938) reported on a large institutional outbreak of food poisoning attributed to beef croquettes which were found, five days after the outbreak, to contain enormous numbers of alpha-hemolytic streptococci of two biochemical types. (From the published tests it is not possible to identify the species concerned.) Seven volunteers were fed filtrate from cultures of these streptococci and showed no ill effects, but the ingestion of living broth cultures resulted in symptoms in five of seven subjects after an incubation period of about 12 hours. This

Table 17.2. Outbreaks of Food Poisoning Attributed to Fecal Streptococci

Reference	No. of Cases	No. at Risk	Food Implicated	Author's Identification	Comments
Linden, Turner and Thom (1926)	9 22	? ?	Albanian cheese American cheese	Not identified	Identified as S. faecalis by Sherman, Smiley and Niven (1943).
Cary, Dack and Myers (1931)	75	182	Vienna sausage	"Green hemolytic strep."	2 biochemical types; one might have been an "enterococcus."
Jordan and Burrows (1934)	?	?	Coconut cream pie	"Str. viridans"	Not possible to name. Cell-free filtrate produced symptoms in susceptible monkeys.
Cary, Dack and Davison (1938)	117	208	Beef croquettes	α hemolytic streptococci; 2 biochemical types	Not possible to name these from the tests recorded. Feeding tests did not show a toxin.
Dack (1943)	294	393	Turkey dressing	α hemolytic streptococci	Identified as S. faecalis by Buchbinder et al.
Buchbinder, Osler and Steffen (1948)	74	161	Evaporated milk	S. faecalis	Physiological reactions were atypical.
	3	3	Charlotte russe	S. faecalis and S. liquefaciens	Both present in large numbers.
	74	87	Barbecued beef	S. faecalis and S. liquefaciens	17 cultures S. faecalis 1 culture S. liquefaciens.
	9	9	Ham bologna	S. faecalis and S. liquefaciens	72 cultures S. faecalis 4 cultures S. liquefaciens.
Dack, Niven, Kirsner and Marshall (1949)	94	98	Turkey à la king	S. liquefaciens	

is similar to the incubation period in the Salmonella type of food poisoning, rather than in the staphylococcal type associated with toxin formation. These experiments indicated that living cultures of certain alpha-hemolytic streptococci and not filtrates from them were necessary to produce symptoms of food poisoning in man.

Osler, Buchbinder and Steffen (1948) also carried out feeding tests on human volunteers with four strains of S. faecalis. Three of these strains had been isolated from human feces within two months of using and one had been isolated about one year previously from a can of evaporated milk which had been implicated in an outbreak of gastroenteritis. The cultures when grown for five hours in milk, custard or egg salad produced symptoms of gastric or intestinal disturbance or both in six out of 26 volunteers but the same strains when grown for 20 hours did not produce symptoms.

The idea that a metabolic product of microbial growth might be responsible for the symptoms was put forward by Sherman, Gunsalus and Bellamy (1944) who examined some half-dozen strains of S. faecalis from outbreaks of food poisoning. They suggested that the decarboxylation of tyrosine to tyramine was the probable cause of the poisoning, but later investigations have shown that tyramine is not a likely cause of poisoning; one gram amounts of tyramine monohydrochloride have been ingested by human volunteers without effect (Dack et al., 1949). The decarboxylation of tyrosine is a general property of S. faecalis (Sharpe, 1948) and not a function of some strains only. There have been numerous accounts of the occurrence of Group D streptococci in food products, e.g., from an investigation of the flora of English hard cheese, Mattick and Shattock (1943) reported that all 30 cheeses examined contained Group D streptococci in numbers ranging from 10,000 to more than 10,000,000 per gram of cheese. Shattock (1949b) recorded that one of the most active tyrosine decarboxylase producing strains of S. faecalis she had encountered had been isolated in large numbers from cheese consumed without any ill effects, and expressed the opinion that because S. faecalis occurred frequently in food it seemed improbable that tyramine could be the toxic agent per se in outbreaks of food poisoning. This was amply confirmed by the work of Dack et al. (1949) who investigated whether a particular strain of S. faecalis used as cheese starter might be a potential source of food poisoning. In experiments involving 52 feeding tests on 37 human volunteers this strain of S. faecalis and cultures of three strains of "enterococci" which had been implicated in outbreaks of food poisoning were studied. Cheese made with the starter strain of S. faecalis contained very large numbers of viable streptococci and very appreciable amounts of tyramine but caused no ill effects when fed to volunteers. Likewise tyramine monohydrochloride in as much as one gram amounts fed in milk to human volunteers caused no ill effects.

If Group D streptococci are enteropathogenic it is surprising that outbreaks of streptococcal food poisoning do not occur more frequently, because these organisms often occur in foodstuffs (e.g. in cheese, Mattick and Shattock, 1943; in commercially frozen fruits, fruit juice concentrates and vegetables, Larkin Litsky and Fuller, 1955a; in meat- and fish-curing brines, Buttiaux and Moriamez, 1957; in frozen sea food, Raj, Wiebe and Liston, 1961) and are well adapted to grow under a wide range of environmental conditions.

More recently doubt has been cast upon the specific role of Group D streptococci in food poisoning. There have been several reports of human volunteers who consumed very large numbers of living S. faecalis without showing symptoms of food poisoning. Buttiaux (1956) reported the consumption of ham containing more than 3 million S. faecalis per gram without ill effects; the inconclusive results of feeding experiments of Osler et al. (1948) and Dack et al. (1949) have already been referred to. Recently Silliker and Deibel (1960) reported an unsuccessful attempt to demonstrate food poisoning potentialities in 24 strains of S. faecalis and S. faecium, including 9 cultures which had been implicated in previous outbreaks of food poisoning. They were unable to demonstrate any symptoms of food poisoning in human volunteers fed pure cultures of these strains grown in skim milk, ham slices and various laboratory media. This study did not find cultural conditions necessary for the production of a noxious agent; none of the strains tested produced any gastro-intestinal symptoms.

From all the evidence one is forced to conclude either that the environmental conditions for producing a pathogenic principle for Group D streptococci are only very occasionally encountered, or that only exceptional strains are potentially enteropathogenic.

The question of the relation, if any, of Group D streptococci to food poisoning will be answered only when strains carefully identified physiologically and more precisely by serological type are incriminated unequivocally in an outbreak on circumstantial evidence. This should include the demonstration that the isolates are capable of causing a disease in human volunteers.

The type serology of Group D streptococci has already been mentioned; perhaps strains incriminated in outbreaks of food poisoning may prove to be of particular antigenic types. There is no suggestion however, that any Group D streptococcus has an antigen analogous to the M antigen of Group A streptococci which is related to virulence. There may, however, be an analogy with another condition; namely, enteritis in infants and young animals caused by certain strains of <u>Escherichia coli</u>. Both <u>E. coli</u> and Group D streptococci are normal inhabitants of the intestinal tract of man and animals, and outside the digestive tract they are both capable of causing certain pathological conditions. Antigenically <u>E. coli</u> is exceedingly diverse and exists as some 140 serogroups and almost innumerable serotypes and although apparently all groups of <u>E. coli</u> are potentially pathogenic outside the intestinal tract less than 10 per cent are etiologically related to infantile enteritis (Neter, 1959). As we have seen, Group D streptococci similarly have many serotypes; perhaps only exceptional types may cause gastroenteritis.

In my opinion, however, Group D streptococci have not conclusively been proved to cause food poisoning. In epidemiological investigations of this nature there must inevitably be a considerable delay between the consumption of food and its examination. Under these circumstances Group D streptococci could grow profusely and mask the real cause of poisoning.

In conclusion, Group D streptococci do not per se present a serious hazard in food. There is no doubt, however, that they provide a very useful indicator of fecal pollution. Moreover the identification of species by physiological tests can provide valuable help in tracing sources of contamination. The identification of serological types will give very precise information on the identity of strains which may be required in some circumstances.

LITERATURE CITED

Andrewes, F. W. and Horder, T. J. 1906. A study of the streptococci pathogenic for man. Lancet ii:708-13, 775-82, 852-55.

Barnes, E. M. 1956a. Tetrazolium reduction as a means of differentiating Streptococcus faecalis from Streptococcus faecium. J. Gen. Microbiol. 14:57-68.

_____. 1956b. Methods for the isolation of faecal streptococci (Lancefield Group D) from bacon factories. J. Applied Bact. 19:193-203.

_____. 1959. Differential and selective media for the faecal streptococci. J. Sci. Food Agric. 12:656-62.

_____, Ingram, M., and Ingram, G. C. 1956. The distribution and significance of different species of faecal streptococci in bacon factories. J. Applied Bact. 19:204-11.

Bartley, C. H. and Slanetz, L. W. 1960. Types and sanitary

significance of fecal streptococci isolated from feces, sewage and water. Am. J. Publ. Hlth. 50:1545-52.

Buchbinder, L., Osler, A. G., and Steffen, G. I. 1948. Studies in enterococcal food poisoning. I. The isolation of Enterococci from foods implicated in several outbreaks of food poisoning. Publ. Hlth. Rep., Wash. 63:109-18.

Buttiaux, R. 1956. Sur quelques faits nouveaux concernant les toxi-infections alimentaires. Rev. Médicale de Liège, XI: 521-33.

_____. 1958. Les streptocoques fécaux des intestins humains et animaux. Ann. Inst. Pasteur 94:778-82.

_____. 1959. The value of the association Escherichieae-Group D streptococci in the diagnosis of contamination in foods. J. Applied Bact. 22:153-58.

_____, and Moriamez, J. 1957. Le comportement des germes tests de contamination fécale dans les saumures de viandes. H. M. Station. Off. 247-62.

_____, and Mossel, D. A. A. 1961. The significance of various organisms of faecal origin in foods and drinking water. J. Applied Bact. 24:353.

Cary, W. E., Dack, G. M., and Davison, E. 1938. Alpha type streptococci in food poisoning. J. Infect. Dis. 62:88-91.

_____, and Myers, E. 1931. An institutional outbreak of food poisoning possibly due to a streptococcus. Proc. Soc. Exp. Biol., N. Y. 29:214-15.

Cooper, K. E. and Ramadan, F. M. 1955. Studies in the differentiation between human and animal pollution by means of faecal streptococci. J. Gen. Microbiol. 12:180-90.

Dack, G. M. 1956. Food Poisoning. Chicago. Univ. of Chicago Press.

_____, Niven, C. F., Kirsner, J. B., and Marshall, H. 1949. Feeding tests on human volunteers with enterococci and tyramine. J. Infect. Dis. 85:131-38.

Dunican, L. K. and Seeley, H. W. 1962. Starch hydrolysis by Streptococcus equinus. J. Bact. 83:264-69.

Elliott, S. D. 1960. Type and group polysaccharides of Group D streptococci. J. Exp. Med. 111:621-30.

Gunsalus, I. C. 1947. Products of anaerobic glycerol fermentation by Streptococcus faecalis. J. Bact. 54:239-44.

Guthof, O. 1957. Streptokokken und Dysbakterie-Problem. Zbl. Bakt. 1. Orig. 70:327-33.

Hartsell, S. E. and Caldwell, J. H. 1961. Lysozyme and the differentiation of Group D streptococci. Proc. 2nd Int. Symp. Fleming's Lysozyme, Milan, Italy.

Hobbs, B. C. 1961. The public health significance of Salmonella carriers in livestock and birds. J. Applied Bact. 24:340-52.

Jones, D. 1958. Physiological and serological studies of Group D streptococci. Univ. Reading: Thesis.

_____, and Shattock, P. M. F. 1960. The location of group antigen of Group D streptococcus. J. Gen. Microbiol. 23:335-43.

Jordan, E. O. and Burrows, W. 1934. Streptococcus food poisoning. J. Infect. Dis. 55:363-67.

Kenner, B. A., Clark, H. F., and Kabler, P. W. 1960. Fecal streptococci. II. Quantification of streptococci in feces. Am. J. Publ. Hlth. 50:1553-59.

Kjellander, J. 1960. Enteric streptococci as indicators of fecal contamination of water. Acta Path. Microbiol. Scand. suppl. 136, 48:1-133.

Kereluk, K. 1959. Studies on the bacteriological quality of frozen meat pies. III. Identification of enterococci isolated from frozen meat pies. Applied Microbiol. 7:324-26.

_____, and Gunderson, M. F. 1959. Studies on the bacteriological quality of frozen meat pies. I. Bacteriological survey of some commercially frozen meat pies. Applied Microbiol. 7:320-23.

Lake, D. E., Deibel, R. H., and Niven, C. F. 1957. The identity of Streptococcus faecium. Bact. Proc. 13.

Lancefield, R. C. 1933. A serological differentiation of human and other groups of hemolytic streptococci. J. Exp. Med. 57:571-95.

Larkin, E. P., Litsky, W., and Fuller, J. E. 1955a. Fecal streptococci in frozen foods. I. A bacteriological survey of some commercially frozen foods. Applied Microbiol. 3:98-101.

_____. 1955b. Fecal streptococci in frozen foods. II. Effect of freezing storage on Escherichia coli and some fecal streptococci inoculated onto green beans. Applied Microbiol. 3:102-4.

_____. 1955c. Fecal streptococci in frozen foods. III. Effect of freezing storage on Escherichia coli, Streptococcus faecalis and Streptococcus liquefaciens inoculated into orange concentrate. Applied Microbiol. 3:104-6.

Linden, B. A., Turner, W. R., and Thom, C. 1926. Food poisoning from a streptococcus in cheese. Publ. Hlth. Rep., Wash. 41:1647-52.

Mattick, A. T. R. and Shattock, P. M. F. 1943. Group D streptococci in English hard cheese. Mon. Bul. Emerg. Publ. Hlth. Lab. Serv. 2:73.

McCarty, M. 1952a. The lysis of Group A hemolytic streptococci by extracellular enzymes of Streptomyces albus. I. Production and fractionation of the lytic enzymes. J. Exp. Med. 96: 555-68.

_____. 1952b. The lysis of Group A hemolytic streptococci by extracellular enzymes of Streptomyces albus. II. Nature o the cellular substrate attacked by lytic enzymes. J. Exp. Med. 96:569-80.

Medrek, T. F. and Barnes, E. M. 1958. A note of the growth of Streptococcus bovis in thallous acetate and tetrazolium media. J. Applied Bact. 21:79.

_____. 1962a. The influence of the growth medium on the demonstration of a Group D antigen in faecal streptococci. J. Gen. Microbiol. 28:701-709.

_____. 1962b. The physiological and serological properties of Streptococcus bovis and related organisms isolated from cattle and sheep. J. Applied Bact. 25:169-79.

Moore, B. 1955. Streptococci and food poisoning. J. Applied Bact. 18:606-18.

Mossel, D. A. A., van Diepen, H. M. J., and de Bruin, A. S. 1957. The enumeration of faecal streptococci in foods, using Packer's crystal violet sodium azide blood agar. J. Applied Bact. 20:265-72.

Neter, E. 1959. Enteritis due to enteropathogenic Escherichia coli. Present-day status and unsolved problems. J. Pediat. 55:223-39.

Orla-Jensen, S. 1919. The Lactic Acid Bacteria. Copenhagen: A. F. Høst and Son.

Osler, A. G., Buchbinder, L., and Steffen, G. I. 1948. Experimental enterococcal food poisoning in man. Proc. Soc. Exp. Biol. N. Y. 67:456-59.

Raibaud, P., Caulet, M., Galpin, J. V., and Mocquot, G. 1961. Studies on the bacterial flora of the alimentary tract of pigs. II. Streptococci: selective enumeration and differentiation of the dominant group. J. Applied Bact. 24:285-306.

Raj, H., Wiebe, W. J., and Liston, J. 1961. Detection and enumeration of fecal indicator organisms in frozen sea foods. II. Enterococci. Applied Microbiol. 9:295-303.

Sharpe, M. E. 1948. Some biochemical characteristics of Group D streptococci isolated from infants' feces, with special reference to their tyrosine decarboxylase activity. Proc. Soc. Applied Bact. p. 13-17.

_____. 1952. Group D streptococci in the faeces of healthy infants and of infants with neonatal diarrhea. J. Hyg. 50:209-28.

Sharpe, M. E. 1962. (Personal communication.)

_____, and Fewins, B. G. 1960. Serological typing of strains of Streptococcus faecium and unclassified Group D streptococci isolated from canned hams and pig intestines. J. Gen. Microbiol. 23:621-30.

_____, and Shattock, P. M. F. 1952. The serological typing of Group D streptocci associated with outbreaks of neonatal diarrhea. J. Gen. Microbiol. 6:150-65.

Shattock, P. M. F. 1949a. The streptococci of Group D; the serological grouping of Streptococcus bovis and observations on serologically refractory Group D strains. J. Gen. Microbiol. 3:80-92.

_____. 1949b. The faecal streptococci. 12th Int. Dairy Congress. Stockholm. 2:598-604.

_____. 1955. The identification and classification of Streptococcus faecalis and some associated streptococci. Ann. Inst. Pasteur Lille 7:95-100.

Sherman, J. M. 1937. The streptococci. Bact. Rev. 1:3-97.

_____. 1938. The enterococci and related streptococci. J. Bact. 35:81-93.

_____, Smiley, K. L., and Niven, C. F. 1943. The identity of a streptococcus associated with food poisoning from cheese. J. Dairy Sci. 26:321-23.

_____, Gunsalus, I. C., and Bellamy, W. D. 1944. 57th Ann. Rep., New York State Coll. Agr., Cornell Univ. Agr. Exper. Stat. 116.

Skadhauge, K. 1950. Studies of enterococci. Copenhagen: Einar Munksgaards.

Silliker, J. H., and Deibel, R. H. 1960. On the association of enterococci with food poisoning. Bact. Proc. 48.

Smith, D. G. and Shattock, P. M. F. 1962. The serological grouping of Streptococcus equinus. J. Gen Microbiol. 29:731-36.

Zaborowski, H., Huber, D. A., and Rayman, M. M. 1958. Evaluation of microbiological methods used for the examination of precooked frozen foods. Applied Microbiol. 6:97-104.

18

Staphylococcal Enterotoxin

G. M. DACK

UNIVERSITY OF CHICAGO

PRODUCTION IN FOODS

Strains of Staphylococci-Producing Enterotoxin

NO SIMPLE LABORATORY PROCEDURE will identify strains of staphylococci capable of producing enterotoxin. In laboratory experience at the Food Research Institute and at the American Meat Institute Foundation (Evans and Niven, 1950) all of the strains of staphylococci from food poisoning outbreaks have been found to produce coagulase. However, many coagulase-producing strains of staphylococci do not produce enterotoxin. With the first food poisoning strain of staphylococcus isolated in our laboratories from a Christmas cake, the production of enterotoxin in veal infusion broth cultures was variable when tested by feeding human volunteers.

Sugiyama, Bergdoll and Dack (1960) prepared an antiserum to S-6 enterotoxin from which nonenterotoxigenic components had been absorbed. When this antiserum was mixed in agar medium in Petri plates and the plates were streaked with S-6 organisms, not all of the cells from this culture produced enterotoxin, as evidenced by failure of a halo precipitate to develop around many of the isolated colonies.

The S-6 strain from which the S-6 enterotoxin has been purified was originally isolated by Evans (1948) from frozen shrimps which were not involved in food poisoning. However, this strain produced a very potent enterotoxin and for this reason was selected for Bergdoll's early work in the purification of enterotoxin. At this time the relationship of S-6 enterotoxin to enterotoxin from staphylococci isolated from food poisoning outbreaks was unknown. As the work has progressed, S-6 type of enterotoxin has not been found to be commonly associated with food poisoning, but is the type of enterotoxin produced by strains of staphylococci involved in pseudomembranous enterocolitis (Surgalla and Dack, 1955). This enterotoxin apparently is produced in lesions other than those of the intestinal tract, e.g., a 23-year-old woman died from gastrointestinal symptoms and shock 5 days following a Cesarean section. A staphylococcus isolated from a local area of peritonitis was found to produce a potent enterotoxin when fed to monkeys. The gel-diffusion test showed that it was related to our S-6 enterotoxin.

Bergdoll's present enterotoxin purification studies are being done with strain 196E which was isolated from food involved in a food poisoning outbreak and its enterotoxin is antigenically like that produced by most other food poisoning strains of staphylococci. There is evidence that a third enterotoxin exists for which future purification studies are contemplated. Perhaps a polyvalent antitoxin to enterotoxin will be developed which will be free of nonenterotoxin staphylococcal antibodies. When such an antiserum is available, testing large numbers of strains of staphylococci for their ability to produce enterotoxin as well as to evaluate conditions necessary for enterotoxin production in food processing will be simplified.

Types of Food in Which Enterotoxin is Produced

There are many types of foods in which staphylococci have been found to produce enterotoxin. In reviewing the outbreaks listed in Morbidity and Mortality Weekly Reports for the years 1956-1961, 137 cases were selected in which the incubation period following the ingestion of the incriminated food was characteristic for staphylococcal food poisoning. In most, but not all, of these outbreaks some laboratory tests were made. Fifty-seven of the outbreaks were due to meat; 45 of these were due to ham. Egg salad accounted for two outbreaks; poultry, for 23, 14 of which were turkey and 9 chicken. Five outbreaks were associated with fish; milk and cheese accounted for 8; bakery goods were involved in 25; potato salad, 10; and 11 were miscellaneous items such as vegetables.

Thatcher et al. (1959) have called attention to the wide dissemination of staphylococci having phage patterns commonly associated, respectively, with cattle and human infections. They have shown that Canadian cheese, made from unpasteurized milk, is one vehicle in the spread of food poisoning staphylococci. They demonstrated enterotoxin in 8 of 149 specimens tested from individual vats of cheese from a single factory selected because cheese from this factory was contaminated with large numbers of staphylococci. The toxic cheeses contained up to 1,500,000 staphylococci per gram.

The fate of food poisoning staphylococci in contaminated foods is dependent upon many factors; e.g., in rapidly cured ham the salt content in the curing solution may make this product more of a selective medium for staphylococci by inhibiting the non-halophilic bacteria. Sodium chloride is not inhibitory to the growth of staphylococci when it has been added to make up 10 per cent of the medium. On the other hand, certain precooked frozen foods such as pot pies containing meat, vegetables and gravy are contaminated with staphylococci in their preparation. If they are mishandled and subjected to times and temperatures favoring the growth of staphylococci, the staphylococci may not reach numbers sufficient to produce enough enterotoxin to cause illness (Peterson et al., 1962). Dack and Lippitz (submitted for publication) inoculated slurries of frozen pot pies with decimal dilutions of staphylococci. In the presence of the natural flora, staphylococci usually did not grow sufficiently to suggest a hazard from staphylococcus food poisoning when the inoculated slurry was incubated for 18 hours at 35° C. The predominant organism, a lactobacillus, produced considerable acid in the medium, but this organism inoculated into sterilized slurry with food poisoning staphylococci did not duplicate the inhibitory effect of the natural flora. Staphylococci grew very well in the slurry at pH 5 and higher. At pH 4.5 they grew slightly and failed to grow at pH's 4.0, 4.1 and 4.3.

Walker et al. (1961), in the case of the manufacture of Colby cheese from raw milk inoculated with food poisoning staphylococci, found that a maximum count was obtained at the time of cutting and hooping the curd. During the curing process at 10 °C over a period of 120 days the staphylococci diminished in numbers. These authors did not carry out studies to determine the cause of reduction in staphylococci during storage. It is, therefore, important in evaluating the safety hazards of a particular food item to staphylococcus food poisoning to consider not only the composition of the food but also the antagonistic effect of the natural flora on food poisoning staphylococci.

Time and Temperature Relationships
for Production of Enterotoxin

Barber (1914) found that cream which contained staphylococci was harmless when refrigerated but was toxic when left at room

temperature for 5 hours. Hauge (1951) found that mashed potatoes made with raw milk and kept at a warm temperature for
6 1/2 hours before serving became toxic. Wain and Blackstone
(1956) reported an outbreak of staphylococcus food poisoning
from ham. When the hams were removed from the oven, the internal temperature of the hams was 165°F. They were sliced between 7:15 and 8 a.m. At the time of slicing, two workers allegedly ate slices of the ham without developing any ill effects. At
8 a.m. the slices of ham were placed in two deep trays in a solid
mass as they came off the slicer. The trays were wrapped in
aluminum foil and placed in an oven at a temperature of about
200°F until 1:30 p.m. to 2 p.m. when they were picked up and
taken to a picnic ground. The outside temperature was approximately 90°F. At 3 p.m. one worker at the picnic grounds ate
some of the ham upon arrival and was taken ill 2 1/2 hours later.
Evidently, the ham was wholesome at 8 in the morning but at 3 in
the afternoon it was toxic. This involved an incubation period of
7 hours.

Segalove and Dack (1941) under experimental laboratory conditions demonstrated enterotoxin production in a culture grown
for 3 days at 18°C and one grown for 12 hours at 37°C. Cultures
grown for shorter periods did not contain enterotoxin. At 9° and
15°C enterotoxin was not produced in 7 and 3 days, respectively.
Enterotoxin was not formed after 4 weeks at 4° and 6.7°C.

Angelotti et al. (1959) found that staphylococci grew in custard and chicken á la king at temperatures of 44°F and above;
whereas, in ham salad no growth occurred at 40-50°F. These
workers suggested that food should be maintained at or below
42°F in order to prevent the growth of staphylococci. No studies
were conducted for the production of enterotoxin under these conditions. Angelotti et al. (1960) found in custard that staphylococci
grew at 114°F but decreased in numbers at 116-120°F. In
chicken á la king, staphylococci grew at 112°F and were killed at
all higher temperatures. In ham salad, staphylococci decreased
in numbers at temperatures of 112°-120°F. Therefore, depending
upon the type of food, staphylococci may grow over a range of
from 44°F to 114°F. No assays were made for enterotoxin in
these foods at the temperatures studied.

Resistance to Heat

The literature concerning the resistance of enterotoxin as it
occurs in food or culture medium has been reviewed (Dack, 1956).
Enterotoxin has been shown to withstand boiling and higher temperatures. Outbreaks of staphylococcus food poisoning in man

have furthermore been demonstrated to occur in products heated
and free from viable staphylococci at the time of consumption.
There is some evidence that purified enterotoxin is less resistant
to heat than the crude enterotoxin, and also, that heating a medium containing enterotoxin may make it ineffective for oral injection into monkeys but not for intraperitoneal or intravenous
injections into cats or monkeys. More work is needed to determine the precise effect of heat on enterotoxin.

EXPERIMENTAL PRODUCTION

Media Required

Bergdoll and his associates have made extensive studies on
suitable mediums for enterotoxin production (Bergdoll, 1962).
They have found an enzyme hydrolyzed casein (Protein Hydrolysate Powder, supplied by Mead Johnson and Co., Evansville, In-
diana) to be satisfactory as a medium for enterotoxin production
provided it is supplemented with niacin and thiamin. Enzyme hydrolyzed caseins obtained from other sources were also satisfactory for enterotoxin production provided they were supplemented
with dipotassium hydrogen phosphate or with protein hydrolysate
powder. The difficulty, however, with the supplemented enzyme
hydrolyzed caseins is the recovery of S-6 enterotoxin from the
culture supernatant by the ion exchange procedures. In the case
of enterotoxin of two strains other than the S-6, the more economical enzyme hydrolysate casein with supplementation may be
used because trichloroacetic acid precipitation is the method of
choice for these enterotoxins.

The amount of enterotoxin produced by the S-6 strain has
been variable. Studies with single isolated colonies have shown
that enterotoxin production varied greatly from colony to colony
when tested by using a method of incorporating antiserum in the
culture plates for selection of colonies producing a specific enterotoxin. Strains of high enterotoxin production have been maintained in the laboratory by drying selected cultures on porcelain
beads and storing them at 4°C. Using this method of storing the
S-6 culture enterotoxin production has been consistently good
over an 18-month period.

Chemical Nature

A number of chemical procedures have been investigated by
Bergdoll and his associates in the course of purifying S-6 and

partially purifying 196E enterotoxin. The S-6 enterotoxin has been purified in the following manner: 1) precipitation at pH 3.5 with phosphoric acid (H_3PO_4), 2) alumina adsorption, 3) ethanol precipitation, 4) cation exchange resin, Amberlite XE-64 chromatography, 5) starch electrophoresis, 6) ethanol precipitation and 7) freeze-drying (Bergdoll, Sugiyama and Dack, 1959).

The 196E enterotoxin has been prepared in approximately 70 per cent purity by: 1) removal from the culture supernatants by precipitation with trichloroacetic acid (pH 3.0); 2) adsorption of the enterotoxin by the cation exchange resin, Amberlite XE-64, pretreated with 0.02 M sodium phosphate, pH 6.2, 3) precipitation from 40 per cent ethanol at -13°C, and 4) chromatography on carboxymethyl cellulose employing sodium acetate buffers.

The S-6 enterotoxin is a protein with a molecular weight of approximately 23,000 having an isoelectric point at pH 8.5 (Hibnick and Bergdoll, 1959). S-6 enterotoxin is soluble and heat coagulable and is made up of 17 amino acids which have been determined by microbiological techniques. Aspartic acid, lysine, and tyrosine are found in greater amounts than any of the other 17 amino acids. The properties of the 196E enterotoxin are currently under investigation.

Number of Kinds

Enterotoxins produced by different strains of staphylococci appear to give identical symptoms in man and animals, but they differ when tested by immunological methods. Enterotoxin produced by strain 196E was found not to react with the antibody prepared with the purified enterotoxin from strain S-6 (Bergdoll, Surgalla and Dack, 1959). Specific antibodies to the 196E enterotoxin have been reported by Casman (1960) who used absorption techniques. Most enterotoxin producing strains that have been examined using two antiserums, S-6 and 196E, have been found to produce either one or the other enterotoxin. Strains isolated from foods implicated in food poisoning outbreaks appear to produce only 196E enterotoxin. However, Fujiwara (1961) reported isolating five strains of staphylococci from foods implicated in food poisoning outbreaks that produced the S-6 type of enterotoxin. In Bergdoll's experience the majority of strains that produce S-6 enterotoxin also produce 196E-type enterotoxin but in much smaller amounts. He has found that foods inoculated with S-6 strain and incubated resulted only in the production of 196E-type enterotoxin. When foods inoculated with the S-6 strain were mixed with water in a Waring Blendor and agitated during incubation, S-6 enterotoxin was produced. This enterotoxin was

absent in foods naturally inoculated with the same strain and incubated without agitation. The reason for this difference has not
been determined.

Specific Antiserum for Purified Enterotoxin

In the absence of a specific antiserum to 196E enterotoxin, a
feeding test for monkeys has been devised for measuring differences in enterotoxin. Monkeys which became resistant to 196E
enterotoxin were assumed to have developed specific antibodies
to this enterotoxin. Therefore, if these animals were fed S-6 enterotoxin to which they had no antibodies, they should react when
an emetic dose is given. Using the resistance of monkeys to specific enterotoxins at least three enterotoxic types have been identified, namely, S-6, 196E and a third enterotoxin from strain 137.
The latter enterotoxin is to be studied chemically at a later time.

MODE OF ACTION OF STAPHYLOCOCCAL ENTEROTOXIN IN THE BODY

Sugiyama and his associates have attempted to identify enterotoxin with biological phenomena with the hope of finding a relationship which may prove helpful in the assay of enterotoxin. A
transitory elevation of the serum glutamic-oxalacetic transaminase
serum levels occurs in monkeys fed enterotoxin (Sugiyama, Bergdoll and Dack, 1958b). They have suggested that this may result
from some tissue damage from the action of enterotoxin.

The incidence of vomiting in monkeys treated with nonemetic
doses of dihydroergotamine methanesulfonate (DHE-45) and fed
staphylococcal enterotoxin is twice as great as in monkeys not
given the drug (Sugiyama, Bergdoll and Dack, 1958a). These results suggested a possible central nervous system site of action
of enterotoxin, since DHE-45 has been closely related to dihydro-
genated ergot alkaloids known to cause emesis by acting at the
chemoreceptor trigger zone of dogs. Perphenazine (Trilafon) and
reserpine (Serpasil) in doses which do not significantly depress
the vomiting center were effective antiemetics against enterotoxin induced vomiting in monkeys (Sugiyama, Bergdoll and
Wilkerson, 1960). Perphenazine is a potent antiemetic on drugs
known to act at the chemoreceptor trigger zone of dogs. Bilateral ablation of the area postrema (chemoreceptor trigger zone)
of monkeys rendered them highly resistant to the emetic action of
enterotoxin. Bilateral transection of the vagus nerve at the

diaphragm level of monkeys also resulted in a high degree of resistance to emetic action of enterotoxin (Sugiyama, Chow and
Dragstedt, 1961). Since the chemoreceptor trigger zone in the
area postrema is adjacent to the dorsal sensory nucleus of the
vagus, serial histological sections of the brain stems were made.
No injury to the dorsal sensory nucleus of the vagus was observed. Both these surgical procedures gave protection against
the S-6 and 196E types of enterotoxin. The protection of monkeys
against enterotoxin following either vagotomy or chemoreceptor
trigger zone ablation is similar to that observed with the early
emesis caused by a single 1200 r X-irradiation. Species difference may exist as to the site of emetic action of enterotoxin.
Cats with chronic ablation of the chemoreceptor trigger zone
challenged by intravenous enterotoxin still vomited but decere-
bration resulted in protection (Clark, Vanderhooft and Borison,
1961).

It is obvious that much knowledge has been gained in a study
of staphylococcus enterotoxin. As the various types of staphylococcal enterotoxins are chemically purified, knowledge should be
extended much further. It is hoped that this knowledge will eventually lead to a satisfactory and simple assay procedure for the
detection of enterotoxin.

LITERATURE CITED

Angelotti, R., Foter, M. J., and Lewis, K. H. 1960. Time-
    temperature effects on salmonellae and staphylococci in
    foods. II. Behavior at warm holding temperatures.
    Thermal-death-time studies. Robert A. Taft Sanitary Engi-
    neering Center Technical Report F60-5.

_____, Wilson, E., Foter, M. J. and Lewis, K. H. 1959.
    Time-temperature effects on salmonellae and staphylococci
    in foods. I. Behavior in broth cultures and refrigerated
    foods. Robert A. Taft Sanitary Engineering Center Technical
    Report F59-2.

Barber, M. A. 1914. Milk poisoning due to a type of Staphylo-
    coccus albus occurring in the udder of a healthy cow. Phil-
    ippine J. Science , 9(6), Sec. B., Trop. Med., Nov. p. 515-19.

Bergdoll, M. S. 1962. Chemistry and detection of staphylococcal
    enterotoxin. Proceedings of the Fourteenth Research Con-
    ference, American Meat Institute Foundation, University of
    Chicago, March 22-23 (in press).

_____, Sugiyama, H., and Dack, G. M. 1959. Staphylococcal
    enterotoxin. I. Purification. Arch. Biochem. Biophys. 85:
    62-69.

Bergdoll, M. S., Surgalla, M. J., and Dack, G. M. 1959. Staphylo-coccal enterotoxin. Identification of a specific precipitating antibody with enterotoxin-neutralizing property. J. Immunol. 83:334-38.

Casman, E. P. 1960. Further serological studies of staphylo-coccal enterotoxin. J. Bact. 79:849-56.

Clark, W. G., Vanderhooft, G. F., and Borison, H. L. 1961. Emetic and pyrogenic effects of staphylococcal enterotoxin (S.E.) in cats. Fed. Proc. 20:230.

Dack, G. M. 1956. Food Poisoning. Chicago. Univ. of Chicago Press, 3rd ed. Pp. 147-50.

_____, and Lippitz, G. 1962. Fate of food poisoning micro-organisms introduced into slurry of frozen pot pies. Appl. Microbiol. (Submitted for publication).

Evans, J. B. 1948. Studies of staphylococci with special refer-ence to the coagulase-positive types. J. Bact. 55:793-800.

_____, and Niven, C. F., Jr. 1950. A comparative study of known food poisoning staphylococci and related varieties. J. Bact. 59:545-50.

Fujiwara, K. 1961. Annual Report Inst. Food Microbiol., Chiba University, Japan 14:1.

Hauge, S. 1951. Staphylococcus food poisoning caused by entero-toxic staphylococci in milk. Nordisk Veterinaermedicin 3: 931-56.

Hibnick, H. E. and Bergdoll, M. S. 1959. Staphylococcal entero-toxin. II. Chemistry. Arch. Biochem. and Biophys. 85:70-73.

Peterson, A. C., Black, J. J., and Gunderson, M. F. 1962. Staph-ylococci in competition. II. Effect of total numbers and pro-portion of staphylococci in mixed cultures on growth in arti-ficial medium. Appl. Microbiol. 10:23-30.

Segalove, M. and Dack, G. M. 1941. Relation of time and tem-perature to growth and enterotoxin production of staphylo-cocci. Food Research 6(2):127-33.

Sugiyama, H., Bergdoll, M. S., and Dack, G. M. 1958a. Staphylo-coccal enterotoxin: Increased vomiting incidence in monkeys following subemetic doses of dihydroergotamine. Proc. Soc. Exptl. Biol. Med. 97:900-903.

_____. 1958b. Increased serum glutamic-oxalacetic tran-saminase of monkeys following oral administration of staph-ylococcal enterotoxin. Proc. Soc. Exptl. Biol. Med. 98:33-36.

_____. 1960. In vitro studies on staphylococcal enterotoxin production. J. Bact. 80(2):265-70.

Sugiyama, H., Bergdoll, M. S., and Wilkerson, R. G. 1960. Perphenazine and reserpine as antiemetics for staphylococcal enterotoxin. Proc. Soc. Exptl. Biol. Med. 103:168-72.

Sugiyama, H., Chow, K. L., and Dragstedt, L. R., II. 1961. Study of emetic receptor sites for staphylococcal enterotoxin in monkeys. Proc. Soc. Exptl. Biol. Med. 108:92-95.

Surgalla, M. J. and Dack, G. M. 1955. Enterotoxin production by micrococci from cases of enteritis after antibiotic therapy. J. Am. Med. Assoc. 158:649-50.

Thatcher, F. S., Comtois, R. D., Ross, D., and Erdman, I. E. 1959. Staphylococci in cheese: Some public health aspects. Canada J. Publ. Health 497-503.

Wain, H. and Blackstone, P. A. 1956. Staphylococcal gastroenteritis. Report of a major outbreak. Am. J. Digestive Dis. 1(10):424-29.

Walker, G. C., Harmon, L. G., and Stine, C. M. 1961. Staphylococci in Colby cheese. J. Dairy Science 44:1272-82.

19

Microbial Toxins:
Commentary and Discussion

C. F. NIVEN, JR.

AMERICAN MEAT INSTITUTE FOUNDATION

COMMENTARY

AMONG THE THOUSANDS of bacterial species, and the hundreds of metabolites produced by them, relatively few bacteria are known to be capable of producing substances that are toxic to man and animals. Even among these few microorganisms, most of them must invade and infect the host in order to produce illness as the result of toxin production; e.g., erythrogenic, tetanus, and diphtheria toxins. Fortuitously only two bacterial species are recognized which are capable of producing extracellular metabolites when growing in foods, and which are toxic when introduced via the oral route; namely, Staphylococcus aureus and Clostridium botulinum.

Several similarities between botulinum toxins and the staphylococcal enterotoxins have been pointed out. Both are so-called extracellular, antigenic proteins of extremely high toxicity. Several serologically distinct types of each toxin are known to exist, and the human or animal hosts can be actively immunized against the individual toxin types. Although information concerning the modes of action of each class of toxin leaves much to be desired, both appear to be neurotoxins.

Of course, there are several differences between the two classes of toxins with respect to the conditions under which they are produced, relative toxicity, time interval required for onset of symptoms, and the clinical manifestations noted in the hosts. In contrast with the staphylococcal enterotoxins, the botulinum toxins are much less stable to heat as they exist in foods, possess greater potency and lethal effect, and have a much wider host range.

Because of the drastic consequences of food-borne botulism

outbreaks, the world's commercial food producers and public health officials are extremely conscious of this hazard. As a result of the precautionary measures taken by the food producers for the past four decades, the world has enjoyed a remarkable record of freedom from botulism episodes originating from commercially canned foods. Such episodes have been confined largely to those resulting from home-canned foods where we encounter the pride of the homemaker in her culinary and canning arts handed down from mother and grandmother. However, we are witnessing an increasing dependence upon precooked, commercially packed foods with an attendant disappearance of home canning. We should, therefore, anticipate a gradual disappearance of botulism episodes confined to individual families.

Fortunately, many foods which have been spoiled by the more common Cl. botulinum types possess an obnoxious odor and may show swelling of the container due to gas production. No doubt, the rejection of foods due to the disagreeable odor has spared untold thousands of lives.

As pointed out by Riemann, several commercially canned foods are of proven safety, not as a result of scientific investigations but from long-standing usage without incidence. Some commonly accepted foods are not thermally processed sufficiently to kill all Cl. botulinum spores that may be present. We think we know most or all the reasons for their safety. Nevertheless, any changes in formulation, processing or subsequent handling of these foods, even though seemingly subtle, should be viewed with a critical eye.

As pointed out by Dack, staphylococcus food poisoning is the most common type of food intoxication of bacterial origin in the United States. It is also common in other parts of the world. Yet, the only reasons for its existence are laxity and ignorance among those who prepare foods for human consumption, and in some instances, the food manufacturer. Today, our knowledge is sufficient to completely eliminate this type of food poisoning from the face of the earth. To accomplish this goal, there is a need for an intensive educational program concerning the proper manner of using facilities already available, and directed mainly to the housewife and others who prepare foods for human consumption. Without this educational program, we shall continue to have the occasional explosive outbreak of staphylococcus food poisoning in spite of any scientific advances yet to come.

In addition to an educational campaign, there is a specific need for methods to detect enterotoxin in foods. Such methods would provide valuable tools for more complete epidemiological investigations of food poisoning episodes. Also, they might become useful as routine tests of safety for foods prepared in large quantities (e.g., precooked, frozen meals) to prove their safety prior to consumption.

Unfortunately, the food poisoning staphylococcus may grow extensively and produce enterotoxin in a food without yielding any indications of spoilage with respect to taste, odor, texture or appearance. Satisfactory methods are available for detecting viable Staphylococcus aureus in foods, but they are rather slow. Their value lies mainly in epidemiological investigations of food poisoning outbreaks, as well as in bacteriological control of foods produced commercially in large quantity and held for relatively long periods of time prior to eating. A specific chemical or serological test for detecting enterotoxin would have complementary value, especially in investigating foods that might have had enterotoxin production prior to a terminal heating.

Although of worldwide occurrence, neither botulism nor staphylococcus food poisoning is an important international problem because few foods are distributed internationally which are likely to be incriminated in such outbreaks. Of course, there are exceptions such as cheese and dried milk, as well as other foods packed in defective cans. However, with the anticipated increase in international distribution of processed dried foods, especially to the underdeveloped countries, we must be increasingly aware of the possibility of some batches of these foods being unwholesome and hazardous to human health. Unlike salmonellosis, botulism and staphylococcal food poisoning are self-limiting and are confined only to the original incriminated foods. Other foods probably would not become contaminated and become involved secondarily in the spread of the disease. It is entirely possible, however, that once the dried foods have been moistened or mixed with other moist foods, they may serve as inocula for subsequent toxin production.

As indicated by Shattock, toxin production by the enterococci has never been demonstrated unequivocally. The need is great to resolve the question as to whether enterococci should be considered food poisoning microorganisms, and if so, under what environmental growth conditions any of the strains will produce symptoms.

Yearly, a number of food poisoning outbreaks are reported in which fecal streptococci are incriminated. Invariably, the only evidence reported to incriminate the enterococci is that they were present in large numbers in the suspected food. Until there is additional evidence, the mere presence of enterococci in foods definitely is not sufficient to conclude that they were the cause of a specific food poisoning outbreak. No doubt, if wholesome foods were examined more often, a high percentage would be found to contain enterococci, some in large numbers.

The enterococci are versatile with respect to the environmental conditions under which they can grow. They survive heat processing to which many foods are subjected, and they can grow at temperatures normally considered to be adequate refrigeration. Many foods, especially cheeses and fermented sausages, normally may contain large numbers of enterococci. Actually, they have been used successfully as starter organisms for these foods.

Evidence is preponderant that the habitat of the enterococci is the intestine of warm blooded animals. Herein lies their value as an index of fecal pollution among some foods. However because of their versatility the enterococci can establish themselves as a normal food plant contaminant and grow under conditions far removed from their original habitat. If the stigma of food poisoning could be removed, these microorganisms could be put to good use in several food fermentations.

Short term gastrointestinal episodes of unknown etiology occur far too frequently. Unfortunately, the absence of any satisfactory reporting system in the United States prevents us from making an accurate estimate as to their frequency. Furthermore, our inadequate systems of investigating such outbreaks do not allow an intelligent guess as to the proportion which originated from foods. It is highly desirable that more thorough investigation of gastrointestinal episodes be conducted, and that a systematic mode of reporting be instituted so that we can deal with them more effectively. It is entirely possible that many such episodes are caused by agents not now recognized as being capable of causing infections or intoxications. A collaborative spirit among physicians, public health and control officials, and the food industries in achieving these goals will certainly result in an even higher level of public health than we enjoy today.

DISCUSSION

Question: Is there any relationship between enterotoxin and other toxins such as the α or β hemolysins and necrotoxins?

Dack: There is no apparent relation between enterotoxin and these other toxins. It is quite a job when you are working with enterotoxin to pick the right one and to be sure you have it; but we have found no correlation.

Question: What is your idea as to the mode of liberation of

botulinus toxin? Is it released from lysing cells or is it actually an extracellular toxin produced by living cells? Can it be part of the spore?

Riemann: It seems that the toxin is liberated rather late in the growth phase. Some people have the opinion that a pro-toxin is formed during logarithmic growth and is transformed by enzymes released from the cell at the time the cells lyse. There have been some reports that resting cells will form toxin but it seems rather to be a question of liberation of the toxin from the cells in the resting stage. There doesn't seem to be any de novo synthesis of toxin in resting cells.

Perry, G. A.: Streptococcus faecalis has been reported to occur in salt whey. Also, at the A.S.M. meeting in Kansas City it was reported as commonly occurring on plants. What is the significance of these organisms in processed fruits and vegetables? Do they represent any health hazard?

Niven: Yes, that was reported by Dr. Mundt (University of Tennessee, Knoxville), I believe. I didn't hear the paper but I did talk with Dr. Mundt. He inoculated plant materials and got extensive growth. How are we going to resolve the enterococcus food poisoning question? Dr. Dack and I have requested the food producing world and public health officials that when a food poisoning outbreak occurs attributable to enterococci, send us the cultures and we will complete the Koch's postulates. This means human feeding. We have already done some of this, but keep in mind that one positive test may bear more weight than a thousand negatives; one negative test means very little.

Ordal, Z. J. (to Shattock or Niven): From the discussions this morning I gather that there is considerable favor for the use of fecal streptococci as index organisms. In relation to your remarks of a minute ago concerning their ability to develop on plants, which Dr. Mundt has been demonstrating, how do you resolve this relationship, especially if you are going to put a numerical value on them?

Niven: And could we also say the same thing about coliforms?

Shattock: It brings out the fact that we just can't use the same yardstick for all foods. We must know what we are doing and something about the history of the food. We can't have the same standards for every kind of food.

Perry: Dr. Mundt says that certain plants produce antibiotics.

Appleman: Yes, these include such things as cabbage, cirtus oil, rape seed, and mustard.

Raj, H.: Dr. Buttiaux has reported that the incidence of E. coli in the human gut is not the same today as it was 10 or 15 years ago because of dietary habits and use of antibiotics. I would like to know your comments on that.

Shattock: This is one of the reasons Dr. Buttiaux suggested the use of Group D streptococci for indicator organisms.

El-Bisi, H. M.: Recently in Kansas City it was reported that when massive populations of wet, detoxified spores of Clostridium botulinum were irradiated, they demonstrated activity similar to that of botulinum toxin. Can clean, detoxified spores produce toxin?

Grecz, N.: Yes, botulinum spores "detoxified" by heat-shocking at 80°C for 10 minutes are capable of inducing typical botulinum toxemia when injected into white mice. Since extracellular botulinum toxin is normally extremely heat sensitive, we are probably confronted with some kind of a biologically stabilized form of toxin in the spore. This matter is of considerable theoretical interest because there is possibly a basic implication as to the mechanism of preserving the normally heat sensitive toxin within the spore, i.e. a way in which the spore offers protection to the toxin — as well as to other normally heat sensitive compounds.

We have done some preliminary experiments showing that unheated Type A spores which have been thoroughly cleaned by enzymes and ultrasonics possessed the same order of toxicity as spores which have been "detoxified" by heat-shocking at 80°C for 10 minutes. We cited this finding as support for the cleanliness of our spores. In this connection — I think — it is perfectly all right to use the lowest, i.e., the basic, toxicity of botulinum spores as a criterion of spore cleanliness.

The question has been raised if toxin is really present within the spores or if it is synthesized after we inject the spores into the mouse; it is conceivable that spores may germinate, grow and produce toxin in the mouse. The answer to this question is not entirely known; however, available observations indicate that some toxin is present in the spore per se. First of all, we need massive doses in order to produce a lethal toxemia; this indicates that it is not a synthesis since very few growing organisms would be needed to synthesize enough toxin to kill a mouse. Secondly, previous experiments have been done by us and other investigators in which botulinum spores were disintegrated and extracted. The spore free extract when injected into mice poisoned the mice in the same fashion. In fact, Dr. Graikoski (University of Michigan, Ann Arbor) told me that there is a ten-fold increase in toxicity of the extract as compared to the original spores.

We would like to point out, however, that in our laboratory only Type A spores have been found to be toxic; heat-shocked Type B or E spores had no effect on the mice even when used in 100-fold higher concentration. A check on published reports on toxicity of botulinum spores will show that all investigators were dealing exclusively with Type A. However, these are just preliminary experiments. We have used only a very limited number of strains, and so no general conclusion can be made as yet.

The pathogenicity of spores has also been investigated in England by Dr. J. Keppie and the results were that actually spores of C. botulinum are not pathogenic in the sense that they do not grow out in healthy animal tissue, but they will grow out and produce toxin in damaged tissue which provides anaerobic conditions. The conclusion was that, in human food poisoning cases, spores of C. botulinum found in spoiled foodstuffs are not of importance compared with the toxin produced by the initial growth of the contaminating organisms.

As to the toxicity of spores in experimentally inoculated food products, the whole matter seems to be an experimental artifact resulting from the use of massive spore loads. Consequently, the spore toxicity appears to be of more academic interest than of practical concern.

The toxicity of spores has been particularly emphasized in connection with food radiation preservation experiments, since botulinum toxin is relatively resistant to ionizing radiations. Two aspects have been of concern: (1) the toxicity of botulinum spores per se experimentally added to inoculated, irradiated food packs, and (2) the hypothetical possibility that radiation killed spores may perhaps still possess sufficient active enzymes to synthesize significant amounts of toxin under "resting culture" conditions. This second possibility may become important during prolonged storage at favorable temperatures.

The former hazard can be ruled out because of the extremely rare natural incidence of botulinum spores in food products. The latter hypothesis is being presently examined by us. The data available at this time indicate that no significant amounts of toxin were synthesized by radiation killed spores in a resting culture situation even when massive spore populations were employed.

Niven: Didn't you report that you required 10^6 spores injected interperitoneally into the mouse to produce toxic symptoms?

Grecz: With Type A spores, 10^6 spores have consistently produced a lethal toxemia. The effect could be neutralized by a specific antitoxin. The people from the University of Michigan disagree with us to the extent that they say it takes 10^7 spores for a lethal effect in a mouse. This may be due to the use of different botulinum strains or to the use of mice of different weight. We consistently use 16-20 g. mice. As many as 10^8 spores of Type B did not cause a lethal toxemia in white mice.

Raj: You mentioned the relationship of Streptococcus bovis to Strep. equinus. We examined stools from all kinds of animals — lions, tigers, elephants and from children. We found Streptococcus bovis had more relation to faecalis than to equinus. Strep. bovis could be taken out of the viridans group. You said equinus is also Group D. Do you have any comment on this?

Shattock: Yes, we have made Group D antisera from Streptococcus equinus which react quite specifically with antigen extracts of all other species in Group D. We have found Streptococcus equinus to be similar in many respects to Strep. bovis. In the physiological tests carried out on your strains did you examine the results with a computer? Did you use the tests I mentioned this morning?

Raj: Yes.

Shattock: Where did you get your Group D antiserum? Did you make it?

Raj: No, it was Difco's.

Shattock: It is possible that the antiserum you used may have been deficient in Group D antibody. In a recent paper from France strains of Strep. equinus isolated from pig feces were reported on and compared with other Group D streptococci. Antigen extracts precipitated with a Group D antiserum produced by an English institute but not with an antiserum prepared by Difco. The production of potent grouping sera for Group D is usually more troublesome than for most other groups of streptococci. Commercial antisera for Group D have often proved to be poor in group antibody.

Raj: We actually tried Cappell's and Difco's. We found Group A was better with Difco.

Niven: The term "enterococcus" is being abused in the literature to such an extent that a good term may be destined to be eliminated from our literature.

Grecz: Dr. Niven, you listed only two organisms which produce a toxin in food: Clostridium botulinum and Staphylococcus aureus.

Niven: No, I said two bacteria.

Grecz: What about Clostridium perfringens and Bacillus cereus?

Niven: Would Dr. Hobbs care to comment on this?

Hobbs: We have no evidence that food poisoning from Cl. per-fringens is caused by a toxin. We can't produce the symptoms. I wonder if it isn't a mechanical effect.

Grecz: Did you get the symptoms when you fed the cultures?

Hobbs: No, not with washed cultures; we got symptoms only when cultures were actively growing in meats.

Grecz: Did you work with Bacillus cereus?

Hobbs: No.

Perry: Are mold toxins important as far as human beings are concerned?

Niven: Dr. Dack, do you have any information about mold toxins that have been reported? It was mentioned briefly last night.

Dack: No, I have no comments.

Niven: I assume it is something we must watch carefully in addition to the possibility of Pseudomonas enteritidis being a food poisoning microorganism, as reported from Japan.

PART SIX
Summary

20

General Conclusions

GEORGE F. STEWART

UNIVERSITY OF CALIFORNIA

T HIS HAS BEEN a remarkable symposium, both in its breadth of the subject matter covered and in its depth. Never before has Food Protection received such close scrutiny by such a competent and diverse group of scientific experts. It is to be hoped that this will be the forerunner of a continuing series of symposia on this most important topic.

In order to produce meaningful and integrating conclusions, it seems appropriate to use somewhat different titles for the symposium and its subdivisions, than appeared on the program leaflet. For the conference as a whole, the title "Protecting our food supply — problems and opportunities" would seem to be a good paraphrase. The following subdivisions seem pertinent to it: (1) "Problems of producing the raw materials," (2) "Problems associated with processing, manufacturing and packaging foods," and (3) "Problems associated with food distribution, marketing, and ultimate use."

Protecting our food supply is a part of an overall program by scientists to provide technical information on how to assure an abundant food supply for peoples everywhere. Not only should there be an ample quantity of foods, but the foods should be of good quality. By the latter we mean: nutritionally adequate, free from harmful added chemicals, toxins and pathogenic microorganisms. Finally, food should be aesthetically enjoyable.

Providing good food for mankind is one of the great challenges of our time, and the scientist has a key role to play in this effort. To provide the necessary knowledge requires a wide variety of scientific talent. Eleven disciplines were represented in discussions here: animal physiology, biochemistry, chemistry, engineering, entomology, food science and microbiology, nutrition, parasitology, pharmacology, plant physiology, and toxicology.

Undoubtedly, there are still others which are required to assure finding scientific solutions to the technical problems to be met.

One should not get the idea that all of the technical problems of food protection have been discussed at this symposium. Actually, they are only a sample of the many problems facing us, although, of course, those discussed are very important and timely. Left for the future are such serious other problems as viral infections of man associated with food, nutritional inhibitors occurring in raw agricultural products and processed foods, intentional additives besides nutritional supplements, antioxidants, flavoring agents, coloring agents, microbial inhibitors, and incidental additives besides pesticides, plant growth regulators and animal growth regulators.

Cannon's, Mrak's, and Mossel's papers provided excellent orientation for the general problems discussed at this conference. They deserve careful study by all serious students of Food Protection.

USE OF CHEMICALS IN FOOD PRODUCTION AND PROCESSING

The papers on intentional additives summarized knowledge of the present status concerning nutritional supplements, antioxidants, flavors, and coloring agents. It is clear that much more information is needed, not only about the fate, metabolism and possible toxicity of the compounds in present use, but also regarding prospects for still better compounds.

Since hundreds of flavoring materials are used and the methods of measuring their effects are complex, tedious, and subjective, these agents present an especially complicated picture. It seems obvious that coloring and flavoring agents will continue to require special and intensive study regarding their safe use in foods. Not only synthetic chemicals but also naturally occurring compounds sometimes present special toxicity problems. It would appear that much more detailed studies on their pathology are needed. Also, specific information is needed concerning their absorption from the gastro-intestinal tract, metabolism, detoxification, and elimination from the body. Only in these ways can the toxicity problem be dealt with realistically.

As mentioned, above, intentional additives are used for many purposes other than those discussed. These, too, require more study.

CHEMICALS USED IN AGRICULTURAL PRODUCTION

The speakers did a masterful job of delineating the problems associated with the use of chemicals in agricultural production. Hundreds of natural and synthetic compounds are being used to destroy insects, nematodes, microbiological pests and weeds, and to control physiological processes in plants and animals. While many have proven extremely useful in controlling pests and improving production efficiency, there remain many more problems to solve. Harmful side effects, resistant strains, and relative ineffectiveness are problems in the use of many of these chemicals. When effective agents have been found, their benefits have been enormous. Decreased crop and animal losses and quality improvement benefits amount to hundreds of millions of dollars a year.

There are many problems of toxicity with these chemicals, although many can be used without much concern or restriction. As with some of the intentional additives, much more needs to be known concerning their absorption and fate in the body. There is little doubt that until this knowledge is gained, a conservative attitude regarding their use must be taken by regulatory bodies.

PROTECTING FOOD FROM PATHOGENIC MICROORGANISMS AND MICROBIAL TOXINS

Excellent presentations and discussions were given on the food-borne salmonellas and certain parasites of man. One was almost overwhelmed by the complexity of the problems associated with isolation and identification and enumeration of these organisms.

It is easy to become alarmed about these matters when one visualizes what is needed in order to implement effective control programs. (In fact, some aspects of the programs for the control of salmonella in egg products seem unrealistic and completely uneconomic to industry.) Nonetheless, food-borne salmonella infections in man are real and, somehow, satisfactory control (but possibly not elimination) programs must be sought and put into force.

Superb presentations were made of the status of knowledge concerning the bacterial toxins associated with food: botulinum toxin and staphylococcal enterotoxin. Also discussed was the possibility of a toxin from the enterococci. There appears to be a real controversy among participants concerning the latter, but it was heartening to learn that steps are being taken to establish unequivocally whether enterococci can produce toxins which affect man.

FINAL REMARKS

On the whole, this was a stimulating and productive symposium. Worldwide authorities discussed a number of topics of great importance to the safety of our food supply. However, there was one thing missing — a serious lack in the opinion of this writer. There was too little discussion of the over-all problems of providing a safe food supply. A great deal of what was said concerned technicalities of chemical, microbiological, and toxicological problems, and too little concerned the practical application of our knowledge to the solution of food protection problems.

There are too few scientists concerning themselves with devising ways and means of assuring the public of safe foods. There are many very serious problems associated with drafting and putting into force adequate but fair laws and regulations. Standardized, reproducible and sensitive sampling, cultural and counting methods are urgently needed. In addition, there are many other technical problems associated with implementing such programs that need solutions before we can have really satisfactory programs for protecting our food supply. It is hoped that other symposia will be organized to deal with some of these problems.

REGISTRANTS

Acosta, Socorro O.	Iowa State University	Ames, Iowa
Alford, John A.	Meat Lab, EURDD	Beltsville, Md.
Allen, L. P.	Campbell Soup Co.	Chicago, Ill.
Amick, Georgia	University of Missouri	Columbia, Mo.
Amundson, C. H.	University of Wisconsin	Madison, Wis.
Andre, Floyd	Iowa State University	Ames, Iowa
Anellis, Abe	QMF&C Institute	Chicago, Ill.
Ansari, Mrs. M. F.	Iowa State University	Ames, Iowa
Ansari, M. F.	Iowa State University	Ames, Iowa
Appl, R. C.	Corn Products Co.	Argo, Ill.
Appleman, M. D.	University of Southern California	Los Angeles, Calif.
Arnrich, Lottie	Iowa State University	Ames, Iowa
Arthur, B. Wayne	Auburn University	Auburn, Ala.
Aylward, J. J.	Canada Dry Corp.	Greenwich, Conn.
Ayres, John C.	Iowa State University	Ames, Iowa
Bakal, T. E.	Libby, McNeill & Libby	Chicago, Ill.
Baker, M. P.	Iowa State University	Ames, Iowa
Baker, Mrs. R. J.	S. Dakota State College	Brookings, S. Dak.
Baker, R. J.	S. Dakota State College	Brookings, S. Dak.
Baldwin, Ruth	University of Missouri	Columbia, Mo.
Banwart, Geo. J.	Purdue University	Lafayette, Ind.
Bartram, M. T.	Food & Drug Admin.	Washington, D. C.
Bates, Mrs. Judy	Iowa State University	Ames, Iowa
Bates, P. K.	Carnation Research Labs.	Van Nuys, Calif.
Baughman, R. W.	Iowa State University	Ames, Iowa
Baumann, D.	Iowa State University	Ames, Iowa
Beers, R.	Iowa State University	Ames, Iowa
Beliles, R.	Iowa State University	Ames, Iowa
Binkerd, E. F.	Armour & Co.	Chicago, Ill.
Bird, E. W.	Iowa State University	Ames, Iowa
Bird, H. D.	Canada Packers Ltd.	Toronto, Canada

345

Bixby, J. N.	Ovaltine Food Products	Villa Park, Ill.
Black, Billy	Iowa State University	Ames, Iowa
Blanken, R. M.	Borden Foods Co.	Syracuse, N. Y.
Blood, Frank R.	Vanderbilt University	Nashville, Tenn.
Boeck, R. N.	International Milling Co.	Minneapolis, Minn.
Bowery, T. G.	University of North Carolina	Raleigh, N. C.
Brandly, P. J.	Meat Inspection Div., U.S.D.A.	Washington, D. C.
Brewer, Wilma	Iowa State University	Ames, Iowa
Brill, Grace D.	Iowa State University	Ames, Iowa
Brissey, G. E.	Swift & Co.	Chicago, Ill.
Brockington, S. F.	Quaker Oats Co., Res. Lab.	Barrington, Ill.
Brockmann, M. C.	QMF&C Institute	Chicago, Ill.
Brown, Margaret	Iowa State University	Ames, Iowa
Brown, R. W.	1212 Clark Ave.	Ames, Iowa
Browning, G. M.	Iowa State University	Ames, Iowa
Bryan, F. L.	Communicable Disease Center	Atlanta, Ga.
Bryant, W. D.	Birds Eye Div., Gen. Foods Corp.	Waseca, Minn.
Buchanan, J. J.	1350 W. 76th St.	Chicago, Ill.
Buchanan, R. E.	Iowa State University	Ames, Iowa
Buettner, L. G.	Accent International	Skokie, Ill.
Cannon, Mrs. Paul R.	R. 2, Box 56	Yorkville, Ill.
Cannon, P. R.	R. 2, Box 56	Yorkville, Ill.
Carlin, Frances	Iowa State University	Ames, Iowa
Carter, J. E.	Campbell Soup Co.	Chicago, Ill.
Clark, Warren	Iowa State University	Ames, Iowa
Cole, J. H.	Gen. Foods Corp.	Battle Creek, Mich.
Cone, J. F.	Penn State University	University Park, Pa.
Coon, J. M.	University of Pennsylvania	Philadelphia, Pa.
Crane, Anatole	Quaker Oats Co.	Barrington, Ill.
Cromwell, T. H.	Rogers Bros. Co.	Idaho Falls, Idaho
Crosby, D. G.	University of Calif.	Davis, Calif.
Dack, G. M.	University of Chicago	Chicago, Ill.
Dahm, Paul	Iowa State University	Ames, Iowa
Dean, R. W.	Durkee Famous Foods	Chicago, Ill.
Danner, F. W.	625 Cleveland Ave.	Columbus, Ohio
Davidson, Lois	Iowa State University	Ames, Iowa
Davis, J. L.	Igleheart Bros.	Evansville, Ind.
DeCelles, G.	Iowa State University	Ames, Iowa
Delo, Hal	117 So. Water Market	Chicago, Ill.
DeNosaquo, N.	A.M.A.	Chicago, Ill.
Devitt, A. B.	Robert A. Johnston Co.	Milwaukee, Wis.
Dodds, W. S.	Chemtronics Research Inc.	Minneapolis, Minn.

Dooling, G. S.	National Confectioner's Assn.	Chicago, Ill.
DuBois, K. P.	University of Chicago	Chicago, Ill.
El Bisi, Hamed M.	University of Massachusetts	Amherst, Mass.
Ellickson, B. E.	National Dairy Products Corp.	Chicago, Ill.
El Negoumy, A. M.	Iowa State University	Ames, Iowa
Eppright, Mrs. Ercel	Iowa State University	Ames, Iowa
Epstein, J.	2100 N. George St.	Melrose Park, Ill.
Eusebio, J. S.	Iowa State University	Ames, Iowa
Fenton, Faith	707 Beech Ave.	Ames, Iowa
Ferguson, R. B.	Pillsbury Co.	Minneapolis, Minn.
Fischer, E. E.	206 E. 4th	Ankeny, Iowa
Forsythe, R. H.	Iowa State University	Ames, Iowa
Foster, E. M.	University of Wisconsin	Madison, Wis.
Galton, Mildred M.	Vet. Public Health Lab., CDC	Atlanta, Ga.
Garcia, Pilar A.	Iowa State University	Ames, Iowa
Gardner, F. A.	Texas A. & M. College	College Station, Tex.
Garlock, E. A.	Hazleton Labs. Inc.	Falls Church, Va.
Gentry, Lelia	University of Wisconsin	Madison, Wis.
Gericke, C. E.	2073 N. Dixie Highway	Ft. Lauderdale, Fla.
Glick, D. P.	U.S. Army Chem. Corps Biol. Labs.	Fort Detrick, Frederick, Md.
Goss, E. F.	Iowa State University	Ames, Iowa
Gowen, J. W.	Iowa State University	Ames, Iowa
Graham, Jewel	Iowa State University	Ames, Iowa
Grecz, Nicholas	QMF&C Institute	Chicago, Ill.
Greenberg, R. A.	Swift & Co.	Elmhurst, Ill.
Goresline, H. E.	QMF&C Institute	Chicago, Ill.
Gunther, F. A.	University of Calif.	Riverside, Calif.
Guthrie, Elsie Ann	Iowa State University	Ames, Iowa
Haddad, S. F.	Box 60	Ontario, Oreg.
Haglund, J. R.	General Mills, Inc.	Minneapolis, Minn.
Halleck, F. E.	Pillsbury Co.	Minneapolis, Minn.
Haman, R. W.		Cannon Falls, Minn.
Hammond, E. G.	Iowa State University	Ames, Iowa
Hammond, P.	University of Minnesota	Minneapolis, Minn.
Harmon, L. G.	Michigan State University	East Lansing, Mich.
Harrold, C. M., Jr.	215 Rosemont Ave.	St. Louis, Mo.
Hartman, P. A.	Iowa State University	Ames, Iowa
Harp, G. L.	Central Div. Lab., Am. Can	Maywood, Ill.
Hayes, W. L., Jr.	Toxicology Section, CDC	Atlanta, Ga.

Hays, H. M.	117 So. Water Market	Chicago, Ill.
Heck, J. G.	Armour & Co.	Chicago, Ill.
Hetrick, J. H.	Dean Milk Co.	Rockford, Ill.
Heeren, R. H.	Iowa State Dept. of Health	Des Moines, Iowa
Hendricks, L. L.	Iowa State Dept. of Health, DVM	Des Moines, Iowa
Hobbs, Betty	Central Public Health Lab.	Colindale, London, England
Hoffman, Mrs. Ayfer	Iowa State University	Ames, Iowa
Hollen, Evelyn	Iowa State University	Ames, Iowa
Hougham, D. F.	Iowa State University	Ames, Iowa
House, D. L.	General Foods Corp.	Kaukakee, Ill.
Huber, Miss D. A.	QMF&C Institute	Chicago, Ill.
Hudson, J. D.	222 Naupin St.	Trenton, Mo.
Hunnell, J. W.	Pillsbury Co.	New Albany, Ind.
Hunter, J. E.	Procter & Gamble	Cincinnati, Ohio
Hurley, N. A.	Gerber Products Co.	Oakland, Calif.
Jackson, D.	Purdue University	Lafayette, Ind.
Jacobs, Evan	Maple Leaf Milling Co., Ltd.	Toronto, Canada
Jacobs, Leon	NIAID, N.I.H.	Bethesda, Md.
Jacobson, Katherine		Story City, Iowa
Janita, R. J.	901 N. Foster St.	Mitchell, S. Dak.
Jansen, C. D.	Pillsbury Co.	Minneapolis, Minn.
Jerome, Miss Norge W.	University of Wisconsin	Madison, Wis.
Jezeski, J.	University of Minnesota	Minneapolis, Minn.
Johnson, E.	415 W. 11th St.	Mitchell, S. Dak.
Johnston, M. R.	University of Tennessee	Knoxville, Tenn.
Kahlenberg, O. J.	1221 Western	Topeka, Kans.
Kastelic, Mrs. Joe	University of Illinois	Urbana, Ill.
Kastelic, Joe	University of Illinois	Urbana, Ill.
Kaufmann, Adrian W.	Continental Can Co.	Kansas City, Mo.
Kemp, G.	Ag. Div., Am. Cyanamid Co.	Princeton, N. J.
Kennedy, D. W.	T. J. Lipton Inc.	Hoboken, N. J.
Kenney, Mary Alice	Iowa State University	Ames, Iowa
Khan, Paul	DCA Food Industries, Inc.	New York, N. Y.
Kirchoff, Arthur	Iowa State University	Ames, Iowa
Kittel, I. D.	M. A. Gedney Co.	Chaska, Minn.
Klusmeyer, P.	Henningsen, Inc.	Springfield, Mo.
Korslund, Mary	Iowa State University	Ames, Iowa
Krabbenhoft, K.	Mankato State College	Mankato, Minn.
Kraft, A. A.	Iowa State University	Ames, Iowa
Kruse, J. F.	Iowa State University	Ames, Iowa
Kueck, D. R.	Rath Packing Co.	Waterloo, Iowa

Lawton, W. C.	2424 Territorial Road	St. Paul, Minn.
Lechowich, R. V.	Continental Can Co., Inc.	Chicago, Ill.
Lenz, R. J.	University of Wisconsin	Madison, Wis.
Lewis, A. A.	General Foods Res. Center	Tarrytown, N. Y.
Lewis, K. H.	4676 Columbia Parkway	Cincinnati, Ohio
Lehman, A. J.	Food & Drug Admin.	Washington, D. C.
Lindsay, Dale	Div. of Res. Grants, N.I.H.	Bethesda, Md.
Lindwall, W. R.	508 W. 32nd St.	Minneapolis, Minn.
Liston, John	College of Fisheries, Univ. of Wash.	Seattle, Wash.
Lockhart, W. R.	Iowa State University	Ames, Iowa
Loy, W. C.	Wilson & Co., Inc.	Chicago, Ill.
McFarren, E. F.	Sanitary Engineering Center	Cincinnati, Ohio
McGrath, F. P.	Div. of Res. Grants, N.I.H.	Bethesda, Md.
McKelvey, C. E.	Canada Dry Corp.	Greenwich, Conn.
McKim, Elizabeth	Iowa State University	Ames, Iowa
McLean, Ruth A.	ARS, USDA	Beltsville, Md.
McLean, W. H.	Merck Chemical Div.	Rahway, N. J.
McIntosh, Mrs. Elaine	Iowa State University	Ames, Iowa
McMillan, T. J.	Iowa State University	Ames, Iowa
McMillan, Martha	Iowa State University	Ames, Iowa
McPherson, Daniel G.	General Mills Inc.	Minneapolis, Minn.
Macquiddy, Mrs. E. L.		Omaha, Nebr.
Macquiddy, E. L.		Omaha, Nebr.
Mahon, J. H.	Hagan Chemicals & Controls, Inc.	Pittsburgh, Pa.
Mahoney, J. F.	Merck & Co., Inc.	Rahway, N. J.
Malin, B.	Lilly Res. Labs., Eli Lilly & Co.	Indianapolis, Ind.
Mallory, Ralph, Jr.	Mead Johnson & Co.	Evansville, Ind.
Mar, Evelyn	Iowa State University	Ames, Iowa
March, R. B.	University of California	Riverside, Calif.
Marceau, P. A.	Corn Products Co.	Argo, Ill.
Marion, W. W.	Iowa State University	Ames, Iowa
Martin, Mrs. Janet	Iowa State University	Ames, Iowa
Matches, Jack R.	Iowa State University	Ames, Iowa
Mattil, K. F.	Swift & Co.	Chicago, Ill.
Mauer, Carol	Iowa State University	Ames, Iowa
Mericle, R. B.	U.S.D.A.	Nevada, Iowa
Middelem, C. H., van	University of Florida	Gainesville, Fla.
Miller, Miss Madge	Iowa State University	Ames, Iowa
Miller, R. L.	Henningsen, Inc.	Omaha, Nebr.
Miller, W. C.	Milk & Food Program, EEFP, PHS	Washington, D. C.
Mitchell, L. E.	Shell Chemical Co.	New York, N. Y.

Montgomery, W. H.	Penick & Ford, Ltd., Inc.	Cedar Rapids, Iowa
Moore, Ray		Sioux City, Iowa
Moore, Thomas	2811 West	Ames, Iowa
Morgan, B. H.	Continental Can Co.	Chicago
Mossel, D. A. A.	Central Institute for Nutr. & Food Res.	Utrecht, The Netherlands
Movafagh	Iowa State University	Ames, Iowa
Mrak, E. M.	University of California	Davis, Calif.
Murnane, T. G. (Lt. Col.)	Defense Subsistence Testing Lab., QM	Chicago, Ill.
Murphy, J. L., Jr.	Canada Dry Corp.	Greenwich, Conn.
Nelson, F. E.	University of Arizona	Tucson, Ariz.
Nerlin, Virginia	Iowa State University	Ames, Iowa
Nickel, F. Ruth	Iowa State University	Ames, Iowa
Nielsen, Albert	Rath Packing Co.	Waterloo, Iowa
Nielsen, Verner	Iowa State University	Ames, Iowa
Niven, C. F.	Am. Meat Institute Foundation	Chicago, Ill.
Noid, Sylvia	Iowa State University	Ames, Iowa
Norton, Steven	General Foods Corp.	Battle Creek, Mich.
Ogilvy, W. S.	Mead Johnson & Co.	Evansville, Ind.
Olsen, Phyllis	Iowa State University	Ames, Iowa
Olson, J. C.	University of Minnesota	Minneapolis, Minn.
Olson, N. F.	University of Wisconsin	Madison, Wis.
O'Neil, C. J.	209 S. Hazel Ave.	Ames, Iowa
Ordal, Z. J.	University of Illinois	Urbana, Ill.
Oser, B. L.	Food & Drug Res. Labs., Inc.	Maspeth, N. Y.
Papp, Clara	Iowa State University	Ames, Iowa
Parker, D.	Iowa State University	Ames, Iowa
Perry, G. A.	99 Ave. "A"	Bayonne, N. J.
Peterson, O. H.	Dr. Salsbury's Laboratories	Charles City, Iowa
Pier, A. C.	Animal Disease Laboratory	Ames, Iowa
Porter, A. R.	Iowa State University	Ames, Iowa
Pritchard, Wm.	Iowa State University	Ames, Iowa
Putnam, G. W.	Creamery Package Mfg. Co.	Chicago, Ill.
Radkins, A. P.	M & M Candies Div., Food Mfrs. Inc.	Hackettstown, N. J.
Raj, H.	University of Washington	Seattle, Wash.
Rakosky, J., Jr.	Central Soya-Chemurgy Div.	Chicago, Ill.
Reinbold, G.	Iowa State University	Ames, Iowa
Roderuck, C.	Iowa State University	Ames, Iowa
Refle, Norma	Iowa State University	Ames, Iowa

Rice, Frank	Iowa State University	Ames, Iowa
Riemann, Hans	Danish Meat Res. Institute	Roskilde, Denmark
Robinson, H. B.	PHS, Bureau of State Services	Washington, D. C.
Roland, Irene	Iowa State University	Ames, Iowa
Rust, R. F.	Iowa State University	Ames, Iowa
Sajbel, Frances (Mrs.)	1314 Spruce St.	Wausau, Wis.
Sailor, J. C.	Clinton Corn Processing Co.	Clinton, Iowa
Salzer, R. H.	Iowa State University	Ames, Iowa
Sander, E.	Iowa State University	Ames, Iowa
Sanders, R.	Iowa State Dept. of Health	Des Moines, Iowa
Saraswat, Devi	Iowa State University	Ames, Iowa
Scaletti, J. V.	Brewster St.	St. Paul 8, Minn.
Scharpf, Lewis	Iowa State University	Ames, Iowa
Schroeder, C. W.	T. J. Lipton, Inc.	Hoboken, N. J.
Schertz, E.	QMF&C Institute	Chicago, Ill.
Schultz, H. W.	Oregon State University	Corvallis, Oreg.
Schweigert, B. S.	Michigan State University	East Lansing, Mich.
Scrimshaw, N. S.	Mass. Institute of Technology	Cambridge, Mass.
Seidel, Brunhilde	Iowa State University	Ames, Iowa
Shannon, Wm.	1705 S. 91st Ave.	Omaha, Nebraska
Sharf, J. M.	Armstrong Cork Co.	Lancaster, Pa.
Shattock, P. M. F.	Dept. of Microbiology, The University	Reading, England
Shillinglaw, C. A.	Coca-Cola Co.	New York City
Shirk, R. J.	American Cyanamid Co.	Princeton, N. J.
Sieling, D. H.	QM Res. & Engineering Command	Natick, Mass.
Singbuisch, J. R.	Rath Packing Co.	Waterloo, Iowa
Sjöstrom, L. B.	A. D. Little Co.	Cambridge, Mass.
Slosberg, Harry	Henningsen Inc.	New York, N. Y.
Snyder, Harry E.	Iowa State University	Ames, Iowa
Southworth, Jill	University of Wisconsin	Madison, Wis.
Sparling, E. M.	Producers Cry Co.	Springfield, Mo.
Splittstoesser, D.	NYS Agric. Experiment Station	Geneva, N. Y.
Stanislav, Mrs. L. R.	Green Giant Co.	LeSueur, Minn.
Stanislav, L. R.	Green Giant Co.	LeSueur, Minn.
Stedman, R. E.	1444 - 44th	Des Moines, Iowa
Steinberg, Ruth	Univ. of Kansas, Medical Center	Kansas City, Kans.
Stewart, G. F.	University of California	Davis, Calif.
Stocker, C. T.	General Foods	Battle Creek, Mich.
Stukis, J. M.	Pillsbury	Minneapolis, Minn.
Strong, Dorothy H.	University of Wisconsin	Madison, Wis.

Sulzbacher, Wm. L.	Meat Laboratory, EURDD: USDA	Beltsville, Md.
Swanson, Milo H.	University of Minnesota	St. Paul 1, Minn.
Swanson, Pearl	Iowa State University	Ames, Iowa
Tarr, H. L. A.	Fisheries Res. Bd. of Canada	Vancouver, B.C.
Taylor, K. E.	Meat Inspection Div., USDA	Washington, D.C.
Thomas, B. H.	Iowa State University	Ames, Iowa
Thompson, Nelle	Iowa State University	Ames, Iowa
Thompson, H. E.	116 W. 67th Ave.	Kansas City, Mo.
Thompson, Mrs. Victor	1614 Carroll Ave.	Ames, Iowa
Thompson, Victor	1614 Carroll Ave.	Ames, Iowa
Thompson, Wm. D.	American Potato Co.	Blackfoot, Idaho
Thorn, J. A.	Red Star Yeast & Products Co.	Milwaukee, Wis.
Torrey, George	Iowa State University	Ames, Iowa
Vlitos, A. J.	Central Agric. Res. Station	Carapichaima, Trinidad, West Indies
Vinton, Miss Catherine	John Morrell Co.	Ottumwa, Iowa
Walker, H. W.	Iowa State University	Ames, Iowa
Walliker, Miss Catherine	QMF&C Institute (formerly)	Burlington, Iowa
Wells, Mrs. F. E.	Midwest Res. Institute	Kansas City, Mo.
Wells, F. E.	Midwest Res. Institute	Kansas City, Mo.
Wierbicki, Eugen	Rath Packing Co.	Waterloo, Iowa
Wills, Rena	S. Dakota State College	Brookings, S. Dak.
Wilson, Chas.	559 N. Spring	Elgin, Ill.
Wood, Elizabeth A.	Better Homes & Gardens	Des Moines, Iowa
Woodburn, Margy	Purdue University	Lafayette, Ind.
Woolsey, Mary	Iowa State University	Ames, Iowa
Weiss, Karl	Iowa State University	Ames, Iowa
Zottala, E. C.	2231 Knapp St.	St. Paul 8, Minn.
Zweig, Gunter	University of California	Davis, Calif.

INDEX